Food and Energy Security: Current Strategies

Food and Energy Security: Current Strategies

Editor: Oliver Jones

STATES
ACADEMIC PRESS
www.statesacademicpress.com

States Academic Press,
109 South 5th Street,
Brooklyn, NY 11249, USA

Visit us on the World Wide Web at:
www.statesacademicpress.com

ISBN: 978-1-63989-206-8 (Hardback)

Cataloging-in-Publication Data

Food and energy security : current strategies / edited by Oliver Jones.
    p. cm.
Includes bibliographical references and index.
ISBN 978-1-63989-206-8
1. Food security. 2. Food supply. 3. Food--Safety measures.
4. Food service--Safety measures. I. Jones, Oliver.
TX357 .F66 2022
641.3--dc23

Table of Contents

Preface

Food security can be understood as a measure to make food available and accessible to human beings. It involves physical, social, and economic access to sufficient, safe, and healthy food to all the people while meeting their food preferences and nutritional needs for an active and healthy life, at all times. This should be irrespective of their class, gender, or region. Food and energy security is measured in terms of calories to digest out to intake per person per day available on a household budget. It addresses the concepts of availability, accessibility, utilization, stability, climate change and food sovereignty. The book presents researches and studies performed by experts across the globe. Also included herein is a detailed explanation of the various concepts and applications of food and energy security. Scientists and students actively engaged in this field will find this book full of crucial and unexplored concepts.

The information contained in this book is the result of intensive hard work done by researchers in this field. All due efforts have been made to make this book serve as a complete guiding source for students and researchers. The topics in this book have been comprehensively explained to help readers understand the growing trends in the field.

I would like to thank the entire group of writers who made sincere efforts in this book and my family who supported me in my efforts of working on this book. I take this opportunity to thank all those who have been a guiding force throughout my life.

Editor

Sweet sorghum ideotypes: genetic improvement of the biofuel syndrome

Sylvester Elikana Anami[1,2], Li-Min Zhang[1], Yan Xia[1], Yu-Miao Zhang[1], Zhi-Quan Liu[1] & Hai-Chun Jing[1]

[1]Key Laboratory of Plant Resources, Institute of Botany, Chinese Academy of Sciences, Beijing 100093, China
[2]Institute of Biotechnology Research, Jomo Kenyatta University of Agriculture and Technology, Nairobi, Kenya

Keywords
Biofuel syndrome, quantitative trait loci, sorghum (Sorghum bicolour), sweet sorghum ideotypes

Correspondence
Hai-Chun Jing, Key Laboratory of Plant Resources, Institute of Botany, Chinese Academy of Sciences, Beijing 100093, China.

E-mail: hcjing@ibcas.ac.cn

Funding Information
The authors gratefully acknowledge the financial support by the Third World Academy of Sciences for visiting fellowship for researchers from developing countries to Sylvester Anami and National Natural Science Foundation of China and Chinese Academy of Science research grants to H.-C. Jing.

Abstract

Compared to other potential feedstocks such as sugarcane, sugar beet, maize, and watermelon, sweet sorghum possesses higher levels of directly fermentable reducing sugars within the culm and the ability to accumulate high biomass under low-input production systems. In addition, it is tolerant to drought and has more efficient utilization of solar radiation and nitrogen-based fertilizers than maize and sugar cane on marginal lands that are not optimal for food production. These collectively make sweet sorghum to be considered with huge potential as a biofuel crop. Novel phenotypes generated during plant domestication and continued crop improvements via artificial selection constitute the domestication syndrome (Am. J. Bot., 101, 2014, 1711). Here, we draw an analogy and introduce the term the biofuel syndrome to refer to a suite of sweet sorghum traits, such as plant architecture (root, leave, and stem), flowering time and maturity as well as biomass bioconversion efficiency, that are associated with biofuel production and distinguish it from grain and forage sorghum traits. We discuss the biofuel syndrome amenable for targeted genetic modulation and what is currently known about the genetics and genomics of these traits as a potential route to optimize sweet sorghum for biofuel production. Continuous availability of sweet sorghum, transport and storing much mass and minimizing the postharvest loss of fermentable sugars are fundamental to exploiting sweet sorghum as a bioenergy crop. Due to the relatively short history of sweet sorghum breeding, we consider the development of ideotypes adapting to various phenological requirements to maximize the rapid deployment of sweet sorghum for biofuel production.

Introduction

The global carbon dioxide (CO_2) concentrations surpassed 400 parts per million in March 2015 for the first time since record-keeping of greenhouse gas levels began (http://research.noaa.gov). These suggests that burning of fossil fuels by humans have caused global CO_2 concentrations to rise more than 102 parts per million since pre-industrial times. Rising concentrations of atmospheric CO_2 is a key contributor to current global climate changes including shifts in seasons, climbing temperatures, rising sea levels and precipitation extremes, and there is worsening

pollution as a result of fossil exploitation (Wilkinson et al. 2007). As a consequence and in addition to recent high expectation for clean energy has given a new impetus for the production of biofuels from nonfood feedstocks. Biofuel from bioenergy crops (Olson et al. 2012) has large greenhouse gas displacement effects that could help slow down the rise of atmospheric CO_2 levels (Jaggard et al. 2010). It has been shown that corn and switchgrass grown for bioenergy resulted in net greenhouse gas reductions of -29 to -396 grams of CO_2 equivalent emissions per megajoule of ethanol per year as a consequence of direct soil carbon sequestration and from the adoption of integrated biofuel conversion pathways (Schmer et al. 2014). In both the developed and developing world, there is an ambitious target for the increased production of biofuel for economic, political, environmental, and national security reasons (O'Lear 2004). The focus now is to identify and develop improved zero carbon-emission bioenergy crops and to promote low polluting alternatives to fossil fuel.

Sweet sorghum (*Sorghum bicolor* (L.) Moench) is a C_4 monocot annual crop native to tropical and subtropical regions of all continents (Doggett 1988) and can be grown also in temperate and semi-arid regions of the world, suggesting high adaptability to different climates and soils (Davila-Gomez et al. 2011). It has the capacity to produce a crop with high biomass yield per hectare on marginal lands that are not suitable for food and feed production (Lipinsky and Kresovich 1980; Rosenow and Clark 1995; Vermerris et al. 2007; Saballos 2008; Rao et al. 2009; Vermerris and Saballos 2013). This eases competition for agricultural land already managed or harvested to meet human food and fiber needs and those containing natural ecosystems, which need to be conserved and restored to store carbon and combat climate change, to protect freshwater resources, and to preserve the planet's biological diversity (http://www.wri.org/publication/avoiding-bioenergy-competition-food-crops-and-land). Sweet sorghum also has high photosynthetic efficiency (Xu et al. 2011), possesses readily available fermentable sugars within its culm and normally does not produce a large seed head due to the partitioning of photoassimilates into the stem (Mcbee et al. 1988; Sipos et al. 2009). As such, enzymatic conversion of starch to sugar is not necessary which gives sorghum an economical advantage over starch-based crops as much energy is used to depolymerize the starch (McCollum et al. 2005). Taken together, in addition to its ability to utilize low fertilizing rates and short growth period (3–5 months), make sweet sorghum an ideal biofuel crop without impacting food prices and fodder security (Rooney et al. 2007; Carpita and McCann 2008; Vermerris 2011).

The development of sweet sorghum, an annual C_4 grass with tractable genetics, can aid in the optimization of energy crops because new genetic designs can be tested in controlled environments and multiple field locations each year, speeding up the rate of genetic improvement (Mullet et al. 2014a). Many recent reviews have convincingly made the case for sweet sorghum as a bioenergy crop (Vermerris 2011; Calvino and Messing 2012; Mullet et al. 2014a; Prakasham et al. 2014). However, the biofuel syndrome of plant architecture (leaves, root, and stem), flowering time and bioconversion efficiency, their genetic components and how these traits can be modulated as potential routes to optimize sweet sorghum for biofuel production have not been reviewed and are the basis of this review. In addition, due to the relatively short history

of sweet sorghum breeding, we consider the development of regional ideotypes adapted to various phenological requirements in order to optimize production of sweet sorghum with high stem sugars and higher biomass on marginal lands. Although, leaf stay green, abiotic, and biotic stress traits are important biofuel syndromes; they have been discussed in another review (Anami et al. 2015).

The Biofuel Syndrome and their Genetic Determinants

The term domestication syndrome was used in plants to refer to suite of traits in the domesticated crop that distinguish it from its wild ancestors (Renny-Byfield and Wendel 2014). They include a shift from the perennial to annual life cycle, the loss of dormancy and development of a nonshattering phenotype in seeds, increases in yield, improved palatability, and a wider geographical range. We draw an analogy and introduce the term the biofuel syndrome to refer to a suite of traits associated with biofuel production in sweet sorghum that distinguish it from grain and forage sorghum types. They include traits related to plant architecture (leaves, root, and stem), flowering time (maturity) and biomass bioconversion efficiency. These traits are amenable to targeted genetic modulation and constitute a potential route to optimize sweet sorghum for biofuel production.

Survey of the literature for the biofuel syndrome and their genetic components (quantitative trait loci) in sweet sorghum is presented (Table 1). The table summarizes the scientific efforts directed in the past toward the genetic and molecular elucidation of sorghum traits relevant to biofuel production. Eight hundred and fifty-eight (858) genetic loci conditioning biofuel-associated traits related to plant architecture (root, leaves, and stem), flowering time and the rate of biomass conversion into biofuel in sorghum were identified. Overlapping Quantitative Trait Loci (QTLs) affecting biofuel-associated traits were reported in different mapping populations suggesting plasticity of the traits in different environments and limited statistical power of QTL detection. For instance, tillering QTLs detected by (Kong et al. 2014) overlapped with tillering QTLs found in a previous F_2 population (Paterson et al. 1995), and with QTLs found in other sorghum populations (Hart et al. 2001; Shiringani et al. 2010; Mace and Jordan 2011). Such overlapping QTLs were considered similar and reported only once in Table 1. Thus far, genes within these QTL regions could be potential targets for improving the sweet sorghum performance for biomass and biofuel production. A total of 154 QTLs and 319 QTLs mapped on sorghum chromosome were found to have known physical and genetic map positions, respectively. An atlas of these QTLs was generated in order to visualize easily and directly their distribution on sorghum chromosomes and for comparison of the

Table 1. QTLs controlling biofuel syndrome in sorghum.

Traits	Trait category	No. of QTLs	QTL names	Reference
Root structure	Root length	1	qRL4	Fakrudin et al. (2013)
	No. of roots per plant	1	qRN1	
	Root volume	1	qRV1	
		1	qRV4	
	Root fresh weight	1	qRF4	
	Root dry weight	1	qRD4	
		1	qRDW1_2 (also for nodal root angle)	Mace et al. (2012)
		1	qRDW1_5, (also for nodal root angle)	
		1	qRDW1_8 (also for nodal root angle)	
	Shoot/root ratio	1	qRS10	Fakrudin et al. (2013)
		1	qRS10.1	
	No. of brace roots	1	qRT6	Li et al. (2014a,b)
		1	qRT7	
	Nodal root angle	1	qRA1_5,	Mace et al. (2012)
		1	qRA2_5,	
		1	qRA1_8,	
		1	qRA1_10	
		1	qSDW1_1 (Shoot dry weight)	
		1	qSDW1_5 (Shoot dry weight)	
		1	qTLA2_8 (Total leaf area)	
		1	qTLA3_8 (Total leaf area)	
		1	qTLA4_8 (Total leaf area)	
Subtotal		22		
Leaf architecture	Leaf angle	1	QLea.txs-A,	Hart et al. (2001)
		1	QLea.txs-E	
		1	QLea.txs-I	
	Total/Leaf number	1	ln3	Lu et al. (2011)
		1	ln7	
		1	ln8	
		1	ln10	
		1	QTnl-sbi01-1	Srinivas et al. (2009)
		1	QTnl-sbi01-2	
		1	QTnl-sbi03	
		1	QTnl-sbi07	
		1	QTl-dsr03a	Reddy et al. (2013)
		1	QTl-dsr03b	
		1	QTl-dsr03c	
		1	QTl-dsr01-1	
		1	QTl-dsr09-3	
		3	qNL10, qNL10.1, qNL10.2, qNL1	Fakrudin et al. (2013)
	Leaf length	1	u6	Lu et al. (2011)
		1	u6	
		1	u10	
		3	QLln.txs-F, QLln.txs-Ga/Gb, QLln.txsHa/Hb,	Feltus et al. (2006)
		2	QLln.uga-F, QLln.uga-D	
	Leaf width	3	Lw1, Lw4, Lw6	Lu et al. (2011)
		3	QLwd.txs-Ea/Eb, QLwd.txs-F, QLwd.txs-H	Feltus et al. (2006)
		6	QLwd.uga-A1, QLwd.uga-A2, QLwd.uga-B1, QLwd.uga-B2, QLwd.uga-J, QLwd.uga-D	
		1	QLwd.txs-E, QLwd.uga-J(coresspondence QTLs)	
	Leaf Curve	6	QLcv.txs-D1, QLcv.txs-D2, QLcv.txs-G, QLcv.txs-Ha, QLcv.txs-Hb, QLcv.txs-I	
	Leaf pitch	2	QLpt.txs-D, QLpt.txs-G	
	Flag leaf length	4	qFLL10, qFLL2, qFLL3, qFLL7	Zou et al. (2012)
	Flag leaf width	9	qFLW1a, qFLW1b, qFLW2a, qFLW4, qFLW6a, qFLW6b, qFLW1c, qFLW2a, qFLW2b	

(Continued)

Table 1. (Continued)

Traits	Trait category	No. of QTLs	QTL names	Reference
	Leaf composition and yield	15	*Not named*	Murray et al. (2008b)
	Stem/leaf fresh weight	8	*SbAGF08–Xcup25, Xtxp043–Xtxp329, Xtxp329–Xtxp88, Xtxp284-Xtxp06, Xcup20–Xtxp34, SbAGF06–Xcup19, Xtxp321–Xtxp250, Sb5-206–SbAGE0*	Guan et al. (2011)
Subtotal		84		
Stem Architecture	Fiber quality (Neutral detergent fiber, Acid detergent fiber, Acid detergent lignin)	79	*Not named*	Shiringani and Friedt (2011)
	Cellulose content	16	*Not named*	
	Fiber-related traits (Fresh leave mass)	56	*Not named*	
	Stripped stalk mass	15	*Not named*	
	Dry stalk mass	10	*Not named*	
	Fresh biomass	10	*Not named*	
	Dry biomass	16	*Not named*	
	Stem/stalk Juice weight/Fresh stalk weight	1	*12–26 cM*	Shiringani et al. (2010)
		14	*Xcup25-SEST249, Xtxp329-Xtxp088, Xtxp284-Xtxp061, Sb1-10-SbAGG02, SbAGF06–Xcup19, Sb5-206–SbAGE03; (Undhsbm105, xtxp141 and xtxp273) fwp1, fwp3, fwp6, Dwp7, Dwp9*	Guan et al. (2011), Lu et al. (2011), Peng et al.
		22	*Not named*	Felderhoff et al. (2012)
	Stem Brix	5	*Xtxp329–Xtxp88, Xcup74–Xcup29, Xtxp009–Sb5-236, SbAGF06-Xcup19, (Xtxp340)*	Guan et al. (2011), Peng et al. (2013)
		14	*Not named*	Shirigani et al., (2010)
		19	*Not named*	Felderhoff et al. (2012)
	Stem composition	36	*Not named*	Murray et al. (2008a,b)
	Stem glucose content	12	*Not named*	Ritter et al. (2008), Shiringani and Friedt (2011)
	Stem sucrose content	14	*Not named*	
	Stem sugar content	22	*Not named*	
	Fructose content	2	*Not named*	
	Sucrose yield	7	*Not named*	Ritter et al. (2008), Felderhoff et al. (2012)
	Sucrose/sugar ratio	3	*Not named*	
	Vegetative yield	17	*Not named*	
	Dry biomass	45	*Not named*	
	Percent moisture	7	*Not named*	
	Plant height	4	*dw1 (Sb-HT9.1), dw2, dw3,dw4,*	Quinby (1974), Pereira and Lee (1995), Rami et al. (1998), Hart et al. (2001), Kebede et al. (2001), Murray et al. (2008a), Ritter et al. (2008), Guan et al. (2011), Sabadin et al. (2012), Takai et al. (2012), Reddy et al. (2013), Upadhyaya et al. (2013)
		2	*Not named*	Felderhoff et al. (2012)
		17	*Not named*	Guan et al. (2011), Shiringani and Friedt (2011)
		2	*QPhe-sbi07-1, QPhe-sbi06-2*	Srinivas et al. (2009)

(Continued)

Table 1. (Continued)

Traits	Trait category	No. of QTLs	QTL names	Reference
		2	Ph2, Ph8	Lu et al. (2011)
		6	HtAvgD1, HtAvgC1, HtAvgA1, HtAvgG1, HtAvgJ1, HtMG2	Lin et al. (1995), Rami et al. (1998), Kebede et al. (2001), Ritter et al. (2008), Guan et al. (2011)
		1	qSV6	
		2	Pht F.2, Pht G.2	Kebede et al. (2001)
		8	QPh-dsr09-1, QPh-dsr09-3, QPh-dsr03a, QPh-dsr03b, QPh-dsr07-1, QPh-dsr01-2, QPh-dsr04-2, QPh-dsr05	Reddy et al. (2013)
		4	QHtu.txs-Ea/Eb, QHtu.txsF, QHtu.txs-G, QHtu.uga-D	Feltus et al. (2006)
			qPH6a, qPH6b, qPH7, qPH1, qPH6ac	Rami et al. (1998)
		14	QNpb-sbi01-1, QNpb-sbi01-2, QNpb-sbi05, QNpb-sbi07, QNpb-sbi08; QPB-dsr03a, QPB-dsr03b, QPB-dsr03c, QPB-dsr03d, QPB-dsr01-1, QPB-dsr01-2, QPB-dsr05, QPB-dsr07-1, QPB-dsr10-2	Srinivas et al. (2009), Reddy et al. (2013)
	Main Culm height	1	QCuh.txs-C	Hart et al. (2001)
	Culm length	2	QCL6.1/qCL6.2,QCL7	Takai et al. (2012)
	Culm width	2	qCW1, qCW6	
	culm number	2	qCN1, qCN6	
	Number of nodes	8	qNN1b, qNN6, qNN1a, qNN1c,qNN7, qNN6, qNN8a, qNN8b	Zou et al. (2012)
	Stem diameter	23	sd1, Sd3, Sd7, 14 others not named, qSD1, qSD6a, qSD7a, qSD7b, qSD4, qSD6b	Shiringani et al. (2010), Lu et al. (2011), Zou et al. (2012)
	Stem/root ratio in weight	1	sl1	Lu et al. (2011)
	Tillering	4	QTih.uga-C, QTih.uga-C2, QTih.uga-D, QTih.uga-J (Tiller height)	Feltus et al. (2006)
		2	QTih.txs-A, QTih.txs-E (Tallest basal tiller height)	Hart et al. (2001), Kong et al. (2014)
		4	QTina.txs-A1, QTina.txs-A2, QTina.txs-H, QTina.txs-I (No. of basal tillers with heads per plant)	
		3	QTinb.txs-A, QTinb.txs-I1, QTinb-txs-I2 (No. of basal tillers per basal tillered plant)	
		4	pSBJ95-pSBO62, pSBO95-pSB428, pSB510-pSB300b, pSBO67-pSB784 (No. of tillers 8 days after seeding)	Paterson et al. (1995a,b)
		6	pSB614-pSB613, pSBO95-pSB428, pSBJ93-pSB341, pSB510-pSB300b, pSB106-pSB430a, pSBO67-pSB784(Regrowth)	
		2	Not named (Tillering)	Murray et al. (2008a)
		20	tn1a, tn1b, tn2, (17 unnamed QTLs for Tiller number)	Lu et al. (2011), Shiringani et al. (2010)
Subtotal		588		
Flowering time	Flowering time/maturity	2	DFB, DFG	Crasta et al. (1999)
		3	QMa.txs-F1, QMa.txs-F2, QMa.txs-G	Hart et al. (2001)
		6	Ma1, Ma2, Ma3, Ma4, Ma5, Ma6	Kebrom et al. (2006), Takai et al. (2012), Murphy et al. (2014)
		16	Not named	Mace et al. (2013)
		16	Not named	Felderhoff et al. (2012)

(Continued)

Table 1. (Continued)

Traits	Trait category	No. of QTLs	QTL names	Reference
		11	qFT1-1, qFT1-2, qFT2, qFT3, qFT5b, qFT7, qFT8, qFT8b, qFT10, qFT5, qFT6	El Mannai et al. (2011)
		2	Flr F, Flr G	Kebede et al. (2001)
		4	qHD6b, qHD6a, qHD6c, qHD8	Zou et al. (2012)
		9	QDan-sbi01-1, QDan-sbi01-2, QDan-sbi02-1, QDan-sbi02-2, QDan-sbi03, QDan-sbi05, QDan-sbi06, QDan-sbi07, QDan-sbi08	Srinivas et al. (2009)
		3	FlrAvgD1, FlrAvgB1, FlrFstG1	Lin et al. (1995)
		5	Not named	Shiringani et al. (2010)
		18	QDma-sbi01-3, QDan-sbi01-1, QDan-sbi01-2, QDan-sbi02-1, QDan-sbi02-2, QDan-sbi03, QDan-sbi05, QDan-sbi06, QDan-sbi07, QDan-sbi08: QDm-dsr01-1, QDm-dsr01-2, QDm-dsr02-3a, QDm-dsr02-3b, QDm-dsr03, QDm-dsr07-1, QDm-dsr09-3, QDm-dsr10-2	Srinivas et al. (2009), Reddy et al. (2013)
		6	QDf-dsr01-1, QDf-dsr03, QDf-dsr05, QDf-dsr07-1, QDf-dsr09-3, QDf-dsr10-2	Reddy et al. (2013)
		4	Not named	Ritter et al. (2008)
	Photoperiod response (Vegetative Calendar Time)	2	VCT	Chantereau et al. (2001)
	Photoperiod response (Vegetative Thermal Time)	2	VTT	
	Photoperiod response (Vegetative Biological Time)	1	VBT	
	Photoperiod response (Change in vegetative calender Time)	2	ΔVCT	
	Photoperiod response (Change in vegetative thermal Time)	2	ΔVTT	
	Photoperiod response (photoperiod sensitivity slope)	1	PSS	
Subtotal		115		
Bioconversion efficiency	Stem 4-hour sugar release	4	QSt4hs_10_Tv_2A, QSt4hs_10_Tv_5A, QSt4hs_10_Tv_8A, QSt4hs_10_Tv_9A,	Vandenbrink et al. (2013)
		2	QSt4hs_11_Tv_3A, QSt4hs_11_Tv_5A	
	Stem 12-hour sugar release	3	QSt12hs_10_Tv_2A, QSt12hs_10_Tv_7, QSt12hs_10_Tv_9A	
		1	QSt12hs_11_Tv_3, QSt12hs_11_Tv_5A	
	Stem 24-hour sugar release	4	QSt24hs_11_Tv_1A, QSt24hs_11_Tv_3A, QSt24hs_11_Tv_5A QSt24hs_11_Tv_9A	
		3	QSt24hs_10_Tv_3A, QSt24hs_10_Tv_7A, QSt24hrs_10_Tv_9A,	
	Stem 12-hour hydrolysis yield potential	3	QStHYP12_11_Tv_3A, QStHYP12_11_Tv_5A, QStHYP12_11_Tv_9A,	
		3	QStHYP12_10_Tv_5A, QStHYP12_10_Tv_7A, QStHYP 12_10_Tv_9A	
	Stem 24-hour hydrolysis yield potential	3	QStHYP24_11_Tv_3A, QStHYP24_ 11_Tv_5A, QStHYP24_11_Tv_9A	
		2	QStHYP24_10_Tv_7A, QStHYP24_10_Tv_9A	
	Leaf 4-hour sugar release	3	QStHYP24_11_Tv_3A, QStHYP24_11_Tv_5A QStHYP24_11_Tv_9A	
		2	QStHYP24_10_Tv_7A, QStHYP24_10_Tv_9A	
	Leaf 12-hour release	3	QLf12hs_11_Tv_3A, QLf12hs_11_Tv_3B, QLf12hs_11_Tv_8A	
		2	QLf12hs_10_Tv_3A QLf12hs_10_Tv_5A	
	Leaf 24-hour release	2	(QLf24hs_10_Tv_5A, QLf24hs_10_Tv_9A	
		1	QLf24hs_11_Tv_3A	

Table 1. (Continued)

Traits	Trait category	No. of QTLs	QTL names	Reference
	Leaf 12-hour hydrolysis yield potential	3	QLfHYP12_11_Tv_3A, QLfHYP12_11_Tv_3B, QLfHYP12_11_Tv_8A	
	Leaf 24-hour hydrolysis yield potential	2	QLfHYP24_11_Tv_1A QLfHYP24_11_Tv_3A	
		1	QLfHYP24_10_Tv_9A	
	Crystalline index	2	QStCI_10_2A, QStCI_10_3A	
Subtotal		49		
Total		858		

number of QTLs linked to specific biofuel-associated traits in sorghum (Fig. 1). The triangulized outer and circularized inner drawings in the map show QTLs with known physical and genetic map positions on sorghum chromosomes, respectively. Three hundred and eighty-five (385) QTLs could not be integrated into the atlas map in part due to the absence of consensus map of genetic markers. In addition, markers were identified before sorghum genome was sequenced and hence sequence information for the markers is absent from reference papers. Sequence information for some markers was too short to be aligned and locate on sorghum chromosomes. These suggests that markers that were not well aligned are possibly due to less recombination information at the end of the chromosome, sequence assembly errors and multiple amplifications of paralogous loci (Kong et al. 2013a). The availability of sorghum genomes will accelerate the identification of consensus markers and allow integration of the whole genome sequence information with identified QTLs to facilitate a detailed analysis of biofuel related traits together with the development of robust marker-assisted breeding approaches. In addition, these will allow comparative QTL mapping providing means to unify, and thereby simplify, molecular analysis of complex phenotypes. Multiple QTLs controlling biofuel related traits in sorghum have been revealed and only one quantitative trait gene (dw3) underlying plant height has been cloned (Multani et al. 2003). This reflects the relatively poor knowledge of genes underlying biofuel-associated traits in sweet sorghum and justifying the huge potential that exists in further breeding of sorghum for biofuel production through exploitation of its genetics. The following section therefore focusses on dissecting the genetic loci conditioning the biofuel syndrome and the application of this knowledge in sweet sorghum improvement for biomass and biofuel production.

Plant Architecture

Root architecture

Plant root architecture refers to the arrangement of primary and lateral roots in the rhizosphere determined by genetic programs and external signals and plays a major role in yield and overall plant productivity (Herder et al. 2010). It has been shown that root diameter and tissue density traits control the length and surface area of root systems for a given biomass allocated to the root system (Fitter 2002). Small diameter and finer roots increases surface area in contact with soil water, the volume of soil that can be explored for water and root hydraulic conductivity in addition to enhancing root growth rate (Robinson et al. 1999; Comas et al. 2012). Accordingly, breeding for decrease in root diameter has the potential to enhance plants acquisition of water and productivity under drought (Wasson et al. 2012). To achieve optimal growth of biofuel crops on marginal lands and promote carbon sequestration, their adventitious and lateral roots need to be shallow and dispersed, respectively, to forage top soils for diffusion-limited nutrients and reduce runoff on steep grades, whereas deeper roots develop to increase water and soluble nutrient uptake (Hirel et al. 2007).

Among the C_4 plants, sorghum has superior water and nutrient-use efficiency, in particular under drought conditions (Lipinsky and Kresovich 1980; Steduto et al. 1997), in part due to an extensive root system that can penetrate from 1.5 to 2.5 m into the soil and extend 1 m away from the stem (Pellerin and Pagès 1996). A drought tolerant sorghum line possessed roots at least 40 cm deeper than a drought-sensitive one, and deeper rooting of stay-green lines under drought conditions was reported (Salih et al. 1999; Vadez et al. 2005). Many QTLs controlling root traits in sorghum including root length, number of roots per plant, root volume, root fresh weight, and dry weight (Mace et al. 2012; Fakrudin et al. 2013), number of brace roots (Li et al. 2014a), and nodal root angle (Mace et al. 2012) have been identified (Table 1). Cloning of genes associated with these loci and improvements in the genetic design of the root structure of sweet sorghum are expected to contribute to increase in biomass yield and create hybrids with strong plant body stability, a prerequisite for sustainable biofuel production.

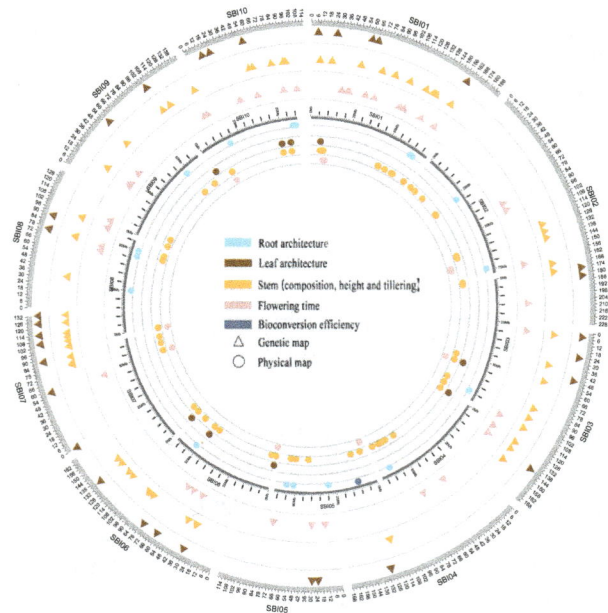

Figure 1. The atlas of QTLs for the biofuel syndrome distributed on sorghum chromosomes. QTLs with known genetic positions are represented on the outer circle with rectangular marks. The circular marks on the inner circles represent QTLs with known physical positions. QTLs controlling specific traits are represented by different colors.

Root traits are developmentally controlled by complex interacting genetic pathways whose effects could change in response to the perception of environmental cues. Indeed, S-phase kinase-associated protein2 (SKP2B) was identified as new early marker for lateral root development and coordinated largely by the phytohormones auxin and cytokinin (Moubayidin et al. 2009; Péret et al. 2009; Manzano et al. 2012). Furthermore, peroxidases activity and reactive oxygen species signaling are specifically required during lateral root emergence (Manzano et al. 2014). Cloning of genes in mutated rice and maize mutants defective in root branching has identified genes involved in polarized auxin transport, cytokinin response pathways, and transcription factors, including auxin-inducible LATERAL ORGAN BOUNDARY transcription factors in maize (Taramino et al. 2007) and rice (Inukai et al. 2005) and a cytokinin-regulated WUSCHEL-related homeobox gene in rice (Zhao et al. 2009). The roles of these genes in sorghum root development have yet to be confirmed and could be critical for shaping pathways for biofuel production.

MicroRNA (MiRNAs) are known to regulate lateral root branching and patterning in Arabidopsis and maize (Marin et al. 2010; Kong et al. 2013b) and a dedicated miRNA microarray approach identified miRNAs in sorghum (Pasini et al. 2014). Recently, efficacy of asymmetric 22-nt amiRNA-directed RNA silencing and associated phasiRNA production and activity, in mediating widespread RNA silencing of an endogenous target gene (CHALCONE SYNTHASE) in

Arabidopsis was demonstrated, providing another avenue to exploit endogenous RNA silencing mechanisms in plants (McHale et al. 2013). In view of these findings, understanding miRNAs in grasses has the potential to develop miRNA as a tool for manipulating root structure in biofuel crops.

Leaf architecture

Leaf architecture refers to the leaf morphological characteristics such as leaf angle, size, shape, weight, number, composition, width, and length. Leaf size and dimension (shape, number, width, and length) are affected by leaf cell number and this relationship suggests that cell division control leaf size. Different sorghum cultivars have tremendous variation in leaf sizes that could affect the plant's capacity for energy capture, conversion of the captured energy into biomass and physiological activity. Smaller leaves are advantageous in hot and dry environments and at high intensities of solar radiation, whereas large leaves with less efficient energy exchange capacity are advantageous in lower irradiance and cooler and moister environments (Ackerly et al. 2002).

Leaves are from 30 to 135 cm long and 1.5 to 13 cm wide, they encircle the stem with their margins overlapping with flat or wavy margins. Midribs are white or yellow in dry pithy cultivars or green in juicy cultivars (http://www.icrisat.org/text/coolstuff/crops/gcrops2.html). The regulation of leaf size and dimension can be disrupted by alteration of genes involved in numerous plant processes such as metabolism, hormone action, cell division, or cell expansion (Tsukaya 2002). Given that photosynthesis of carbohydrate is the primary source of biomass in sweet sorghum, it is important to genetically analyze the morphological characteristics of functional leaves, especially size and dimension, in sorghum biomass improvement. As detailed in Table 1 and Fig. 1, 84 genetic loci controlling leaf morphological diversity such as leaf number (Srinivas et al. 2009; Lu et al. 2011; Reddy et al. 2013), leaf length, width, curve and pitch (Feltus et al. 2006; Lu et al. 2011), flag leaf length and flag leaf width (Zou et al. 2011) and stem and leaf fresh weight (Guan et al. 2011) in sorghum have been resolved from Recombinant Inbred Line (RIL) mapping populations. These studies reveal the molecular mechanisms with regard to leaf architectural traits and indicate that there is a potential to modulate leaf architecture to enhance biomass production is sorghum. Genetic analysis of morphological diversity in sorghum leaves can be helpful to the breeders to enhance crop improvement. Further studies are necessary to fine map and construct single chromosome segment lines for genetic regions of these QTLs controlling diversity in leaf for future marker-assisted breeding. Sorghum has been shown to have increased nitrogen use efficiency, a leaf nonarchitectural trait that adds to sorghum's appeal as a bioenergy feedstock

(Heaton et al. 2008). Higher levels of leaf nitrogen concentration were found to enhance sugar production of sweet sorghum (Serrão et al. 2012), suggesting that the application of nitrogen fertilizers correlates with high sugar and biofuel production. Indeed, leaf lignocellulosic biomass (structural carbohydrates) deliver higher yields of energy per hectare than sugar and starch (nonstructural carbohydrates) (Murray et al. 2008b). Leaf area index, used to quantify vegetative canopy structure, influences photon capture, photosynthesis, assimilate partitioning and growth (Tsialtas and Maslaris 2008) and is an important parameter to simulate growth and development in sorghum (Narayanan et al. 2014).

Upright leaf angle (LA), the angle of the third leaf-blade junction relative to the main culm, (erect leaf) determines planting density and increases light harvesting for photosynthesis, thereby influencing biomass yield and has been demonstrated to drive the continuous yield increases in modern maize hybrids (Mason et al. 2008; Tian et al. 2011). Mutations in the *Oryza sativa BRASSINOSTEROID INSENSITIVE1 (OsBRI1)* gene confers semi dwarf stature and erect leaves and led to 30% more yield than the wildtype at high planting densities (Sakamoto et al. 2005; Morinaka et al. 2006). Many independent erect leaf mutants are available in an induced sorghum mutant population that could facilitate functional genomics (Xin et al. 2009). In addition, these mutants may be useful for improving sorghum biomass-based biomass improvements achieved in maize hybrids through improved leaf angle. Major loci genetically controlling leaf angle variation trait in sorghum have been mapped (Hart et al. 2001), and require further validation for their future application in sweet sorghum breeding programs.

Stem composition

Sweet sorghum stems contain directly fermentable sugars and the secondary cell walls are rich in cellulose, hemicellulose xylan, and lignin. (Murray et al. 2008a; Wang et al. 2009). (Hoffmann-Thoma et al. 1996) investigated the activities of sugar metabolizing enzymes (sucrose phosphate synthases and invertases) that are associated with sugar accumulation in sugarcane during stem development in three sorghum cultivars (NK 405, Keller and Tracy) and found out that none of these enzymes were responsible for the extent of sucrose storage in the stem. In addition, the expression of sucrose phosphate synthase (*SPS2* and *SPS3*) and vacuolar invertase genes in sweet sorghum were found to be lower when compared with grain sorghum. These suggests that the regulatory networks for high sugar content is more complex and that other phenomena such as transport processes within the stern tissue require further investigation. Lower

expression of two sucrose transporters (*SUT1* and *SUT4*) correlated with higher sugar accumulation in sweet sorghum (Qazi et al. 2012).

The inheritance of increased stem sugar depends on introgression and has been shown to be either additive or dominant (Schluhuber 1945; Clark et al. 1981). Many genetic loci controlling sorghum stem composition (Murray et al. 2008a; Ritter et al. 2008; Shiringani et al. 2010; Guan et al. 2011; Lu et al. 2011; Felderhoff et al. 2012) (Table 1) have been identified and they remain to be cloned. Cloning of genes conditioning stem sugar and juice accumulation may create opportunities for enhancing biofuel production in sweet sorghum. Compared to grain sorghum at flowering stage, sweet sorghum transcripts related to cell wall processes were downregulated; suggesting that carbon partitioning in the stem could be a mechanism that contribute to genotypic variation in sugar content (Calvino and Messing 2012).

Plant cell wall contains cellulose, hemicelluloses, and lignin polysaccharides which influence biomass quality (Carpita and McCann 2008) and they can be hydrolyzed to sugars and then fermented to ethanol. The interactions among these polymers prevent facile accessibility and deconstruction by enzymes and chemicals. Plant biomass with increased cellulose content that can with minimal pretreatment be degraded into sugars is required to produce renewable biofuels in a cost-effective manner (Biswal et al. 2015). Sorghum biomass on a dry matter (the portion of biomass that is not water) basis is about 23% cellulose, 14% hemicellulose, and 11% lignin (www.eere.energy.gov/biomass/progs/search1.cgi). Improvement in biomass quality in sorghum is dependent on genetic variability within the species, the heritability of the trait(s), selection intensity, and the ability of plant breeders to understand the genetic architecture controlling these traits. Identifying genes controlling these traits is difficult because of limited amenability to high-resolution genetic mapping. To address this problem, chromosomal regions linked with cell wall content (neutral detergent fiber, acid detergent fiber, acid detergent lignin, cellulose, hemicellulose) and biomass yield traits (fresh leaf mass, stripped stalk mass, dry stalk mass, fresh biomass, and dry biomass) have been identified (Murray et al. 2008a; Shiringani and Friedt 2011) (Table 1). The combination of favorable alleles for high fiber quality with high biomass in single RILs represents interesting new sorghum breeding material for the production of biomass and bioenergy. (Shiringani and Friedt 2011) used an additive x additive model that resulted in detection of more QTL controlling cell wall content and biomass yield with high additive effects that are particularly important for breeding because they can be directly exploited in selection for the respective trait(s). Furthermore, many QTL's for cell wall components in maize and sorghum stalks are co-localized and clustered on particular chromosomes

(Cardinal et al. 2003; Barriere et al. 2008; Shiringani and Friedt 2011) suggesting that genes controlling cell wall components may either be linked in the sorghum genome or act in a pleiotropic manner. Targeted biotechnological approaches hold promise in deconstructing the cell wall components for biofuel production. In Populus, the downregulation of GAUT12.1 lead to a reduction in a population of xylan and pectin during wood formation and to reduced recalcitrance, more easily extractable cell walls, and increased growth (Biswal et al. 2015).This strategy could be applicable in sweet sorghum.

Plant height

Plant height is an important biomass component and its modulation has the potential to enhance biomass production. Sweet sorghum cultivars are often over three meters tall and are able to produce biomass in the order of 58.3–80.5 tons of fresh stems per hectare in semi-arid zones (Wang and Liu 2009). Biomass yields realized in heterosis and sugar content could be enhanced by creating hybrids through crossing grain-type seed parents and sweet-type pollen parents (Hunter and Anderson 1997). Indeed, plant growth regulators including gibberellins and brassinosteroids promote stem elongation and overall plant growth, respectively, and their metabolism and signaling are both crucial for controlling plant height (Li and Jin 2007; Yamaguchi 2008). Gibberellins deficiency resulting from any loss-of-function mutation in four genes (SbCPS1, SbKS1,SbKO1, SbKAO1) involved in the early steps of GA biosynthesis, not only results in severe dwarfism but also in abnormal culm bending in sorghum (Ordonio et al. 2014). Breaking-type lodging resistance was improved in rice plants producing high amounts of GA due to increased lignin accumulation and/or larger culm diameters with concomitant increase in total biomass weight (Okuno et al. 2014). In addition, transgenic switchgrass plants overexpressing PvSUS1 had increases in plant height by up to 37%, biomass by up to 13.6%, and tiller number by up to 79% compared to control plants (Poovaiah et al. 2014). These indicates that the use of sweet sorghum cultivars producing high levels of GA and the expression of sucrose synthase (SUS) would be novel targeted approaches to create higher lodging resistance and biomass yield, respectively, and to stack with other genes to increase biofuel production per land area cultivated. Identification of QTLs of plant height components is crucial in understanding dwarfing mechanisms and efficient use of novel dwarf germplasms for a breeding program.

Four (dw1, dw2, dw3, and dw4) genetic loci have consistently been identified in sorghum across many environments controlling height (Lin et al. 1995; Pereira and Lee 1995; Rami et al. 1998; Hart et al. 2001; Kebede et al. 2001; Klein et al. 2001; Natoli et al. 2002; Murray

et al. 2008b; Ritter et al. 2008; Guan et al. 2011; Sabadin et al. 2012; Takai et al. 2012; Reddy et al. 2013; Upadhyaya et al. 2013), and breeders have introgressed these dwarfing mutations for reducing height into elite cultivars (Quinby 1974). Current sorghum commercial lines contain three combined mutations (dw1, dw2, and dw4). Dw3 is the only dwarfing gene that has been cloned encoding a phosphoglycoprotein involved in auxin transport, it is often included in the combination because of its robust ability to improve the harvest index of sorghum (Multani et al. 2003). However, the only mutant allele of dw3 available is unstable and spontaneously reverts back to the tall type by unequal cross over at a frequency of 0.1–0.5%, depending on the genetic background (Klein et al. 2001). Additional genetic loci have been identified for controlling height and plant uniformity traits in sorghum (Lin et al. 1995; Rami et al. 1998; Kebede et al. 2001; Ritter et al. 2008; Srinivas et al. 2009; Shiringani et al. 2010; Guan et al. 2011; Lu et al. 2011; Felderhoff et al. 2012; Reddy et al. 2013), main culm height (Hart et al. 2001), culm length, width and number (Takai et al. 2012), number of nodes (Zou et al. 2012) and stem number (Shiringani et al. 2010; Lu et al. 2011; Zou et al. 2012) (Table 1), providing important insights for improving molecular breeding strategies for sweet sorghum.

Tillering

In addition to plant height, tillering or the degree of branching is a fundamental component of shoot structure and biomass yield (Conway and Toenniessen 1999) and has a positive impact on sugar accumulation (Jordan et al. 2004). Crop genotypic diversity and the associated growing environment contribute to phenotypic plasticity in tillering. For instance, high-tillering genotypes are better adapted to maximize resource utilization in optimal environments (Borrell et al. 2014). In an environment that does not support optimal growth, low tillering limit plant size and improves postanthesis water availability and grain yield (Hammer et al. 2006). On the other hand, excessive tillering under water limited conditions can lead to high tiller abortion, poor grain set and small panicle size, thereby reducing biomass and grain yield (Kariali and Mohapatra 2007). Recently, it has been shown that tillers with mature panicles and immature secondary branches each show consistent positive correlation with dry biomass (Kong et al. 2014). The understanding of the physiological and genetic control of vegetative branching in sorghum is limited, yet it is essential in deterministic breeding of optimized genotypes for sustainable cellulosic biomass production in both optimal and marginal conditions (Saracutu et al. 2010; Kong et al. 2014).

Tiller angle is another key agronomic trait for achieving ideal plant architecture because a plant with smaller tiller angles or erect growth habits is considered to be a compact plant architecture, which may allow planting at a high density, enhance photosynthesis efficiency and improve biomass yield and is regulated mainly by shoot gravitropism. In rice, strigolactone growth regulator regulate rice tiller angle by attenuating shoot gravitropism through inhibiting auxin biosynthesis mainly by decreasing the local indoleacetic acid content (Sang et al. 2014). Furthermore, by characterizing dwarf27, a classic rice mutant exhibiting increased tillers and reduced plant height suggests that DWARF27 (D27) an iron-containing protein required for the biosynthesis of strigolactones, regulates rice tiller bud outgrowth (Lin et al. 2012). Therefore, strigolactones could be considered as an important metabolite for modulation to achieve ideal sorghum plant architecture in future. Variations in the degree of tillering in sorghum including tiller height (Feltus et al. 2006), tallest basal tiller height, number of basal tillers with heads per plant, number of basal tillers per basal tillered plant (Hart et al. 2001), number of tillers 8 days after seeding, regrowth (Paterson et al. 1995) and number of tillers (Shiringani et al. 2010; Lu et al. 2011) have a genetic basis (Table 1). TAC1, a major gene controlling tiller and leaf angle in rice was shown to control multiple traits associated with biofuel production in Miscanthus sinensis (Zhao et al. 2014), providing information on the molecular mechanisms of tillering and an opportunity for the improvement of plant architecture with regard to tillering and leaf angle in sweet sorghum breeding.

Flowering Time

Breeding sweet sorghum for flowering time, a trait regulated by photoperiod sensitivity, has the potential to enhance biofuel production because the trait significantly impact plant adaptation to agro-ecological environments and biomass accumulation (Rooney et al. 2007). For annual crops such as sweet sorghum, delay on flowering and floral development usually increase plant biomass yield. Thus, we can increase plant biomass yield though targeted regulation of plant flowering and floral development (Zhang and Wang 2015). For instance, energy sorghum hybrids are selected for late flowering to enhance biomass yield. (Rooney 2004; Rooney et al. 2007). In addition, cultivars with a wide range of maturity classes would allow staggered planting dates and extended harvesting periods to better fit the requirements of the processing industry. The degree of photoperiod sensitivity in sorghum, a short day plant, depends in part on alleles of the maturity loci $Ma1$ through $Ma6$. $Ma1$ corresponds to *PSEUDORESPONSE REGULATOR PROTEIN37*

(*SbPRR37*), a repressor of flowering in long days and has been important in more recent breeding efforts to generate lines for biofuel production (Mullet et al. 2014b). SbPRR37 expression is circadian clock regulated in a photoperiod-dependent waveform: in short day conditions, SbPRR37 expression peaks during the morning, while in long day or constant light conditions, a morning and an evening peak of expression are present. The synchronization of biological processes with daily and seasonal environmental conditions enables resource allocation during the most beneficial times of day and year (Bendix et al. 2015). $Ma3$ encodes *PHYTOCHROME B* (*PhyB*), a red-light photoreceptor that plays an important role in photoperiod sensing and repression of flowering by repressing *TEOSINTE BRANCHED1* expression and induces sorghum axillary bud outgrowth in response to light signals (Kebrom et al. 2006). $Ma6$ encodes *SbGhd7* (*Sb06 g000570*), a repressor of *EARLY HEADING DATE 1* (*SbEHD1*) expression and flowering in long days independent of $Ma1$ pathway, thus enhancing biomass accumulation and grain production (Murphy et al. 2014). $Ma2$, $Ma4$, and $Ma5$ are flowering time loci that enhance photoperiod sensitivity in sorghum. Sorghum *CONSTANS* has been characterized as a floral activator that promotes flowering by inducing the expression of *EARLY HEADING DATE 1*(*SbEHD1*) and sorghum orthologs of the maize floral integrator (*FT*) genes *ZCN8* (*SbCN8*) and *ZCN12* (*SbCN12*). *BBX19* and *CONSTANS* co-localize in the nucleus and interacts physically in vivo repressing *FLOWERING LOCUS T* transcription (Wang et al. 2014). The floral repressor *PSEUDORESPONSE REGULATOR PROTEIN 37* (*PRR37*) inhibits sorghum CONSTANS activity and flowering in long days (Yang et al. 2014). More than 100 genetic loci have been identified controlling flowering time in sorghum (Table 1) and may assist in fine mapping, map-based gene isolation and also for sorghum improvement as an energy crop through marker-assisted breeding.

Bioconversion Efficiency

Bioconversion efficiency or fermentation efficiency refers to the degree of conversion of soluble carbohydrates (glucose, fructose, and sucrose) and insoluble carbohydrates (cellulose and hemicellulose) into biofuels. Conversion efficiency in bioenergy grasses is inversely correlated with crystallinity index, the relative amounts of crystalline material in cellulose. A lower crystallinity index enhances enzymatic hydrolysis in forage sorghums by easily transforming crystalline cellulose to amorphous cellulose (Corredor et al. 2009). Lignocellulosic biomass conversion efficiency trait is under genetic control and 49 loci associated with enzymatic biomass conversion efficiency in sorghum leaf and stems and two loci controlling crystallinity index trait have recently been identified (Vandenbrink et al. 2013) (Table 1). These bioprocess-relevant QTLs are necessary to identify genes and develop genotypes that are favorable to realistic industrial processes.

Saccharification is a process by which hydrolytic enzymes break down lignocellulosic materials to fermentable sugars for biofuel production. Mapping and identifying genes underlying saccharification yield is an important first step to genetically improve the plant for higher biofuel productivity. Seven loci are associated with saccharification yield, and β-tubulin and NAC SECONDARY WALL THICKENING PROMOTING FACTOR1 (NST1), are candidate genes within these loci (Wang et al. 2013). Saccharification assays under different pre-treatment conditions and simultaneous saccharification and fermentation assays on Poplar transgenic plants in which cinnamoyl-CoA reductase was downregulated, showed that wood from the most affected transgenic trees (FAS13) yielded 161% more ethanol than wilt type (Van Acker et al. 2014). Saccharification ability of cellulosic biomass for efficient production of biofuels has been enhanced through the expression of cell wall degrading enzymes such as cellulase linked to a senescence-inducible promoter in Arabidopsis (Furukawa et al. 2014). In Switchgrass, the downregulation of caffeic acid O-methyltransferase (COMT) gene through manipulation of microRNA (mRNAs) decreased lignin content modestly, reduced the syringyl:guaiacyl lignin monomer ratio, improved forage quality, and most importantly, increased the ethanol yield by up to 38% using conventional biomass fermentation processes (Fu et al. 2011). Thus, the downregulation of genes in the monolignol-specific branch of the lignin biosynthetic pathway may become a successful strategy to improve biomass processing in sweet sorghum.

Sorghum leaves constitute sucrose known to be translocated and transformed into starch during the development of grain (Smith and Frederiksen 2000). A mutant, named RED for GREEN (RG) displayed increased accumulation of lignin and reduced saccharification efficiency in leaves with concomitant depletion in the stems suggesting that the red leaf coloration of the RG mutant represents a potential marker for improved conversion of stem cellulose to fermentable sugars in the C_4 grass sorghum (Petti et al. 2013).

The brown midrib mutant (bmr) is associated with increased conversion efficiency of sorghum stover to ethanol and reduced lignin content in the cell walls and vascular tissues, which could potentially be advantageous for cellulosic biofuel production (Vermerris et al. 2007). There are about 29 mutants with altered lignin biosynthesis categorized into bmr2, bmr6, bmr12, and bmr19. Bmr6 (Sb04 g005950) located on SBI-04 and bmr12

(Sb07 g003860) located on SBI-07 represent the mutant forms of cinnamoyl alcohol dehydrogenase (CAD) and COMT of the monolignol pathways, respectively (Mace and Jordan 2010). Mutants with impaired CAD or COMT activity have attracted considerable agronomic interest for their altered lignin composition and improved digestibility. Silencing of CAD and COMT in Brachypodium distachyon overexpressing artificial microRNA improved ethanol yield by microbial fermentation (Trabucco et al. 2013), suggesting that modulation of these two genes may result in greater stem biomass yield and bioconversion efficiency. Indeed, valuable insights into the mechanisms of lignin biosynthesis was unearthed from differential gene expression analysis for bmr6 that led to the upregulation of 11 key enzyme genes of monolignols biosynthesis with their promoter having a common MYB sites indicating that a MYB1 transcription factor (Sb02 g031190) could associate with the upregulation of these genes in sorghum (Li et al. 2014b). Search for novel mutants and new alleles of previously known loci provide new genetic resources to improve the conversion efficiency of sorghum stover to ethanol. Suppression of SQUAMOSA PROMOTER BINDING PROTEIN LIKE (SPL) genes by the overexpression of miR156 in switchgrass caused an increase in overall biomass accumulation coupled with an increase in conversion efficiency of 24.2–155.5% in non-pretreated lignocellulosic material and between 40.7 and 72.3% increase in acid-pretreated samples. Transgenic tobacco plants overexpressing Arabidopsis CESA3ixr1-2 gene exhibits improved saccharification with 45% and 25% more sugar being released from transgenic leaf and stem samples, respectively, without chemical or heat pretreatment (Sahoo and Maiti 2014). Thus, from a practical standpoint, similar transgenic strategies could be employed to introduce genes into sweet sorghum to improve the efficiency of biomass conversion.

Harvest Time and Postharvest Storage

The potential of sweet sorghum as a possible ethanol feedstock has found limited use because of poor postharvest storage characteristics and short harvest window limited by freezing in temperate climates. Sorghum is seasonally available and storage is expensive, making it difficult to use infrastructure efficiently and to schedule labor. The delayed fermentation and freezing weather can lead to "souring" of juices characterized by loss of sugar content through production of organic acids and associated reduction in ethanol yield or failed fermentation (Parrish and Cundiff 1985). The rate of carbohydrate degradation is related to the extent of tissue damage, so whole stalks are more stable than chopped stalks, which are more stable than pressed juice. Directly fermentable sugars in sweet sorghum are converted to lactic acids by the natural microflora including bacteria, yeasts, and molds within a relatively short time after harvest and thus, chilling juices has been found to be stable for longer days with little or no deterioration if stored at 4°C (Daeschel et al. 1981).

Minimizing the loss of fermentable sugars after harvest is fundamental to exploiting sweet sorghum as a bioenergy crop. Ensilage has been used to extend the shelf life of the sugars by overcoming the limitations associated with short harvest windows and frost and freeze damage. In addition, cool/cold (no freeze) storage was used successfully to maintain whole stalks of sorghum up to 150 days without significant loss in fermentable carbohydrates (Parrish and Cundiff 1985). The concentration of extracted juice into a stable syrup and use of the syrup as a feedstock for year-round fermentation is another strategy to store fermentable carbohydrates; however, the approaches are associated with relatively high capital requirements (Bennett and Anex 2009). Adjusting the pH to an extreme level to inhibit microbial activity could extend the shelf life of the freshly pressed juices or simply begin the fermentation processes shortly after pressing since the freshly pressed juice is unstable. If yeast is pitched shortly after pressing the juice, conversion to ethanol begins immediately, minimizing sugar loss (Bellmer et al. 2010). To adjust harvesting time, it has been proposed to plant the seeds at different times for the same variety and another is to plant different genotypes with various maturities (Xin and Wang 2011).

Perspective: Sweet Sorghum Ideotypes

Currently, biofuels are produced from crops such maize, vegetable oils, and sugarcane and provide about 2.5% of the world's transportation fuel. *Miscanthus* and *switchgrass* are also potential feedstocks for biofuel production but have the disadvantage that ethanol has to be produced from cellulose, raising the production costs compared to sweet sorghum. Sugarcane is polyploid and so is a less suitable model system for other species to increase stem sugar by translational genetics (Calvino and Messing 2012). The production of biofuel from improved sweet sorghum ideotypes in addition to its inherent characteristics might contribute to the displacement of fossil fuel use and phasing out of crop-based biofuels for a sustainable food future and must therefore be deployed rapidly.

In this regard, the concept of super sweet sorghum ideotypes is proposed to stimulate discussion on how to combine the various biofuel syndromes to design the genomes of sweet sorghum hybrids for the various climatic zones. The super sweet sorghum aims to produce high biomass, higher yields, and stem soluble sugars, under marginal land. Indeed (Murray et al. 2008b) suggested that both traits, high grain yield and soluble sugar content, could be bred into a single sorghum cultivar. In addition, as demonstrated in tobacco, pyramiding genes involved in sucrose metabolism, UDP-glucose pyrophosphorylase, sucrose synthase, and sucrose phosphate synthase directly impacted primary growth and therefore biomass production as seen in an increase in height growth (Coleman et al. 2010), and might appear to be a promising strategy to achieve the proposed super sweet sorghum ideotypes. Depending on the climatic zones, the traits required to combine will be different. While high stem sugar and juice content and high biomass are probably the universal traits needed to be optimized in all conditions, the ways to achieve the same outcomes may be different. For instance, tall plants appear to be a key for high sugar and biomass production per acre, but increase in stem diameters might be more in favor for regions where windy weather or lodging is the major constraint. Furthermore, specific traits may need to be considered in the breeding programmes for different climatic zones. For Northern China and high land Europe, chilling tolerance is a favorable trait for early seed germination and seedling growth. For tropical and subtropical regions, defining the right maturity to achieve three or four crops in a year to increase annual biomass yields and the development of perennial sweet sorghum cultivars that are more resilient to extreme environmental conditions would be the goal of breeding in addition to developing cultivars resistant to *Striga* parasitism (Anami et al. 2015). For arid and semi-arid regions, breeding ideotypes with enhanced water use efficiency and adult plant drought tolerance would be essential.

Targeting the biofuel syndromes through molecular biology approaches to increase biomass yield, sugar and juice contents in sweet sorghum under adverse environmental conditions may contribute toward achieving sustainability in biofuel production. Improved biomass yields in sweet sorghum per unit area of existing farmland are extremely important in the conservation of natural areas that could be converted to agricultural fields (Duke 2014). Genetically engineered sweet sorghum might produce more biomass on less land, sweet sorghum would not require extensive use of prime agricultural lands and they should have a low cost of energy production from biomass.

Acknowledgments

The authors gratefully acknowledge the financial support by the Third World Academy of Sciences for visiting fellowship for researchers from developing countries to Sylvester Anami and National Natural Science Foundation of China and Chinese Academy of Science research grants to H.-C. Jing. We wish to thank X.-B. He, X. Feng for help in literature retrieving and two anonymous reviewers for critically evaluating the manuscript and providing constructive comments to improve its quality.

Conflict of Interest

None declared.

References

Ackerly, D., C. Knight, S. Weiss, K. Barton, and K. Starmer. 2002. Leaf size, specific leaf area and microhabitat distribution of chaparral woody plants: contrasting patterns in species level and community level analyses. Oecologia 130:449–457.

Anami, S. E., L. M. Zhang, Y. Xia, Y. M. Zhang, Z. Q. Liu, and H. C. Jing. 2015. Sweet sorghum ideotypes: genetic improvement of stress tolerance. Food and Energy Security. 4:3–24.

Barriere, Y., J. Thomas, and D. Denoue. 2008. QTL mapping for lignin content, lignin monomeric composition, p-hydroxycinnamate content, and cell wall digestibility in the maize recombinant inbred line progeny F838 × F286. Plant Sci. 175:585–595.

Bellmer, D., R. Huhnke, R. Whiteley, and C. Godsey. 2010. The untapped potential of sweet sorghum as a bioenergy feedstock. Biofuels 1:563–573.

Bendix, C., C. M. Marshall, and F. G. Harmon. 2015. Circadian clock genes universally control key agricultural traits. Mol. Plant DOI: http://dx.doi.org/10.1016/j.molp.2015.03.003

Bennett, A. S., and R. P. Anex. 2009. Production, transportation and milling costs of sweet sorghum as a feedstock for centralized bioethanol production in the upper Midwest. Bioresour. Technol. 100:1595–1607.

Biswal, A. K., Z. Hao, S. Pattathil, X. Yang, K. Winkeler, C. Collins, et al. 2015. Downregulation of GAUT12 in Populus deltoides by RNA silencing results in reduced recalcitrance, increased growth and reduced xylan and pectin in a woody biofuel feedstock. Biotechnol. Biofuels 8:1–26.

Borrell, A. K., E. J. van Oosterom, J. E. Mullet, B. George-Jaeggli, D. R. Jordan, P. E. Klein, et al. 2014. Stay-green alleles individually enhance grain yield in sorghum under drought by modifying canopy development and water uptake patterns. New Phytol. 203:817–830.

Calvino, M., and J. Messing. 2012. Sweet sorghum as a model system for bioenergy crops. Curr. Opin. Biotechnol. 23:323–329.

Cardinal, A., M. Lee, and K. Moore. 2003. Genetic mapping and analysis of quantitative trait loci affecting fiber and lignin content in maize. Theor. Appl. Genet. 106:866–874.

Carpita, N. C., and M. C. McCann. 2008. Maize and sorghum: genetic resources for bioenergy grasses. Trends Plant Sci. 13:415–420.

Clark, R., P. Pier, D. Knudsen, and J. Maranville. 1981. Effect of trace element deficiencies and excesses on mineral nutrients in sorghum. J. Plant Nutr. 3:357–374.

Chanterau, J., Trouche, G., Rami, J. F., Deu M., Barro C., Grivet L. 2001. RFLP mapping of QTLs for photoperiod response in tropical sorghum. Euphytica 120:183–194.

Coleman, H. D., L. Beamish, A. Reid, J.-Y. Park, and S. D. Mansfield. 2010. Altered sucrose metabolism impacts plant biomass production and flower development. Transgenic Res. 19:269–283.

Comas, L., K. Mueller, L. Taylor, P. Midford, H. Callahan, and D. Beerling. 2012. Evolutionary patterns and biogeochemical significance of angiosperm root traits. Int. J. Plant Sci. 173:584–595.

Conway, G., and G. Toenniessen. 1999. Feeding the world in the twenty-first century. Nature 402:C55–C58.

Corredor, D., J. Salazar, K. Hohn, S. Bean, B. Bean, and D. Wang. 2009. Evaluation and characterization of forage sorghum as feedstock for fermentable sugar production. Appl. Biochem. Biotechnol. 158:164–179.

Crasta, O. R., Xu, W. W., Rosenow, D. T., Mullet J, Nguyen H. T. 1999. Mapping of post-flowering drought resistance trait in grain sorghum: association between QTLs influencing premature senescence and maturity. Mol. Gen. Genet. 262:579–588.

Daeschel, M. A., J. O. Mundt, and I. E. McCarty. 1981. Microbial changes in sweet sorghum (Sorghum bicolor) juices. Appl. Environ. Microbiol. 42:381–382.

Davila-Gomez, F., C. Chuck-Hernandez, E. Perez-Carrillo, W. Rooney, and S. Serna-Saldivar. 2011. Evaluation of bioethanol production from five different varieties of sweet and forage sorghums (Sorghum bicolor (L) Moench). Ind. Crops Prod. 33:611–616.

Doggett, H. 1988. Sorghum: Longman Scientific and Technical.

Duke, S. O.. 2014. Perspectives on transgenic, herbicide-resistant crops in the USA almost 20 years after introduction. Pest Manag. Sci. 71:652–657.

El Mannai, Y., T. Shehzad, and K. Okuno, 2012. Mapping of QTLs underlying flowering time in sorghum (Sorghum bicolor (L.) Moench). Breed. Sci. 62:151–9.

Fakrudin, B., S. Kavil, Y. Girma, S. Arun, D. Dadakhalandar, B. Gurusiddesh, et al. 2013. Molecular mapping of genomic regions harbouring QTLs for root and yield traits in sorghum (Sorghum bicolor L. Moench). Physiol. Mol. Biol. Plants 19:409–419.

Felderhoff, T. J., S. C. Murray, P. E. Klein, A. Sharma, M. T. Hamblin, S. Kresovich, et al. 2012. QTLs for energy-related traits in a sweet x grain sorghum Sorghum bicolor (L.) Moench mapping population. Crop Sci. 52:2040–2049.

Feltus, F. A., G. E. Hart, K. F. Schertz, A. M. Casa, S. Kresovich, S. Abraham, et al. 2006. Alignment of genetic maps and QTLs between inter- and intra-specific sorghum populations. Theor. Appl. Genet. 112:1295–1305.

Fitter, A. (2002) Roots as dynamic systems: the developmental ecology of roots and root systems. Pp. 115. Physiological Plant Ecology: 39th Symposium of the British Ecological Society. Cambridge Univ. Press, Pp 115–131.

Fu, C., J. R. Mielenz, X. Xiao, Y. Ge, C. Y. Hamilton, M. Rodriguez, et al. 2011. Genetic manipulation of lignin reduces recalcitrance and improves ethanol production from switchgrass. Proc. Natl Acad. Sci. 108:3803–3808.

Furukawa, K., S. Ichikawa, M. Nigorikawa, T. Sonoki, and Y. Ito. 2014. Enhanced production of reducing sugars from transgenic rice expressing exo-glucanase under the control of a senescence-inducible promoter. Transgenic Res. 23:531–537.

Guan, Y.-A., H.-L. Wang, L. Qin, H.-W. Zhang, Y.-B. Yang, F.-J. Gao, et al. 2011. QTL mapping of bio-energy related traits in Sorghum. Euphytica 182:431–440.

Hammer, G., M. Cooper, F. Tardieu, S. Welch, B. Walsh, F. van Eeuwijk, et al. 2006. Models for navigating biological complexity in breeding improved crop plants. Trends Plant Sci. 11:587–593.

Hart, G., K. Schertz, Y. Peng, and N. Syed. 2001. Genetic mapping of Sorghum bicolor (L.) Moench QTLs that control variation in tillering and other morphological characters. Theor. Appl. Genet. 103:1232–1242.

Heaton, E. A., R. B. Flavell, P. N. Mascia, S. R. Thomas, F. G. Dohleman, and S. P. Long. 2008. Herbaceous energy crop development: recent progress and future prospects. Curr. Opin. Biotechnol. 19:202–209.

Herder, G. D., G. Van Isterdael, T. Beeckman, and I. De Smet. 2010. The roots of a new green revolution. Trends Plant Sci. 15:600–607.

Hirel, B., J. Le Gouis, B. Ney, and A. Gallais. 2007. The challenge of improving nitrogen use efficiency in crop plants: towards a more central role for genetic variability and quantitative genetics within integrated approaches. J. Exp. Bot. 58:2369–2387.

Hoffmann-Thoma, G., K. Hinkel, P. Nicolay, and J. Willenbrink. 1996. Sucrose accumulation in sweet sorghum stem internodes in relation to growth. Physiol. Plant. 97:277–284.

Hunter, E., and I. Anderson. 1997. Sweet sorghum. Hortic. Res. 21:73–104.

Inukai, Y., T. Sakamoto, M. Ueguchi-Tanaka, Y. Shibata, K. Gomi, I. Umemura, et al. 2005. Crown rootless1, which is essential for crown root formation in rice, is a target of an AUXIN RESPONSE FACTOR in auxin signaling. Plant Cell Online 17:1387–1396.

Jaggard, K. W., A. Qi, and E. S. Ober. 2010. Possible changes to arable crop yields by 2050. Philos. Trans. R. Soc. B: Biol. Sci. 365:2835–2851.

Jordan, D., R. Casu, P. Besse, B. Carroll, N. Berding, and C. McIntyre. 2004. Markers associated with stalk number and suckering in sugarcane colocate with tillering and rhizomatousness QTLs in sorghum. Genome 47:988–993.

Kariali, E., and P. K. Mohapatra. 2007. Hormonal regulation of tiller dynamics in differentially-tillering rice cultivars. Plant Growth Regul. 53:215–223.

Kebede, H., P. K. Subudhi, D. T. Rosenow, and H. T. Nguyen. 2001. Quantitative trait loci influencing drought tolerance in grain sorghum (Sorghum bicolor L. Moench). Theor. Appl. Genet. 103:266–276.

Kebrom, T. H., B. L. Burson, and S. A. Finlayson. 2006. Phytochrome B represses Teosinte Branched1 expression and induces sorghum axillary bud outgrowth in response to light signals. Plant Physiol. 140:1109–1117.

Klein, R. R., R. Rodriguez-Herrera, J. A. Schlueter, P. E. Klein, Z. H. Yu, and W. L. Rooney. 2001. Identification of genomic regions that affect grain-mould incidence and other traits of agronomic importance in sorghum. Theor. Appl. Genet. 102:307–319.

Kong, W., H. Jin, C. D. Franks, C. Kim, R. Bandopadhyay, M. K. Rana, et al. 2013a. Genetic analysis of recombinant inbred lines for Sorghum bicolor× Sorghum propinquum. G3 3:101–108.

Kong, Y., Y. Zhu, C. Gao, W. She, W. Lin, Y. Chen, et al. 2013b. Tissue-specific expression of SMALL AUXIN UP RNA41 differentially regulates cell expansion and root meristem patterning in Arabidopsis. Plant Cell Physiol. 54:609–621.

Kong, W., H. Guo, V. H. Goff, T.-H. Lee, C. Kim, and A. H. Paterson. 2014. Genetic analysis of vegetative branching in sorghum. Theor. Appl. Genet. 127:2387–2403.

Li, J., and H. Jin. 2007. Regulation of brassinosteroid signaling. Trends Plant Sci. 12:37–41.

Li, R., Y. Han, P. Lv, R. Du, and G. Liu. 2014a. Molecular mapping of the brace root traits in sorghum (Sorghum bicolor L. Moench). Breed. Sci. 64:193–198.

Li, Z., C. Zhao, Y. Zha, C. Wan, S. Si, F. Liu, et al. 2014b. The minor wall-networks between monolignols and interlinked-phenolics predominantly affect biomass enzymatic digestibility in Miscanthus. PLoS One 9:e105115.

Lin, Y.-R., K. F. Schertz, and A. H. Paterson. 1995. Comparative analysis of QTLs affecting plant height and maturity across the Poaceae, in reference to an interspecific sorghum population. Genetics 141:391.

Lin, Q., D. Wang, H. Dong, S. Gu, Z. Cheng, J. Gong, et al. 2012. Rice APC/CTE controls tillering by mediating the degradation of MONOCULM 1. Nat. Commun. 3:752.

Lipinsky, E., and S. Kresovich 1980. Sorghums as energy crops. Proceedings, Bio-Energy'80, World Congress and

Exposition, April 21-24, 1980, Atlanta, Georgia, USA. The Bio-Energy Council, pp. 91–93.

Lu, X.-P., J.-F. Yun, C.-P. Gao, and S. Acharya. 2011. Quantitative trait loci analysis of economically important traits in Sorghum bicolor x S. sudanense hybrid. Can. J. Plant Sci. 91:81–90.

Mace, E., and D. Jordan. 2010. Location of major effect genes in sorghum (Sorghum bicolor (L.) Moench). Theor. Appl. Genet. 121:1339–1356.

Mace, E. S., and D. R. Jordan. 2011. Integrating sorghum whole genome sequence information with a compendium of sorghum QTL studies reveals uneven distribution of QTL and of gene-rich regions with significant implications for crop improvement. Theor. Appl. Genet. 123:169–191.

Mace, E., V. Singh, E. Van Oosterom, G. Hammer, C. Hunt, and D. Jordan. 2012. QTL for nodal root angle in sorghum (Sorghum bicolor L. Moench) co-locate with QTL for traits associated with drought adaptation. Theor. Appl. Genet. 124:97–109.

Manzano, C., E. Ramirez-Parra, I. Casimiro, S. Otero, B. Desvoyes, B. De Rybel, et al. 2012. Auxin and epigenetic regulation of SKP2B, an F-box that represses lateral root formation. Plant Physiol. 160:749–762.

Manzano, C., M. Pallero, I. Casimiro, B. De Rybel, B. Orman-Ligeza, G. Van Isterdael, et al. 2014. The emerging role of ROS signalling during lateral root development. Plant Physiol. 165:1105–1119

Marin, E., V. Jouannet, A. Herz, A. S. Lokerse, D. Weijers, H. Vaucheret, et al. 2010. miR390, Arabidopsis TAS3 tasiRNAs, and their AUXIN RESPONSE FACTOR targets define an autoregulatory network quantitatively regulating lateral root growth. Plant Cell Online 22:1104–1117.

Mason, S. C., D. Kathol, K. M. Eskridge, and T. D. Galusha. 2008. Yield increase has been more rapid for maize than for grain sorghum. Crop Sci. 48:1560–1568.

Mcbee, G. G., R. A. Creelman, and F. R. Miller. 1988. Ethanol yield and energy potential of stems from a spectrum of sorghum biomass types. Biomass 17:203–211.

McCollum, T., K. McCuistion, and B. Bean 2005. Brown midrib and photoperiod sensitive forage sorghums. Proceedings of the 2005 plains nutrition council spring conference.

McHale, M., A. L. Eamens, E. J. Finnegan, and P. M. Waterhouse. 2013. A 22-nt artificial microRNA mediates widespread RNA silencing in Arabidopsis. Plant J. 76:519–529.

Morinaka, Y., T. Sakamoto, Y. Inukai, M. Agetsuma, H. Kitano, M. Ashikari, et al. 2006. Morphological alteration caused by brassinosteroid insensitivity increases the biomass and grain production of rice. Plant Physiol. 141:924–931.

Moubayidin, L., R. Di Mambro, and S. Sabatini. 2009. Cytokinin–auxin crosstalk. Trends Plant Sci. 14:557–562.

Mullet, J., D. Morishige, R. McCormick, S. Truong, J. Hilley, B. McKinley, et al. 2014a. Energy Sorghum—a genetic model for the design of C4 grass bioenergy crops. J. Exp. Bot. doi: 10.1093/jxb/eru229

Mullet, J., D. Morishige, R. McCormick, S. Truong, J. Hilley, B. McKinley, et al. 2014b. Energy Sorghum-a genetic model for the design of C4 grass bioenergy crops. J. Exp. Bot. 65:3479–3489.

Multani, D. S., S. P. Briggs, M. A. Chamberlin, J. J. Blakeslee, A. S. Murphy, and G. S. Johal. 2003. Loss of an MDR transporter in compact stalks of maize br2 and sorghum dw3 mutants. Science 302:81–84.

Murphy, R. L., D. T. Morishige, J. A. Brady, W. L. Rooney, S. Yang, P. E. Klein, et al. 2014. Ghd7 (Ma6) represses sorghum flowering in long days: Ghd7 alleles enhance biomass accumulation and grain production. Plant Gen. 7:10.

Murray, S. C., W. L. Rooney, S. E. Mitchell, A. Sharma, P. E. Klein, J. E. Mullet, et al. 2008a. Genetic improvement of sorghum as a biofuel feedstock: II. QTL for stem and leaf structural carbohydrates. Crop Sci. 48:2180–2193.

Murray, S. C., A. Sharma, W. L. Rooney, P. E. Klein, J. E. Mullet, S. E. Mitchell, et al. 2008b. Genetic improvement of sorghum as a biofuel feedstock: I. QTL for stem sugar and grain nonstructural carbohydrates. Crop Sci. 48:2165–2179.

Narayanan, S., R. M. Aiken, P. Prasad, Z. Xin, G. Paul, and J. Yu. 2014. A simple quantitative model to predict leaf area index in sorghum. Agron. J. 106:219–226.

Natoli, A., C. Gorni, F. Chegdani, P. A. Marsan, C. Colombi, C. Lorenzoni, et al. 2002. Identification of QTLs associated with sweet sorghum quality. Maydica 47:311–322.

Okuno, A., K. Hirano, K. Asano, W. Takase, R. Masuda, Y. Morinaka, et al. 2014. New approach to increasing rice lodging resistance and biomass yield through the use of high gibberellin producing varieties. PLoS One 9:e86870.

O'Lear, S. 2004. Resources and conflict in the Caspian Sea. Geopolitics 9:161–186.

Olson, S. N., K. Ritter, W. Rooney, A. Kemanian, B. A. McCarl, Y. Zhang, et al. 2012. High biomass yield energy sorghum: developing a genetic model for C4 grass bioenergy crops. Biofuels, Bioprod. Biorefin. 6:640–655.

Ordonio, R. L., Y. Ito, A. Hatakeyama, K. Ohmae-Shinohara, S. Kasuga, T. Tokunaga, et al. 2014. Gibberellin deficiency pleiotropically induces culm bending in sorghum: an insight into sorghum semi-dwarf breeding. Sci. Rep. doi:10.1038/srep05287

Parrish, D., and J. Cundiff. 1985. Long-term retention of fermentables during aerobic storage of bulked sweet sorghum. Proceedings of 5th Annual Solar and Biomass Workshop, Atlanta, GA. pp. 137–140.

Pasini, L., M. Bergonti, A. Fracasso, A. Marocco, and S. Amaducci. 2014. Microarray analysis of differentially expressed mRNAs and miRNAs in young leaves of sorghum under dry-down conditions. J. Plant Physiol. 171:537–548.

Paterson, A. H., K. F. Schertz, Y.-R. Lin, S.-C. Liu, and Y.-L. Chang. 1995. The weediness of wild plants: molecular analysis of genes influencing dispersal and persistence of johnsongrass, Sorghum halepense (L.) Pers. Proc. Natl Acad. Sci. 92:6127–6131.

Pellerin, S., and L. Pagès. 1996. Evaluation in field conditions of a three-dimensional architectural model of the maize root system: comparison of simulated and observed horizontal root maps. Plant Soil 178:101–112.

Pereira, M., and M. Lee. 1995. Identification of genomic regions affecting plant height in sorghum and maize. Theor. Appl. Genet. 90:380–388.

Péret, B., B. De Rybel, I. Casimiro, E. Benková, R. Swarup, L. Laplaze, et al. 2009. Arabidopsis lateral root development: an emerging story. Trends Plant Sci. 14:399–408.

Petti, C., A. E. Harman-Ware, M. Tateno, R. Kushwaha, A. Shearer, A. B. Downie, et al. 2013. Sorghum mutant RG displays antithetic leaf shoot lignin accumulation resulting in improved stem saccharification properties. Biotechnol. Biofuels 6:1–16.

Poovaiah, C. R., M. Mazarei, S. R. Decker, G. B. Turner, R. W. Sykes, M. F. Davis, et al. 2014. Transgenic switchgrass (Panicum virgatum L.) biomass is increased by overexpression of switchgrass sucrose synthase (PvSUS1). Biotechnol. J. 10:552–563.

Prakasham, R. S., D. Nagaiah, K. S. Vinutha, A. Uma, T. Chiranjeevi, A. V. Umakanth, et al. 2014. Sorghum biomass: a novel renewable carbon source for industrial bioproducts. Biofuels 5:159–174.

Qazi, H. A., S. Paranjpe, and S. Bhargava. 2012. Stem sugar accumulation in sweet sorghum–activity and expression of sucrose metabolizing enzymes and sucrose transporters. J. Plant Physiol. 169:605–613.

Quinby, J. R. 1974. Sorghum improvement and the genetics of growth.

Rami, J. F., P. Dufour, G. Trouche, G. Fliedel, C. Mestres, F. Davrieux, et al. 1998. Quantitative trait loci for grain quality, productivity, morphological and agronomical traits in sorghum (Sorghum bicolor L. Moench). Theor. Appl. Genet. 97:605–616.

Rao, S., S. Rao, N. Seetharama, A. Umakath, P. S. Reddy, B. Reddy, et al. 2009. Sweet sorghum for biofuel and strategies for its improvement: International Crops Research Institute for the Semi-Arid Tropics.

Reddy, R. N., R. Madhusudhana, S. M. Mohan, D. V. N. Chakravarthi, S. P. Mehtre, N. Seetharama, et al. 2013. Mapping QTL for grain yield and other agronomic traits in post-rainy sorghum Sorghum bicolor (L.) Moench. Theor. Appl. Genet. 126:1921–1939.

Renny-Byfield, S., and J. F. Wendel. 2014. Doubling down on genomes: polyploidy and crop plants. Am. J. Bot. 101:1711–1725.

Ritter, K. B., D. R. Jordan, S. C. Chapman, I. D. Godwin, E. S. Mace, and C. L. McIntyre. 2008. Identification of QTL for sugar-related traits in a sweet x grain sorghum (Sorghum bicolor L. Moench) recombinant inbred population. Mol. Breeding 22:367–384.

Robinson, D., A. Hodge, B. S. Griffiths, and A. H. Fitter. 1999. Plant root proliferation in nitrogen–rich patches confers competitive advantage. Proc. R. Soc. Lond. B Biol. Sci. 266:431–435.

Rooney, W. 2004. Sorghum improvement—integrating traditional and new technology to produce improved genotypes. Adv. Agron. 83:37–109.

Rooney, W. L., J. Blumenthal, B. Bean, and J. E. Mullet. 2007. Designing sorghum as a dedicated bioenergy feedstock. Biofuels, Bioprod. Biorefin. 1:147–157.

Rosenow, D., and L. Clark. 1995. In Proceedings of the 50th annual corn and sorghum industry research conference. Chicago, IL. Drought and lodging resistance for a quality sorghum crop. pp. 16.

Sabadin, P. K., M. Malosetti, M. P. Boer, F. D. Tardin, F. G. Santos, C. T. Guimaraes, et al. 2012. Studying the genetic basis of drought tolerance in sorghum by managed stress trials and adjustments for phenological and plant height differences. Theor. Appl. Genet. 124:1389–1402.

Saballos, A.. 2008. Development and utilization of sorghum as a bioenergy crop. Genetic Improvement of Bioenergy Crops. Springer, 211–248.

Sahoo, D. K., and I. B. Maiti. 2014. Biomass derived from transgenic tobacco expressing the Arabidopsis CESA3 ixr1-2 gene exhibits improved saccharification. Acta Biol. Hung. 65:189–204.

Sakamoto, T., Y. Morinaka, T. Ohnishi, H. Sunohara, S. Fujioka, M. Ueguchi-Tanaka, et al. 2005. Erect leaves caused by brassinosteroid deficiency increase biomass production and grain yield in rice. Nat. Biotechnol. 24:105–109.

Salih, A., I. Ali, A. Lux, M. Luxova, Y. Cohen, Y. Sugimoto, et al. 1999. Rooting, water uptake, and xylem structure adaptation to drought of two sorghum cultivars. Crop Sci. 39:168–173.

Sang, D., D. Chen, G. Liu, Y. Liang, L. Huang, X. Meng, et al. 2014. Strigolactones regulate rice tiller angle by attenuating shoot gravitropism through inhibiting auxin biosynthesis. Proc. Natl Acad. Sci. 111:11199–11204.

Saracutu, O., G. Cnops, I. Roldán-Ruiz, and A. Rohde. 2010. Phenotypic assessment of variability in tillering and early development in ryegrass (Lolium spp.). Sustainable use of genetic diversity in forage and turf breeding. Springer, 155–160.

Schluhuber, A. 1945. INHERITANCE OF STEM CHARACTERS in certain sorghum varietes and their hybrids. J. Hered. 36:219–222.

Schmer, M. R., K. P. Vogel, G. E. Varvel, R. F. Follett, R. B. Mitchell, and V. L. Jin. 2014. Energy potential and greenhouse gas emissions from bioenergy cropping systems on marginally productive cropland. PLoS One 9:e89501.

Serrão, M., M. Menino, J. Martins, N. Castanheira, M. Lourenço, I. Januário, et al. (2012) Mineral leaf composition of sweet sorghum in relation to biomass and sugar yields under different nitrogen and salinity conditions. Commun. Soil Sci. Plant Anal. 43, 2376–2388.

Shiringani, A. L., Frisch, M., Friedt W. 2010. Genetic mapping of QTLs for sugar-related traits in a RIL population of Sorghum bicolor L. Moench. Theor. Appl. Genet. 121:323–336.

Shiringani, A. L., and W. Friedt. 2011. QTL for fibre-related traits in grain x sweet sorghum as a tool for the enhancement of sorghum as a biomass crop. Theor. Appl. Genet. 123:999–1011.

Shiringani, A. L., M. Frisch, and W. Friedt. 2010. Genetic mapping of QTLs for sugar-related traits in a RIL population of Sorghum bicolor L. Moench. Theor. Appl. Genet. 121:323–336.

Sipos, B., J. Reczey, Z. Somorai, Z. Kadar, D. Dienes, and K. Reczey. 2009. Sweet sorghum as feedstock for ethanol production: enzymatic hydrolysis of steam-pretreated bagasse. Appl. Biochem. Biotechnol. 153:151–162.

Smith, C. W., and R. A. Frederiksen. 2000. Sorghum: origin, history, technology, and production. John Wiley & Sons, Inc., New York, USA.

Srinivas, G., K. Satish, R. Madhusudhana, R. N. Reddy, S. M. Mohan, and N. Seetharama. 2009. Identification of quantitative trait loci for agronomically important traits and their association with genic-microsatellite markers in sorghum. Theor. Appl. Genet. 118:1439–1454.

Steduto, P., N. Katerji, H. Puertos-Molina, M. Mastrorilli, and G. Rana. 1997. Water-use efficiency of sweet sorghum under water stress conditions Gas-exchange investigations at leaf and canopy scales. Field. Crop. Res. 54:221–234.

Takai, T., J.-I. Yonemaru, H. Kaidai, and S. Kasuga. 2012. Quantitative trait locus analysis for days-to-heading and morphological traits in an RIL population derived from an extremely late flowering F-1 hybrid of sorghum. Euphytica 187:411–420.

Taramino, G., M. Sauer, J. L. Stauffer, D. Multani, X. Niu, H. Sakai, et al. 2007. The maize (Zea mays L.) RTCS gene encodes a LOB domain protein that is a key regulator of embryonic seminal and post-embryonic shoot-borne root initiation. Plant J. 50:649–659.

Tian, F., P. J. Bradbury, P. J. Brown, H. Hung, Q. Sun, S. Flint-Garcia, et al. 2011. Genome-wide association study of leaf architecture in the maize nested association mapping population. Nat. Genet. 43:159–162.

Trabucco, G. M., D. A. Matos, S. J. Lee, A. J. Saathoff, H. D. Priest, T. C. Mockler, et al. 2013. Functional characterization of cinnamyl alcohol dehydrogenase and

caffeic acid O-methyltransferase in Brachypodium distachyon. BMC Biotechnol. 13:61.

Tsialtas, J., and N. Maslaris. 2008. Leaf area prediction model for sugar beet (Beta vulgaris L.) cultivars. Photosynthetica 46:291–293.

Tsukaya, H. 2002. Interpretation of mutants in leaf morphology: genetic evidence for a compensatory system in leaf morphogenesis that provides a new link between cell and organismal theories. Int. Rev. Cytol. 217:1–39.

Upadhyaya, H. D., Y.-H. Wang, C. Gowda, and S. Sharma. 2013. Association mapping of maturity and plant height using SNP markers with the sorghum mini core collection. Theor. Appl. Genet. 126:2003–2015.

Vadez, V., J. Kashiwagi, L. Krishnamurthy, R. Serraj, K. Sharma, J. Devi, et al. 2005. Pp. 24–28. Recent advances in drought research at ICRISAT: using root traits and rd29a: DREB1A to increase water use and water use efficiency in drought-prone areas. Poster presented at the Interdrought II conference. Italy, Rome.

Van Acker, R., J.-C. Leplé, D. Aerts, V. Storme, G. Goeminne, B. Ivens, et al. 2014. Improved saccharification and ethanol yield from field-grown transgenic poplar deficient in cinnamoyl-CoA reductase. Proc. Natl Acad. Sci. 111:845–850.

Vandenbrink, J. P., R. N. Hilten, K. Das, A. H. Paterson, and F. Alex Feltus. 2013. Quantitative models of hydrolysis conversion efficiency and biomass crystallinity index for plant breeding. Plant Breeding 132:252–258.

Vermerris, W. 2011. Survey of genomics approaches to improve bioenergy traits in maize, sorghum and sugarcanefree access. J. Integr. Plant Biol. 53:105–119.

Vermerris, W., and A. Saballos. 2013. Genetic enhancement of sorghum for biomass utilization. Genomics of the Saccharinae. Springer, 391–425.

Vermerris, W., A. Saballos, G. Ejeta, N. S. Mosier, M. R. Ladisch, and N. C. Carpita. 2007. Molecular breeding to enhance ethanol production from corn and sorghum stover. Crop Sci. 47:S-142–S-153.

Wang, F., and C.-Z. Liu. 2009. Development of an economic refining strategy of sweet sorghum in the inner Mongolia region of China. Energy Fuels 23:4137–4142.

Wang, M. L., C. Zhu, N. A. Barkley, Z. Chen, J. E. Erpelding, S. C. Murray, et al. 2009. Genetic diversity and population structure analysis of accessions in the US historic sweet sorghum collection. Theor. Appl. Genet. 120:13–23.

Wang, Y.-H., A. Acharya, A. M. Burrell, R. R. Klein, P. E. Klein, K. H. Hasenstein, et al. 2013. Mapping and candidate genes associated with saccharification yield in sorghum. Genome 56:659–665.

Wang, C.-Q., C. Guthrie, M. K. Sarmast, and K. Dehesh. 2014. BBX19 interacts with CONSTANS to repress FLOWERING LOCUS T transcription, defining a flowering time checkpoint in Arabidopsis. Plant Cell 26:3589–602.

Wasson, A., R. Richards, R. Chatrath, S. Misra, S. S. Prasad, G. Rebetzke, et al. 2012. Traits and selection strategies to improve root systems and water uptake in water-limited wheat crops. J. Exp. Bot. 63:3485–3498.

Wilkinson, P., K. R. Smith, M. Joffe, and A. Haines. 2007. A global perspective on energy: health effects and injustices. Lancet 370:965–978.

Xin, Z., and M. L. Wang. 2011. Sorghum as a versatile feedstock for bioenergy production. Biofuels 2:577–588.

Xin, Z., M. L. Wang, G. Burow, and J. Burke. 2009. An induced sorghum mutant population suitable for bioenergy research. Bioenergy Res. 2:10–16.

Xu, F., Y.-C. Shi, X. Wu, K. Theerarattananoon, S. Staggenborg, and D. Wang. 2011. Sulfuric acid pretreatment and enzymatic hydrolysis of photoperiod sensitive sorghum for ethanol production. Bioprocess Biosyst. Eng. 34:485–492.

Yamaguchi, S. 2008. Gibberellin metabolism and its regulation. Annu. Rev. Plant Biol. 59:225–251.

Yang, S., B. D. Weers, D. T. Morishige, and J. E. Mullet. 2014. CONSTANS is a photoperiod regulated activator of flowering in sorghum. BMC Plant Biol. 14:148.

Zhang, B., and Q. Wang. 2015. MicroRNA-based biotechnology for plant improvement. J. Cell. Physiol. 230:1–15.

Zhao, Y., Y. Hu, M. Dai, L. Huang, and D.-X. Zhou. 2009. The WUSCHEL-related homeobox gene WOX11 is required to activate shoot-borne crown root development in rice. Plant Cell Online 21:736–748.

Zhao, H., Z. Huai, Y. Xiao, X. Wang, J. Yu, G. Ding, et al. 2014. Natural variation and genetic analysis of the tiller angle gene MsTAC1 in Miscanthus sinensis. Planta 240:161–75.

Zou, G., S. Yan, G. Zhai, Z. Zhang, J. Zou, and Y. Tao. 2011. Genetic variability and correlation of stalk yield-related traits and sugar concentration of stalk juice in a sweet sorghum (Sorghum bicolor L. Moench) population. Aust. J. Crop Sci. 5:1232–1238.

Zou, G., G. Zhai, Q. Feng, S. Yan, A. Wang, Q. Zhao, et al. 2012. Identification of QTLs for eight agronomically important traits using an ultra-high-density map based on SNPs generated from high-throughput sequencing in sorghum under contrasting photoperiods. J. Exp. Bot. 63:5451–5462.

Impact of biofuel development on Malaysian agriculture: a comparative statics, multicommodity, multistage production and partial equilibrium approach

Yaghoob Jafari[1] & Jamal Othman[2]

[1]Institute for Food and Resource Economics, University of Bonn, Bonn, Germany
[2]National University of Malaysia, Bangi, Selangor, Malaysia

Keywords
Consumer preferences, Malaysian agriculture, multicommodity partial equilibrium model, palm oil-based biofuel

Correspondence
Yaghoob Jafari, Institute for Food and Resource Economics, University of Bonn, Nussallee 21 53115 Bonn, Germany.

E-mail: yaghoob.jafari@ilr.uni-bonn.de

Funding Information
No funding information provided.

Abstract

In light of the move toward a low carbon economy, Malaysia is currently promoting the production of palm oil-based biodiesel. This study develops a multicommodity, multistage, comparative static, partial equilibrium model for the Malaysian agricultural sector to be used to examine the impact of shifts in domestic demand for palm oil-based biofuel on the Malaysian agricultural sector. The model links explicitly factor markets, related outputs, and consumer preferences for biofuels. Model results demonstrate a clear converse relationship between the variables representing the competing activities within the same sector. An increase in domestic biofuel demand leads to an increase in biofuel output and primary factor uses while the outputs of competing uses, especially food and other non-food products decline markedly.

Introduction

The biofuel industry in Malaysia is synonymous with the oil palm crop, the country's most important agro-industrial commodity. The country considers biofuel as a new growth sector in light of its contribution toward improving energy security and increased use of renewable and low-carbon source of energy.

The issue on whether biodiesel actually contributes to a net reduction in carbon emission is rather contentious. Palm biodiesels are a low-carbon energy source if increases in demand are not associated with oil palm expansion via conversion of peatlands or unsustainable farming system which may release more CO_2 than the reductions than these biofuels may provide by displacing fossil fuels. It has been argued that Malaysian palm oil is produced in a sustainable manner and therefore Malaysian palm biodiesel is deemed a sustainable energy (MPOB 2016).

Malaysia is currently promoting the production of palm biodiesel to capitalize on the growing demand for biofuels.

The country is set to implement nationwide the B5 Program which calls for the blending of 5% biodiesel with 95% petroleum diesel. Malaysian biodiesel players are expecting to see increases in domestic biodiesel blend, from B5 to B10 (blending of 10% biodiesel with 90% petroleum diesel) and eventually to B20. The government is also expected to provide a range of economic incentives to enhance the downstream palm biodiesel industries (Hanim 2012). The mandatory blending targets along with biofuel consumption subsidies and other production incentives may have positive impact on demand for biofuels which is unlikely to exist otherwise.

Given that Malaysia's biofuel industry is synonymous with oil palm, there are concerns that there will be increased competition in the use of the same palm oil input for the manufacture of traditional products, namely food and inedible use products in various industries. Therefore, the implementation of policy measures such as biofuel mandates may induce a shift in consumer demand and have considerable impacts on related markets including food and non-food products, which utilize the same processed palm oil (PPO) intermediate input as biofuel. Consequently, this may lead to changes in agricultural outputs in related subsectors in the long run, as the competing agricultural subsectors in Malaysia are firmly linked through resource constraints, especially land and labor resources.

The paper develops a comparative statics, partial equilibrium model to be used to examine the inter subsectoral effects of a shift in domestic demand for palm biodiesel on related factor markets, output, and trade within the agricultural sector in Malaysia.

Sections Literature Review and Conceptual Framework present the development of the model, data requirements, and operational issues. Section Solution Method and Database presents the empirical application of the model and its findings. The final section discusses the policy implication and concluding remarks.

Literature Review

Models of agricultural policy analysis

Extensive bodies of literature exist on the appraisals of agricultural policies. The most common approaches are econometric and market equilibrium models including partial equilibrium (PE) and computable general equilibrium (CGE) models.

The principal characteristic of econometric models is the use of historical or cross-sectional data to calculate the underlying model's parameters through a variety of estimation methods. Pollitt et al. (2007) argues that econometric models are often very resource-intensive and for this reason alone, their use tends to be somewhat limited.

Another criticism of econometric models is that they are subject to the Lucas Critique. This states that it is a simplicity to predict the effect of a policy experiment based on the relationships estimated from historical data. This is especially crucial for highly disaggregated bottom-up models, where agents' behavior in a particular sector could change considerably in a relatively short period of time.

Market equilibrium models, on the other hand, often require just a single year data for model calibration. They are also not generally subject to the Lucas Critique as the outcome tends to be shaped by the underlying microeconomic behavioral foundations.

The CGE models have the unique advantage to model all sectoral linkages within the economy either based on a static or dynamic framework. In contrast, partial equilibrium models generally consider a single sector in isolation of all other sectors. They typically describe the various agricultural subsectors with detailed treatment of supply and demand, including policy-price linkages, interdependency of inputs and outputs between different subsectoral outputs, etc. The idea behind the models is the neoclassical approach in which the supply and demand attain their equilibrium while producers and consumers, respectively, maximize profits and utility. Owing to the emphasis on a single sector, they can provide a focused and tractable analysis of how a limited number of variables are affected by the imposition of any policy changes or restrictions.

Conceptual Framework

This section illustrates the development of the comparative static, partial equilibrium model for multiple subsectoral

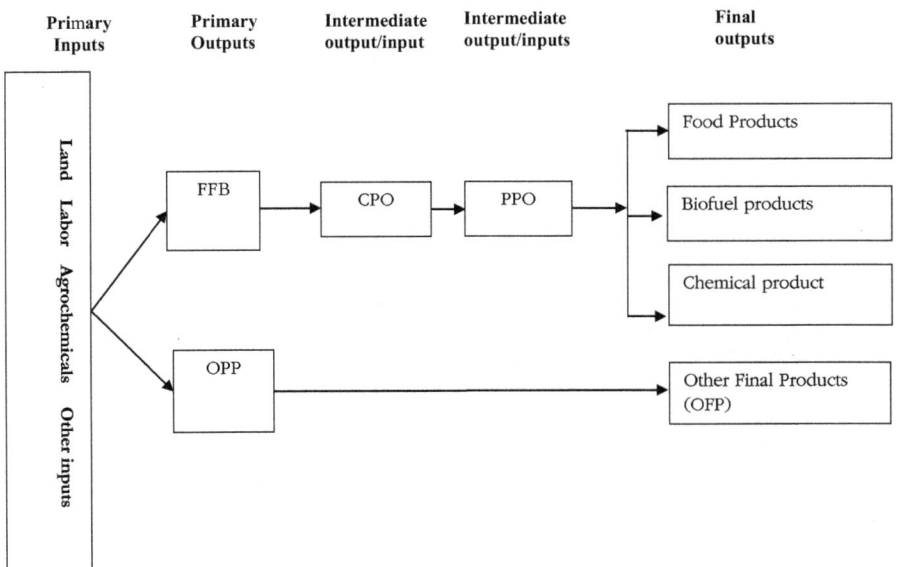

Figure 1. Schematic model of the partial equilibrium, multicommodity model for the Malaysian agricultural sector.

outputs with two subsequent stages of production. Figure 1 depicts the conceptual framework.

Representation of the Malaysian agricultural sector

The Malaysian agricultural sector is represented by two competing subsectors. The first subsector represents the oil palm subsector which is Malaysia's most important agricultural subsector. The other is an aggregate of all other subsectors that compete for the pre-existing resources including land, labor, agrochemicals, and other inputs. As noted in the figure, the primary outputs of the oil palm subsector and other subsectors are Fresh Fruit Bunches (FFB), and Other Primary Products (OPP), respectively. The first level intermediate output in the oil palm subsector (CPO) can be further processed into Processed Palm Oil (PPO), which itself is used to produce palm oil-based food, palm oil-based chemical, and biodiesel products. The OPP is intended to produce Other Final Products in aggregate (OFP). Both intermediate products and final products are tradable in the world marketplace. Since both outputs utilize the same input base, any policy shocks or exogenous changes affecting either subsector will have repercussions in all related markets – primary inputs, primary and final outputs as well as trade.

Mathematical framework of the partial equilibrium model

This subsection presents the general form of the model while the more technical, differentiated form which is actually used in the simulation exercise is shown in Appendix A.

The model is specified for a long-run partial equilibrium scenario under the assumption of perfect competition in the market. The following discusses the different modules of the model.

Commodity market demand module

There are four commodities at the final stage of production, food, biofuel, chemicals, OFP, which are produced by the firms in domestic market and are consumed by the consumers in both domestic and foreign markets. It is assumed that consumers have homothetic preferences, implying that the budget share that the consumers allocate to the commodities is independent of their total expenditure. Provided that market demand is the summation of domestic and export demand, the demand function for commodity y is:

$$Q_y^M = Q_y^D \left(P_y^D, \ldots, P_x^D \right) + Q_y^E (P_y^E, \ldots, P_x^E) \qquad (1)$$

Here, notations Q and P refer to demand quantity and price, respectively. Superscripts D and E denote domestic

and export markets while M refers to the market demand. Subscripts y and x reflect final commodities, and $y, x \in$ {Food, Biofuel, Chemical, and OFP}. Thus, a variable such as Q_y^M will be defined as quantity (Q) of market demand (M) for commodity y. By total differentiation of Equation 1 and manipulating it to obtain the elasticities and market shares, and allowing for shifts in consumer preferences, the market demand function for commodity y is defined as Equation A1 in Appendix A. Interested readers are advised to refer to Hertel (1989), Jamal Othman (2003) and Jafari and Jamal (2015) for detailed mathematical procedure in driving the differentiated form.

Demand for intermediate inputs (PPO and CPO)

The model presumes that PPO is being exported from Malaysia on a net basis, hence the market demand for PPO (Q_{ppo}^M) is the sum of domestic demand (Q_{ppo}^D) and export demand (Q_{ppo}^E). Based on a linear demand function, the market demand function is written;

$$Q_{ppo}^M = Q_{ppo}^D + Q_{ppo}^E \qquad (2)$$

This study disregards the role of primary factors as an input in the production process of final output from intermediate inputs. Conditional demand for inputs can be derived under assumptions of cost minimization and competitive markets. The general specification of cost function begins with the general form:

$$\text{Cost} = C(\text{input prices, producible output}) \qquad (3)$$

where $C(')$ refers to cost function. Assuming that cost function is mathematically well- behaved and twice differentiable, one can derive the conditional factor demand functions using Shephard lemma and the Envelope theorem. Domestic demand for PPO is derived from the demand for their associated final output. Accordingly, the derived demand for PPO to be used in production of y ($Q_{ppo,y}^D$) is defined:

$$Q_{ppo,y}^D = f\left(P_{ppo,y}^D, Q_y^S \right) \text{ for } y \in \{\text{Food, Biofuel, Chemicals}\} \qquad (4)$$

Here, $f(')$ represents the derived demand function. $P_{ppo,y}^D$ refers to the price of domestic demand for PPO to be used in production of y while Q_y^S is the quantity of supply for output y. Note that superscript S refers to domestic supply market.

Having defined the domestic demand for PPO as:

$$Q_{ppo}^D = \sum_y Q_{ppo,y}^D \text{ for } y \in \{\text{Food, Biofuel, Chemicals}\}; \qquad (5)$$

and export demand for PPO as:

$$Q_{ppo}^E = e\left(P_{ppo}^E \right) \qquad (6)$$

where $e(')$ is the export demand function, the market demand function for PPO is specified;

$$Q_{ppo}^M = \sum_y f\left(P_{ppo,y}^D, Q_y^S\right) + e(P_{ppo}^E) \qquad (7)$$

The first right hand side component of Equation 7 is especially formulated to capture the competition of food, biofuel, and chemical products.

In addition, CPO is used an input for PPO, and the model presume that CPO is being exported on a net basis. Accordingly, the derived demand for CPO to be used in production of PPO (Q_{cpo}^D) is defined:

$$Q_{cpo}^D = f\left(P_{cpo}^D, Q_{ppo}^S\right) \qquad (8)$$

Here, P_{cpo}^D refers to the price of domestic demand for CPO to be used in production of PPO while Q_{PPO}^S is the quantity of supply for PPO.

Having defined the export demand for CPO as:

$$Q_{cpo}^E = e\left(P_{cpo}^E\right) \qquad (9)$$

the market demand function for CPO is specified;

$$Q_{cpo}^M = f\left(P_{cpo}^D, Q_{ppo}^S\right) + e(P_{cpo}^E) \qquad (10)$$

Algebraic manipulation of equations 5 and 10 and expressed them in percentage change, and assuming that firms operate under locally constant return to scale condition, the market demand function for PPO and CPO is written as Equation A2 and A3 in Appendix A. It should be noted that in order to obtain the export and domestic demand quantities of PPO, Equations 4 through 6 and to achieve the export and domestic demand quantities of CPO, Equations 8 and 9 are turned into the differentiated form, separately. However, in order to save the space, these equations are not shown in the appendix.

Derived demand for primary outputs (FFBs and OPPs) and primary factors

Using similar concept as above, the demand for FFB to be used in production of CPO (Q_{ffb}^D) and demand for OPP to be used in production of OFP (Q_{opp}^D) is expressed as;

$$Q_{ffb}^D = f\left(P_{ffb}^D, Q_{cpo}^S\right) \qquad (11)$$

$$Q_{opp}^D = f\left(P_{opp}^D, Q_{ofp}^S\right) \qquad (12)$$

where P_{ffb}^D and P_{opp}^D are price of FFB and OPP to be used in production of CPO and OFPs, respectively. Further, Q_{cpo}^S and Q_{ofp}^S signify the domestic supply of CPO and OFPs.

Total differential of the above equations and assuming that firms exhibit constant returns to scale technology results in Equations A4 and A5.

Similarly, the derived demand function for primary inputs to be used in the production of primary outputs ($k \in \{FFB, OPP\}$) is defined as:

$$Q_{i,k}^D = D_{i,k}\left(P_{i,k}^D, \dots, P_{j,k}^D, Q_k^S\right) \forall$$
$$\text{for } k \in \{FFB, OPP\}, \text{ and } i = 1, \dots, j \qquad (13)$$

Here, $Q_{i,k}^D$ and $P_{i,k}^D$ are quantity and price of ith input in production of kth primary output, respectively; while S_k^D denotes the supply of primary output.

Again, total differentiation of the above equations and given constant returns to scale technology, we obtain Equation A6.

Zero profit conditions

Under the assumption of perfect competition and constant returns to scale, firm's profit in the long run is equal to zero. To ensure this condition holds, unit cost function for each output must be equal to its respective prices. This is shown below;

$$P_{kk}^S = C_k\left(P_{i,k}^D, \dots, P_{j,k}^D\right)$$
$$\text{for } k, kk \in \{FFB, OPP\}, \text{ and } i = 1, \dots, j \qquad (14)$$

$$P_{cpo}^S = C_{cpo}\left(P_{ffb}^D\right) \qquad (15)$$

$$P_y^S = C_y\left(P_{cpo,y}^D\right) \text{ for } y \in \{Food, Biofuel, Chemicals\} \qquad (16)$$

$$P_{ofp}^S = C_{ofp}\left(P_{opp}^D\right) \qquad (17)$$

Here, C_k, $C_{k,cpo}$, C_y, and C_{ofp} signify the associated unit cost production of commodity k, CPO, y, and OFP, respectively. The left hand side variables denote the supply price of each output while the right hand side variables show the demand price for each input.

Total differential of the above equations and employing the *Young theorem*, the zero profit condition for production of each commodity is defined as in Equations A7 through A10.

Factor supply equations

The supply of ith input for use in production of FFB and OPP ($Q_{i,ffb}$, $Q_{i,opp}$) is function of the supply price of the input in each subsector ($P_{i,ffb}^S$, $P_{i,opp}^S$). Therefore, the supply function of ith input for producing FFB and OPP is written as:

$$Q_{i,k}^S = g_k(P_{i,k}^S, P_{i,kk}^S) \text{ for } k, kk \in \{FFB, OPP\}, \text{and } i = 1, \dots, j \quad (18)$$

Here, g_k refers to the primary supply function of FFB and OPP activities, respectively.

Taking the total differential of Equation 18, and using some algebraic manipulation to obtain elasticities yield Equation A11.

Output and input clearing condition

This condition requires that prices must adjust to equate supply with demand for each commodity and input, thus there will be no surpluses or deficits in inventory of outputs and inputs. The output and input clearing conditions in differentiated form are presented in Equations A12 through A15.

Solution method and database

Mathematically, all the equations in Appendix A form a linear system that can be solved given the nonsingularity of coefficients matrix condition. The necessary and sufficient condition for nonsingularity is that the matrix shall satisfy the squareness and linear independence equations. A convenient way of solving a linear equation system is by using the well-known Cramer's rule. The system of equations in the model can be written in a matrix form, so that the general system of algebraic equations can be represented as follows;

$$AX = C$$

Here, A is the Jacobean matrix (coefficient of the endogenous variables of the model), X represents the matrix of endogenous variables (prices and quantities) while the right hand side matrix denotes the exogenous variables (policy shocks). Thereafter, we can apply Cramer's rule to solve for the endogenous variables.

Before any simulation is performed, the baseline parameters or coefficients for the endogenous variables must be established. Likewise, many partial and general equilibrium models which contain a large number of parameters, the required parameters are either calibrated, guesstimated, or assumed (See Salhofer 2000). Most of the parameters used in this study are based on Jafari and Jamal (2015). The rest are obtained from various sources or determined based on guided assumptions (Tables 1–5 .

The domestic demand elasticity for PPO products to be used in production of food, chemical, and biofuel products (−0.43, −0.43 and −0.13, respectively) is sourced from the online FAPRI elasticities database. Elasticity of

export demand for PPO product is assigned a value of −0.457 based on Basri and Darawi (2002). The demand elasticity values for food and biofuel products (−0.38 and −0.1, respectively) are obtained from FAPRI elasticities database, while chemical demand elasticity is taken from Beckman and Hertel (2008).

Table 1. Distribution share of primary inputs in different subsectors.

Primary inputs	Oil Palm	Other crops in aggregate
Land	0.758	0.242
Labor	0.8878	0.1122
Capital	0.83	0.17
Agrochemicals	0.841	0.159

Source: Obtained from Jafari and Jamal (2015)

Table 2. Allen elasticities of substitution between primary inputs in oil palm plantation.

	Land	Labor	Agrochemicals	Capital
Land	−0.3	0.078	−0.042	0.645
Labor	0.078	−0.79	0.492	0.895
Agrochemicals	−0.042	0.492	−1.007	0.378
Capital	0.645	0.895	0.378	−4.147
Factor Cost Share	0.36	0.31	0.19	0.14

Source: Jafari and Jamal (2015).

Table 3. Allen elasticities of substitution between primary inputs for other crops in aggregate.

	Land	Labor	Capital	Agrochemicals
Land	−4.2	0.3	0.1	2.7
Labor	0.3	−7.35	0.4	1.3
Capital	0.1	0.4	−2.27	0.6
Agrochemicals	2.7	1.3	0.6	−1.322
Factor Cost Share	0.3	0.1	0.15	0.45

Source: Jafari and Jamal (2015)

Table 4. Input supply elasticities.

Parameters	Values
Land input supply elasticity	0.6
Non-land input supply elasticities	1

Source: Jafari and Jamal (2015)

Table 5. Domestic and export demand shares and elasticities.

Parameters	Values
CPO export elasticity	−0.39
CPO-derived demand elasticity	−0.43
OFP-derived demand elasticity	−0.19
OFP export demand elasticity	−0.19

Source: Jafari and Jamal (2015)

The value of OFP own price elasticities (−0.19) is taken from GTAP database (Betina et al. 2006). The aggregated elasticity is assumed to be normally distributed between foreign and domestic market, hence the value of −0.19 is assigned to own export and domestic demand elasticities. Table 6 depicts the domestic demand share of intermediate and final products. Note that the entire FFBs and OPPs are utilized in the domestic market.

Given the baseline data, the model is validated by its ability to replicate the initial value of endogenous variables upon implementation of the baseline policies.

It is important to note here that due to the unavailability of reliable econometric estimates, most of the coefficients for the endogenous variables are obtained or determined based on guided assumptions. Nevertheless, provided that meaningful sign and magnitudes are accorded to the various parameters, appraisals of the direction and relative order of impacts which remains the focus of our analysis would still be meaningful and useful. More indicative insights and implications of the results shall be derived by conducting sensitivity analyses to examine the effects of varying levels of exogenous parameters and baseline coefficients of the endogenous variables in the model.

Policy simulation – A 10% shift in domestic demand for biodiesel

The constructed model is employed to appraise the effects of shifts in domestic demand schedule due to some exogenous factors (such as changes in consumer preference due to increasing concerns on environmental issues). Since the parameters of the model are assumed to be constant, the differentiated version model (see Appendix A) can be used to appraise the impact of small policy perturbations (Hertel 1989). This paper focuses on simulating a 10% increase in domestic demand for biodiesel in Malaysia. This is modeled by shifting upwards the relevant inverse demand schedule for biodiesel products by 10% (Equation A1). The effects of the policy change on the endogenous variables are listed in Table 7. It shall be recalled here that the focus of this type of appraisals is on the direction and relative order of magnitudes/impacts. Given the uncertain nature of the various baseline values, examination of fine-tune numbers will be immaterial.

Simulation results and discussion

The findings generally show an inverse relation of long-run impacts among the endogenous variables representing each subsector. The results for the baseline scenario where the elasticity of domestic demand for biodiesel is −0.13 is shown in the third column of Table 7.

As indicated, an upward shift in domestic demand for biodiesel by 10% increases the equilibrium domestic demand for the commodity by 0.99%. Price of biodiesel increases very marginally by 0.0002% while export demand may remain unchanged. Market demand for biodiesel may see a 0.27% rise. Consequently, the derived demand (PPO and CPO) for biodiesel use will also increase. This in turn reduces the outputs of competing products, i.e., food and biochemicals as they are now faced with relatively higher prices of inputs.

The projected decline in export demand for PPO (−0.002%) is substantially smaller relative to the increase in PPO domestic demand (0.0104%). Despite the smaller domestic demand share, the market demand for PPO increases, resulting in an increase in demand for CPO and FFBs, which are the intermediate inputs for PPO production. The increase in demand for FFBs provokes a rise in the demand and market prices for primary factors of production in the oil palm subsector through the derived factor demand functions.

Higher returns for primary factors of production give rise to two direct impacts: First, changes in the price of primary factor inputs lead to changes in input mix in each subsector; these linkages are provided through the Allen elasticities of substitution. Second, factor owners will reallocate their inputs to the subsector that yield higher returns; the reallocation is captured through input transformation elasticities which are used in the calculation of cross-supply elasticities. Note that the model assumes that changes in land supply are a result of changes in relative input prices within the two subsectors that are

Table 6. Domestic and export demand share of intermediate and final products.

Outputs (intermediate and final)	Share	Sources
CPO domestic demand share	0.878	Ministry of Plantation Industries and Commodities, Malaysia (2010b)
OFP domestic demand share	0.867	Ministry of Plantation Industries and Commodities, Malaysia (2010a)
PPO-based food products domestic demand share	0.576	MPOB(2008a,b)
PPO-based chemical products domestic demand share	0.271	MPOB (2008a)
Biodiesel domestic demand share	0.268	MPOB (2008a)
PPO export share	0.52	MPOB(2008a,b)
PPO domestic share	0.48	MPOB(2008a,b)

Table 7. Effect of a shift in biofuel domestic demand and varying domestic elasticities and shares on the variables of interest.

Notation	Definition of variables	Percent changes (Ed = 0.1)	Percent changes (Ed = 0.7)	Percent changes (Ed = 1)	Percent changes (Ed = 1.2)	Percent changes (Ed = 1.2, Biofuel share = 0.2)
Primary output – Fresh Fruit Bunches (FFB)						
Q^M_{FFB}	Market Demand for FFB	0.0020	0.0143	0.0205	0.0246	0.1014
P^M_{FFB}	Market Price of FFBs	0.0044	0.0308	0.0440	0.0528	0.2171
$Q_{land,FFB}$	Demand for Land in Production of FFB	0.0019	0.0131	0.0187	0.0224	0.0924
$Q_{che,FFB}$	Demand for Agro chemical in Production of FFB	0.0023	0.0163	0.0233	0.0279	0.1149
$Q_{lab,FFB}$	Demand for Labor in Production of FFB	0.0021	0.0148	0.0211	0.0253	0.1043
$Q_{oth,FFB}$	Demand for Other inputs in Production of FFB	0.0020	0.0141	0.0201	0.0242	0.0995
$P^M_{land,FFB}$	Market Price of Land in Production of FFBs	0.0061	0.0429	0.0613	0.0736	0.3028
$P^M_{che,FFB}$	Market Price of Agro chemical in Production of FFBs	0.0022	0.0157	0.0224	0.0269	0.1106
$P^M_{lab,FFB}$	Market Price of Labor in Production of FFBs	0.0036	0.0255	0.0364	0.0437	0.1798
$P^M_{oth,FFB}$	Market Price of Other Inputs in Production of FFBs	0.0045	0.0317	0.0454	0.0545	0.2240
Other primary output (OPP)						
Q^M_{OPP}	Market Demand for OPP	−0.0051	−0.0364	−0.0521	−0.0625	−0.2569
P^M_{OPP}	Market Price of OPP	0.0064	0.0450	0.0643	0.0771	0.3172
$Q_{land,OPP}$	Demand for Land in Production of OPP	−0.0058	−0.0410	−0.0586	−0.0704	−0.2895
$Q_{che,OPP}$	Demand for Agro chemical in Production of OPP	−0.0073	−0.0511	−0.0730	−0.0875	−0.3600
$Q_{lab,OPP}$	Demand for Labor in Production of OPP	−0.0066	−0.0463	−0.0662	−0.0794	−0.3267
$Q^D_{oth,OPP}$	Demand for Other inputs in Production of OPP	−0.0063	−0.0442	−0.0631	−0.0758	−0.3117
$P^M_{land,OPP}$	Market Price of Land in Production of OPP	−0.0067	−0.0473	−0.0676	−0.0811	−0.3337
$P^M_{che,OPP}$	Market Price of Agro chemical in Production of OPP	−0.0073	−0.0517	−0.0738	−0.0886	−0.3643
$P^M_{lab,OPP}$	Market Price of Labor in Production of OPP	−0.0050	−0.0356	−0.0509	−0.0611	−0.2512
$P^M_{oth,OPP}$	Market Price of Other Inputs in Production of OPP	−0.0037	−0.0265	−0.0379	−0.0455	−0.1871
Total demand for primary inputs						
$Q_{land,T}$	Total Demand for Land	−0.0000	−0.0000	−0.0000	−0.0000	−0.0000
$Q_{che,T}$	Total Demand for Agro Chemical	0.0007	0.0055	0.0079	0.0095	0.0394
$Q_{lab,T}$	Total Demand for labor	0.0011	0.0079	0.0113	0.0136	0.0559
$Q_{oth,T}$	Total Demand for Other inputs	0.0006	0.0042	0.0060	0.0072	0.0296
Crude palm oil (CPO)						
Q^E_{CPO}	Export Demand for CPO	−0.0017	−0.0120	−0.0171	−0.0206	−0.0846
Q^D_{CPO}	Domestic Demand for CPO	0.0025	0.0180	0.0258	0.0309	0.1272
Q^M_{CPO}	Market Demand for CPO	0.0020	0.0143	0.0205	0.0246	0.1014
P^M_{CPO}	Market Price of CPO	0.0044	0.0308	0.0440	0.0528	0.2171
P^E_{CPO}	Export Price of CPO	0.0044	0.0308	0.0440	0.0528	0.2171
Processed palm oil (PPO)						
$Q^D_{PPO,T}$	Domestic Demand for PPO	0.0104	0.0732	0.1046	0.1255	0.5159
Q^E_{PPO}	Export Demand for PPO	−0.0020	−0.0140	−0.0201	−0.0241	−0.0992
Q^M_{PPO}	Market Demand for PPO	0.0044	0.0313	0.0447	0.0536	0.2206
P^E_{PPO}	Export Price of PPO	0.0044	0.0308	0.0440	0.0528	0.2171
P^M_{PPO}	Market Price of PPO	0.0044	0.0308	0.0440	0.0528	0.2171
$Q^D_{PPO,FOOD}$	Demand for PPO in Production of Food Products	−0.0032	−0.0225	−0.0323	−0.0387	−0.1593
$Q^D_{PPO,OleoChem}$	Demand for PPO in Production of OleoChemical Products	−0.0021	−0.0150	−0.0214	−0.0257	−0.1057
$Q^D_{PPO,Biofuel}$	Demand for PPO in Production of Biodiesel Products	0.2674	1.8715	2.6735	3.2080	3.1834
Palm oil-based food products (food)						
P^E_{Food}	Export Price of Food Products	0.0035	0.0246	0.0352	0.0422	0.1737
P^M_{Food}	Market Price of Food Products	0.0035	0.0246	0.0352	0.0422	0.1737
Q^D_{Food}	Domestic Demand for Food Products	−0.0013	−0.0093	−0.0133	−0.0160	−0.0660
Q^E_{Food}	Export Demand for Food Products	−0.0013	−0.0093	−0.0133	−0.0160	−0.0660
Q^M_{Food}	Market Demand for Food Products	−0.0013	−0.0093	−0.0133	−0.0160	−0.0660
Biodiesel products						
$P^M_{Biofuel}$	Market Price of Biodiesel Products	0.0002	0.0015	0.0022	0.0026	0.0108
$Q^D_{Biofuel}$	Domestic Demand for Biodiesel Products	0.9999	6.9989	9.9978	11.9960	11.9871

(Continued)

Table 7. Continued.

Notation	Definition of variables	Percent changes (Ed = 0.1)	Percent changes (Ed = 0.7)	Percent changes (Ed = 1)	Percent changes (Ed = 1.2)	Percent changes (Ed = 1.2, Biofuel share = 0.2)
$Q^E_{Biofuel}$	Export Demand for Biodiesel Products	−0.0000	−0.0001	−0.0002	−0.0002	−0.0010
$Q^M_{Biofuel}$	Market Demand for Biodiesel Products	0.2679	1.8756	2.6792	3.2149	3.2117
Oleochemicals						
$P^M_{Oleo-Chem}$	Market Price of OleoChemical Products	0.0006	0.0046	0.0066	0.0079	0.0325
$P^E_{Oleo-Chem}$	Export Price of OleoChemical Products	0.0006	0.0046	0.0066	0.0079	0.0325
$Q^D_{Oleo-Chem}$	Domestic Demand for OleoChemical Products	−0.0002	−0.0017	−0.0020	−0.0030	−0.0123
$Q^E_{Oleo-Chem}$	Export Demand for OleoChemical Products	−0.0002	−0.0017	−0.0025	−0.0030	−0.0123
$Q^M_{Oleo-Chem}$	Market Demand for OleoChemical Products	−0.0002	−0.0017	−0.0025	−0.0030	−0.0123
Other food products						
Q^M_{OFP}	Market Demand for OFP	−0.0012	−0.0085	−0.0122	−0.0146	−0.0602
P^M_{OFP}	Market Price of OFP	0.0064	0.0450	0.0643	0.0771	0.3172
P^E_{OFP}	Export Price of OFP	0.0064	0.0450	0.0643	0.0771	0.3172

Source: Simulation Results

modeled. Hence, no inference can be made with respect to the possibility of land supply emanating from deforestation or other sources. The sum of the two above-mentioned effects is shown in Table 7. Total demand for agrochemicals, labor, and other inputs are projected to increase very marginally while land demand is expected to remain unchanged.

The increase in demand for primary inputs in the oil palm subsubsector will affect the use of primary inputs in other competing subsectors. The relatively higher price of primary factors in the oil palm subsector compels resource owners to reallocate these factors from the OPP subsector to oil palm in the long run. Thus, the demand for primary factors is anticipated to decline in the OPP subsector. Consequently, OFP outputs will also decline. However, as indicated in Table 7, the decline in the production of these products is smaller than that of biochemicals and palm-oil based food, as the OPP and oil palm subsector are not close substitute in terms of land use.

Total demand for inputs going into the whole agricultural sector can also be appraised. The total use of primary inputs in agricultural sector except for land is estimated to rise as oil palm subsector has the highest share of each primary input use in Malaysian agriculture. Demand for land going into the oil palm subsector increases by 0.0018%, but declines in the competing subsector (0.0058%) leading to a net decline in total land demand, albeit very minutely (close to zero).

Generally, the findings of the study demonstrate clear opposite relationships of impacts among variables across subsectors. Given policy shocks, inputs may be reallocated away from the relatively "unattractive" subsectors, thereby impacting its outputs. The results also demonstrate a clear

converse relationship between the variables representing the competing activities within the same oil palm subsector. For example, while an increase in domestic biodiesel demand leads to an increase in output and its associated resources, the outputs of competing uses, i.e., food and chemical products decline. Such results are consistent with Chen and Khanna (2013) who found that biofuel policies in the US may have substantial impacts on food prices. Gal et al. (2013) based on country level data also observed that the corn-ethanol biofuel mandate may provoke substantial increases in the price of corn.

Drawing upon the direction and order of impacts affecting the different endogenous variables, one may draw useful insights on the usefulness of the modeling framework that is used in this study. It especially provides tractable endogenous results for a set of closely related and/or competing commodities given a particular policy shock. Nevertheless, more meaningful insights and implications of the results may be derived by conducting sensitivity analyses on the economic effects of varying assumptions of the baseline coefficients.

Simulation results using the presumed baseline parameters and coefficients have given mainly very minute results which may be less appealing to analysts and policymakers. As earlier noted, more indicative insights shall be obtained by examining the effects of varying levels of exogenous parameters and baseline coefficients on the endogenous variables. It will be important to recall that the focus of the model is on the direction and order of impacts, rather than on fine-tune numbers. Nevertheless, the levels of exogenous variables, parameter values, and elasticities are undoubtedly instrumental in determining the impact of alternative policies.

The predetermined base share of biodiesel product in the domestic use of CPO as employed in the preceding simulation is only 0.05 while the presumed domestic demand elasticity is −0.13. Such small share of CPO use for biodiesel and the highly inelastic demand may explain much the reason why the impact of biodiesel demand shifts on endogenous variables have been rather negligible. The following section appraises the impact of varying levels of domestic demand elasticity of biodiesel and CPO share in biodiesel on the various endogenous variables. This will provide more insights on the importance of the share of CPO use and greater CPO-fossil fuel substitution effects.

The study simulated a range of domestic demand elasticities; 0.7, 1, and 1.2 under scenario 2 through 4, ceteris paribus. Table 7 provides the simulation results for each elasticity level. With the dynamics of time, as demand becomes more elastic, more pronounced sector-wide impacts can be expected, albeit all other parameters may remain unchanged. We then simulated a change in biofuel share of CPO use from 0.05 to 0.2 along with an elastic domestic biofuel demand of −1.2. The results are listed in the last column of Table 7.

Results clearly show substantial changes in the outcome of each endogenous variable with increases in domestic demand share. Comparing the results from different scenarios, one may notice that most pronounced impacts are seen when biofuel share of CPO use rises to 0.2 and domestic demand for biofuel is highly elastic at −1.2. In this scenario, market demand for biodiesel products and PPO demand for use in biodiesel may rise to 3%.

Given the complexity of the model, varying one or two specific parameters may not be satisfactory to indicate the extent of sensitivity of model results. As Salhofer (2000) pointed out, for such analysis, we usually do not know ex ante what the most crucial parameters are. Future simulation may consider more comprehensive analysis which includes varying the input substitution elasticities, domestic and export market share P^{M}_{FFB}, and input supply rigidity.

Conclusion and policy implications

For many countries, agricultural sub-sectors are inevitably linked due to resource constraints, especially land and labor inputs. Agricultural policies can affect the pattern of production through commodity and input price changes and consequently have implications on primary inputs use and their reallocation between subsectors and economic activities. Changes in agricultural policies in one subsector would affect input use, production, output prices, and exports within and other related subsectors.

This study developed a multicommodity, multistage production, comparative statics, partial equilibrium model to simulate the effects of an exogenous shift in domestic demand for palm oil-based biofuel in the Malaysian agricultural sector. Simulation findings demonstrate a clear converse relationship between the variables representing the competing activities within the same and across subsectors.

Impact of increases in biodiesel demand on competing subsectors depend much on the elasticity of demand for the product. Increases in demand elasticity might be due to demand and supply augmentation measures such as consumption subsidies, biodiesel mandates, and increased consumer preferences for more environmentally friendly or renewable source of energy. Such shifts in supply and consumer demand may induce pronounced impacts on related subsectors, most importantly the food subsectors and oleochemical industry. While increases in biofuel consumption and supply may help improve energy security and emission efficiency, it inevitably leads to loss of welfare for the affected interest groups and stakeholders. Appropriate policies will be necessary to encourage efficient and equitable allocation of palm oil inputs for biodiesel development and other uses, especially food use of palm oil in Malaysia.

For future studies, the model can be further generalized and expanded to consider multiple subsectors including a non-agriculture aggregate and incorporation of primary factors in the production of intermediate inputs and final goods. Policy simulation may also consider rising public concerns on climate change issues, minimum wages, immigration reforms, and more intricate food-fuel issues. Welfare function representing the various interest groups may also be incorporated into the model framework.

Conflict of Interest

None declared.

References

Basri, T., and Z. Darawi. 2002. An economic analysis of the Malaysian palm oil market. Oil Palm Ind. Econ. J. 2:19–27.

Beckman, J. F., and T. W. Hertel. 2008. Validating energy oriented CGE models. GTAP Working Paper No. 54, Centre for Global Trade Analysis, Purdue University.

Betina, V. D., R. A. McDougal, and T. W. Hertel. 2006. GTAP 6 data base documentation - Chapter 20: Behavioral Parameters. Center for Global Trade Analysis, Purdue University.

Chen, X., and M. Khanna. 2013. Food vs fuel: the effect of biofuel policies. Am. J. Agric. Econ. 95.2:289–295.

Food and Agricultural Policy Research Institute (FAPRI). Elasticities database. Available at http://www.fapri.iastate.edu/. (accessed 1 June 2014).

Gal, H., S. Kaplan, and D. Zilberman. 2013. The causes of recent food commodity crises. Paper prepared for presentation at the Agricultural and Applied Economics Association's 2013 AAEA and CAES Joint Annual Meeting, Washington DC, August 4- 6, 2013.

Hanim, A. 2012. The Star, April 17. "Food subsidy should only be for the needy".

Hertel, T. W. 1989. Negotiating reductions in agricultural support: implications of technology and factor mobility. Am. J. Agric. Econ. 71:559–573.

Jafari, Y., and O. Jamal. 2015. A comparative statics, partial equilibrium, multi-commodity model for Malaysian agriculture. Malaysian J. Econ. Studies 52:205–226.

Ministry of Plantation Industries and Commodities, Malaysia. 2010a. Handbook of Agricultural Statistics.

Ministry of Plantation Industries and Commodities, Malaysia. 2010b. Statistics on commodities 2009.

MPOB. 2008a. Review Of Malaysian Oil Palm Industry 2007.

MPOB. 2008b. Malaysian Oil Palm 2007 Statistics.

MPOB. 2016. Biodiesel Solutions. Available at http://www.palmoilworld.org/biodiesel.html. (accessed 12 February 2016).

Othman, J. 2003. Linking agricultural trade, land demand, and environmental externalities: Case of oil palm in Southeast Asia. ASEAN Econ. Bull. 3:244–255.

Pollitt, H., U. Chewpreecha, and P. Summerton. 2007. E3ME: An energy–environment– economy model for Europe: A Technical Description.

Salhofer, K. 2000. Elasticities of substitution and factor supply elasticities in European agriculture: A review of past studies. In OECD, 2001. Markets Effects of Crop Support Measures, pp. 89-110.

Appendix A

Table A1. Multi commodity partial equilibrium model of the agricultural sector.

Equations	Definition of variables
Market demand Equation: $$\hat{Q}_y^M = \sum_x \alpha_y^D \varepsilon_{y,x}^D (\hat{P}_x^M - \hat{U}_x^D) + \sum_x \alpha_y^E \varepsilon_{y,x}^E (\hat{P}_x^E - \hat{U}_x^E)$$ (A1) for $y, x \in$ {Food, Biofuel, Chemicals, OFP}	Here, *hat* notation denotes the percentage changes in the variable (e.g., $\hat{Q}_y^M = \frac{dQ_y^M}{Q_y^M}$), while the notation *d* refers to absolute changes in variables. Here, $\alpha_y^D (= \frac{Q_y^D}{Q_y^M})$ is the share of domestic demand for y (Q_y^D) of the total market demand (Q_y^M), while $\alpha_y^E (= [1 - \alpha_y^D])$ is the export demand share. Parameters $\varepsilon_{y,x}^D$ and $\varepsilon_{y,x}^E$ refer to price elasticity of domestic and export demand for *y*th final output Allowing for the shifts in domestic output and export demand schedules (as the shifts in the direction of price axis), $\hat{U}_x^D$ and $\hat{U}_x^E$ represent percentage shifts in domestic output and export demand schedules for commodity *x*, respectively. $\hat{P}_x^M$ and $\hat{P}_x^E$ represent changes in market, and export price if commodity *x*, respectively; and $\hat{Q}_y^M$ is the percentage change in quantity of *y*.
Market Demand for PPO $$\hat{Q}_{ppo}^M = \sum_y \alpha_{ppo,y}^D * C_{ppo,y} * \sigma_{ppo} * \hat{P}_{ppo,y}^D + \alpha_{ppo}^E \varepsilon_{ppo}^E * \hat{P}_{ppo}^E$$ (A2) $y \in$ {Food, Biofuel, Chemicals}	Here, $\alpha_{ppo,y}^D = \frac{Q_{ppo,y}^D}{Q_{cpo}^M}$ represent the domestic demand share of PPO to be used in production of *y* with respect to market demand for PPO; α_{ppo}^E denotes the export share of PPO with respect to its market demand. Parameter, $C_{ppo,y}$ represent the cost share of PPO in production of *y* and * σ_{ppo} is the own Allen elasticity of substitution for PPO.
Market Demand for CPO $$\hat{Q}_{cpo}^M = \alpha_{cpo}^D * C_{cpo} * \sigma_{cpo} * \hat{P}_{cpo}^D + \alpha_{cpo}^E \varepsilon_{cpo}^E * \hat{P}_{cpo}^E$$ (A3)	Here, α_{cpo}^D represent the domestic demand share of CPO with respect to market demand for CPO; α_{cpo}^E denotes the export share of CPO with respect to its market demand. Parameter, C_{cpo} represent the cost share of CPO in production of *PPO* and * σ_{cpo} is the own Allen elasticity of substitution for CPO.

Table A1. Continued.

Equations	Definition of variables
Derived Demand for Primary Products	c_{ffb}: cost share of FFB in production of CPO
$$\hat{Q}_{ffb}^{D} = c_{ffb} * \sigma_{ffb}\hat{P}_{ffb}^{D} + \hat{Q}_{cpo}^{S} \qquad (A4)$$	σ_{ffb}: own Allen elasticity of FFBs
$$\hat{Q}_{opp}^{D} = c_{opp} * \sigma_{opp}\hat{P}_{opp}^{D} + \hat{Q}_{ofp}^{S} \qquad (A5)$$	c_{opp}: cost share of OPPs in production of OFPs. σ_{opp}: own Allen elasticity of OPPs.
Derived Demand for Primary Factors	$C_{j,k}$: cost share of ith input in production of kth primary output.
$$\hat{Q}_{i,k}^{D} = \sum_{i=1}^{j} C_{i,k}\sigma_{ij,k}\hat{P}_{i,k}^{D} + \hat{Q}_{k}^{S} \, k \in \{FFB, OFP\}\,, \text{and} \, i = 1, \dots j \qquad (A6)$$	$\sigma_{ij,k}$: Own and cross Allen elasticity of ith input in production of kth primary output
Zero Profit Conditions	$C_{cpo,y}$: cost share of CPO in production of yth output
$$\hat{P}_{k}^{S} = \sum_{i=1}^{j} C_{i,k}\hat{P}_{i,k}^{D}\, k \in \{FFB, OFP\}\, \text{and} \, i = 1, \dots, j \qquad (A7)$$	
$$\hat{P}_{cpo}^{S} = C_{ffb}\hat{P}_{ffb}^{D} \qquad (A8)$$	
$$\hat{P}_{y}^{S} = C_{cpo,y}\hat{P}_{cpo,y}^{D} \qquad (A9)$$	
$$\hat{P}_{ofp}^{S} = C_{opp}\hat{P}_{opp}^{D} \qquad (A10)$$	
Input Supply Equations	$\theta_{i,k}$: Share of ith input employed in production of kth primary output
$$\hat{Q}_{i,k}^{S} = \sum_{k=1}^{kk} \theta_{i,kk} * v_{i,kk}\hat{P}_{i,kk}^{S} \qquad (A11)$$	$v_{i,k}$: Supply elasticities of primary inputs for use in production of k.
$k, kk \in \{FFB, OFP\} \text{and} \, i = 1, \dots j$	
Factor Clearing Conditions	
$$\hat{Q}_{i,k}^{D} = \hat{Q}_{i,k}^{S} \, for \, k \in \{FFB, OFP\} \text{and} \, i = 1, \dots, j \qquad (A12)$$	
$$\hat{Q}_{k}^{D} = \hat{Q}_{k}^{S} \, for \, k \in \{FFB, OFP\} \qquad (A13)$$	
$$\hat{Q}_{r}^{M} = \hat{Q}_{r}^{S} \, for \, r \in \{CPO, PPO\} \qquad (A14)$$	
Commodity Market Clearing Condition	The definition of other variables remains as denoted in the text.
$$\hat{Q}_{y}^{M} = \hat{Q}_{y}^{S} \, for \, y \in \{Food, Biofuel, Chemicals, OFP\} \qquad (A15)$$	

3

Soil health and carbon management

Rattan Lal

Carbon Management and Sequestration Center, The Ohio State University, Columbus, Ohio 43210

Keywords
Agriculture, food, soil.

Correspondence
Rattan Lal, Carbon Management and Sequestration Center, The Ohio State University, Columbus, OH 43210.

E-mail: lal.1@osu.edu

Funding Information
No funding information provided.

Abstract

Soil, a natural four-dimensional body at the atmosphere–lithosphere interface, is organic-carbon-mediated realm in which solid, liquid, and gaseous phases interact at a range of scales and generate numerous ecosystem goods and services. Soil organic carbon (SOC) strongly impacts soil quality, functionality and health. Terms soil quality and soil health should not be used interchangeable. Soil quality is related to what it does (functions), whereas soil health treats soil as a living biological entity that affects plant health. Through plant growth, soil health is also connected with the health of animals, humans, and ecosystems within its domain. Through supply of macro- and micronutrients, soil health, mediated by SOC dynamics is a strong determinant of global food and nutritional security. Soil C pool consists of two related but distinct components: SOC and soil inorganic C (SIC). The SIC pool comprises of primary and secondary carbonates, and the latter consists of calcitic (no net sequestration of atmospheric CO_2) and silicatic (net sequestration). While SOC is highly dynamic, its mean residence time depends on the degree of protection (physical, chemical, biological, and ecological) within the soil matrix. Formation of stable microaggregates and of organo–mineral complexes can protect SOC against microbial processes for millennia. In addition to formation of silicatic type of secondary carbonates, leaching of bicarbonates into the subsoil or shallow water table is also an important mechanism of sequestration of CO_2 as SIC. Numerous soil functions and ecosystem services depend on SOC and its dynamics. Improvements in soil health, along with increase in availability of water and nutrients, increases soil's resilience against extreme climate events (e.g., drought, heat wave) and imparts disease-suppressive attributes. Enhancing and sustaining soil health is also pertinent to advancing Sustainable Development Goals of the U.N. such as alleviating poverty, reducing hunger, improving health, and promoting economic development.

Introduction

Soil, a natural body at the atmosphere–lithosphere interphase, is a dynamic entity and teeming with life. It is essential to recycling of dead and decaying organic matter and storing of plant nutrients, denaturing of pollutants and filtering of water, sequestering of carbon (C) and moderating of climate, and storing of germplasm and provisioning of habitat for biodiversity. It is a medium for plant growth, generates the net primary productivity, and supports all terrestrial life through provisioning of the necessary conditions for its well-being. Soil organic C (SOC) concentration, along with its quality and dynamics, is essential to diverse soil functions and ecosystem services. Thus, soil is an organic-C-mediated realm in which solid, liquid, and gaseous phases interact at a scale ranging from nanometers to kilometers and create dynamic environments conducive to growth and development of plants and other biota. Soil organic matter (SOM), comprising about 45–60% of its mass as SOC, is a principal source of energy for soil microorganisms.

Three principal components of SOM are as follows: (1) plant and animal residues and living microbial biomass; (2) active or labile SOM; and (3) relatively stable

SOM. The live biomass in a healthy and dynamic soil may be as much as 5 Mg/ha (Mg = megagram = 10^6 g = 1 metric ton). The objective of this article is to describe the relationship between SOC and soil quality, functionality, and health.

Soil Quality, Functionality and Health

Humanity's interest in soil quality and functionality dates back to the dawn of civilization (Brevik and Sauer 2015). Moses outlined soil functionality or quality by asking his followers as they entered Canaan around circa 1400 BC by stating, "See what the land is like and whether the people who live there are strong or weak, few or many. What kind of land do they live in? Is it good or bad? How is the soil? It is fertile or poor? Are there trees or not? Do your best to bring back some fruits from the land" (Numbers 13:18–20). The Book of Oades, dating back to China's Zhou dynasty from 770 to 476 BC, describes landforms, animals and plants, and explains agricultural practices for their management. In 400 BC, Hippocrates provided a list of things regarding the land or ground: "...whether it be naked or deficient in water or wooded and well watered, and whether it lies in a hollow, confined situation, or is elevated and cold..." (Hippocrates, 400 BC, quoted by Brevik and Sauer 2015). Around 60 BC, Columella related human health to soil conditions and described human diseases that may be contracted from marshes (Sylvia et al. 1998). In a Sanskrit manual "Artha Sastra," Chanukya/Kautilya (4th Century BCE) explains to land managers, techniques of improving soil functions by applying manure and other systems of managing soil fertility and conserving water. Ibn-Al-Awwam, a Moorish philosopher of the 12th Century, wrote in the Book on Agriculture (Kitab-Al-Felaha) "the first step in agriculture is the recognition of soils and how to distinguish that which is of a good quality and that which is of an inferior quality. One must also take into consideration the depth of the soil, for it often happens that its surface layer may be black."

During the modern era, soil quality and functionality have received a growing attention since 1970s because of the increasingly affluent world population (Warkentin and Fletcher 1977; Larson and Pierce 1991; Doran et al. 1994, 1996; Karlen et al. 1994; Parr et al. 1992; Lal 1993, 1994, 1997; Smith et al. 1993; Harris and Bezdicek1994, Doran and Jones1996; Doran and Zeiss 2000). Three separate but interrelated terms emerged since 1970s are as follows: soil quality, soil functionality, and soil health. Soil quality is defined as the "fitness for use" (Larson and Pierce 1991), and "capacity of the soil to function" (Karlen et al. 1997). Thus, soil functions, intricately linked with soil quality, depend on specific land use, and include

sustaining plant and animal productivity (agricultural land use), forest productivity (silviculture land use), air and water quality in relation to human health and habitation (urban land use), contamination with heavy metals (minelands and urban lands) etc. Important soil functions, of relevance to human well-being and nature conservancy, include: retention and cycling of nutrients, formation and stabilization of soil structure (aggregation), retention and transmission of water, aeration and gaseous exchange, buffering of soil reaction, transformation of compounds, and maintenance of biodiversity. These functions are also termed as ecosystem services provided by soil resources (Daily et al. 1997). However, soil functions are difficult to quantify, and are usually measured through assessment of soil quality indicators, which must be flexible and broad-based to be pertinent to a wide range of soil functions (Andrews et al. 2004). As Edward Demmings (1943) pointed out, important things which cannot be measured must be judiciously managed. Therefore, the question is not – what is there in the soil that can be measured, but what it does which must be quantified. What it does is... soil quality, and it can be measured neither precisely nor directly. Therefore, it is measured indirectly by assessing some soil quality indicators (SQI).

Terms soil quality and soil health, while similar, should not be used interchangeably. Soil quality is related to soil functions or what it does, whereas soil health presents the soil as a finite and dynamic living soil resource, and is directly related to plant health. More specifically, soil health is defined as "capacity of soil to function as a vital living system to sustain biological productivity, maintain environment quality and promote plant, animal and human health" (Doran et al. 1996; Doran and Zeiss 2000). In this definition, however, the impact of soil biotic activity and species diversity on climate is overlooked, and yet both are intricately interconnected and this linkage should not be ignored. Soil attributes essential to life include: (1) physical for provisioning of air, water, and gaseous exchange and habitat; (2) chemical for moderating soil reaction and availability and transformation of nutrients; (3) biological for source of energy and food and nutrient cycling; and (4) ecological for hydrological and energy budget, and landscape processes (Fig. 1). These attributes, individually and through interaction, create environments which are conducive to life, and vice versa. Yes, soil health affects life and is in turn affected by its diversity and dynamics.

Soil health is also connected with human health and nutritional security. The soil–human health nexus has also been recognized ever since the dawn of civilization (Brevik and Sauer 2015), and vividly explained by U.S. soil scientists during 1920s (McCarrison 1921) and 1930s (Knight et al. 1938; Albrecht 1945, 1951, 1957). Voisin (1959)

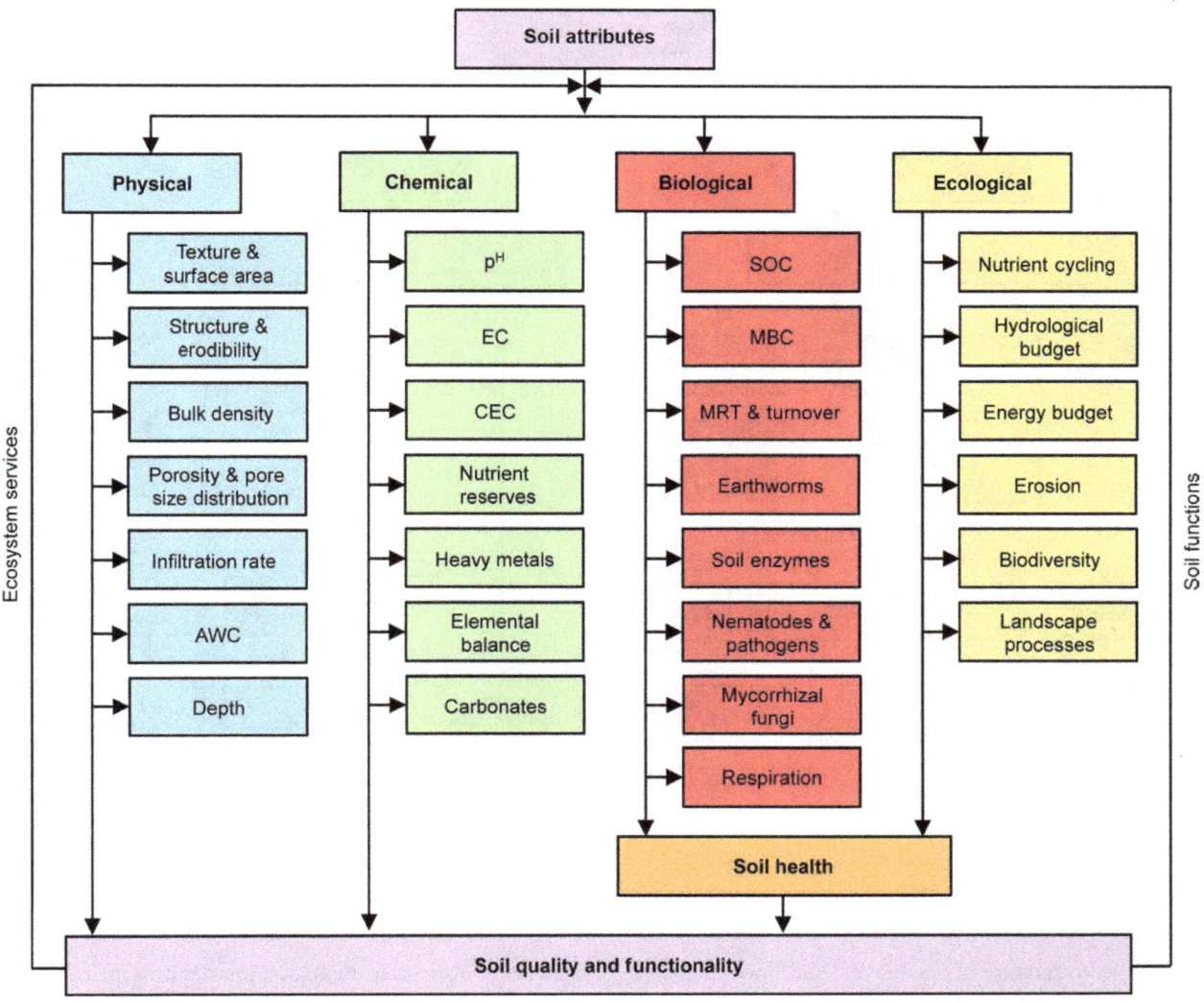

Figure 1. Soil attributes as indicators of soil health (AWC = available water capacity; SOC = soil organic C; CEC = cation exchange capacity; EC = electrical conductivity; MBC = microbial biomass; MRT = mean residence time).

linked cancer to soil health. In British India, Howard (1940, 1947) linked human health to soil health. Being a pioneer in organic agriculture, Sir Albert's work was contemporary to that of Lady Balfour (1943) and that of Rodale (1945). For example, micronutrient deficiencies are an important cause of morbidity and mortality. Children are especially vulnerable to deficiency in Zn, and that of Fe causes anemia in children and nursing mothers. Seventeen micronutrients essential to human health are as follows: Fe, Zn, Cu, Mo, I, F, B, Se, Mu, Ni, Cr, Si, As, Li, Sn, V, and Co. In addition, macronutrients also essential to human health are as follows: N, P, K, Ca, Mg, Na, S, and Cl (Lal 2009). All these micro and macro-nutrients must be supplied through soil. Thus, the widely recognized truism—the health of soil, plants, animals, people, and ecosystems is one and indivisible. Soil health is defined as — soil's capacity, as a biologically active

entity, within natural and managed landscapes, to sustain multiple ecosystem services, including net primary productivity (NPP), food and nutritional security, biodiversity, water purification, and renewability, C sequestration, air quality, and atmospheric chemistry and elemental cycling for human well-being and nature conservancy. This definition of soil health is in accord with that of the Gaia Hypothesis (Lovelock 1979), which states that life creates environment suited to its well-being.

Soil Carbon

Soil carbon (C) consists of two related but distinct components (Fig. 2). Soil organic C (SOC) comprises of the remains of plants and animals at different stages of decomposition and of the microbial biomass and their by-products. As a component of SOM (45–60%), SOC is a

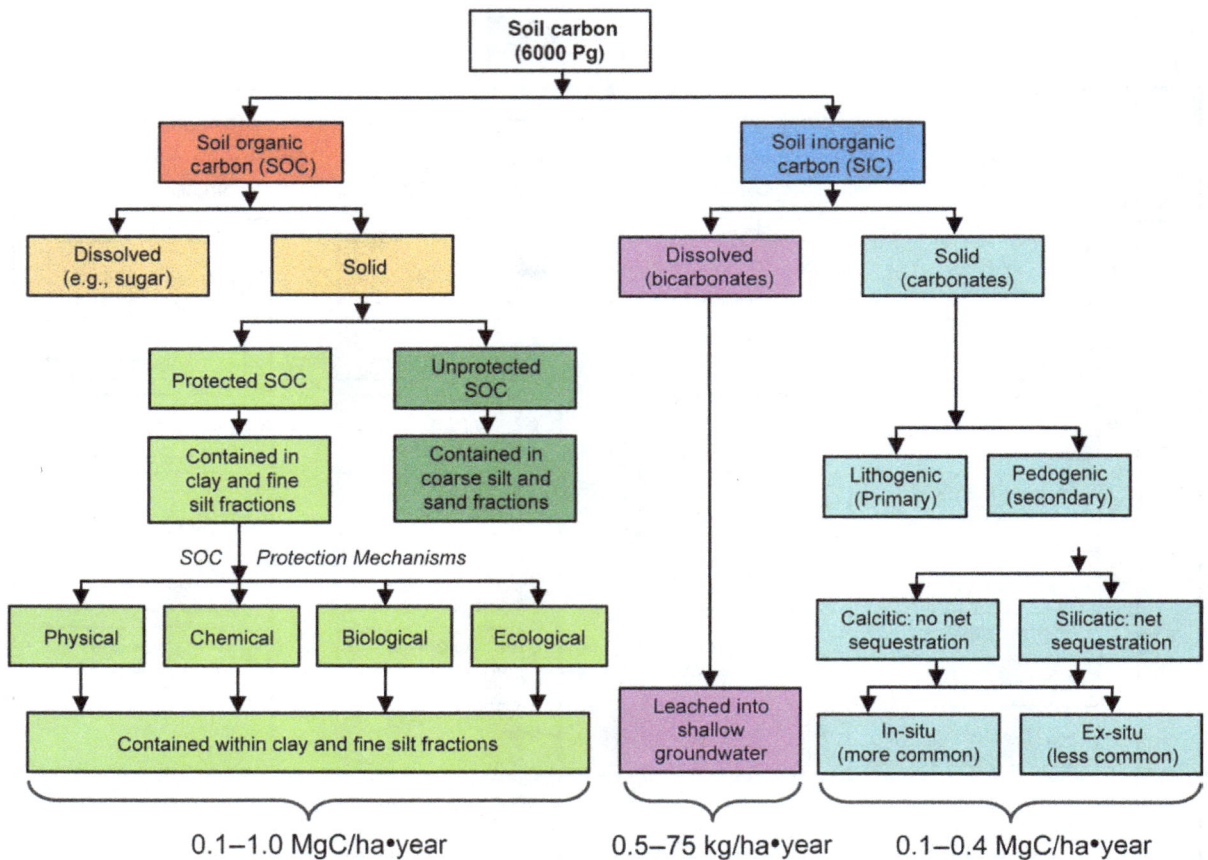

Figure 2. Types of organic and inorganic carbon pools in soil. The numerical values listed on the last line are ranges of sequestration of organic and inorganic carbon in diverse soils and ecoregions. Soil C pool of 6000 Pg is to 3-m depth and comprises of all components.

heterogeneous mixture of organic materials including fresh litter, carbohydrates, and simple sugars, complex organic compounds, some inert materials, and pyrogenic compounds. Dynamics of SOC as a component of the terrestrial C cycle is discussed by Jansson et al. (2010), among others. The SOC is a highly reactive component, and is the basis of numerous pedogenic processes. Because of a high surface area and charge density, it reacts with clay and minerals to form organo–mineral complexes. The mean residence time (MRT) or the rate of its turnover depends on the degree of protection within the soil matrix (Dungait et al. 2012). Among numerous protective mechanisms are physical, chemical, biological, and ecological. Physical mechanisms include encapsulation within stable microaggregates (Six et al. 2000, 2002), formation of organo–mineral complexes, and transfer deep into the subsoil away from the zone of natural and anthropogenic perturbations. Formation of organo–mineral complexes can store SOC for millennia. Chemical protection involves formation of some recalcitrant compounds (von Lützow and Kögel-Knaber 2009), including aromatic and double-bond hydrocarbons and some hydrophobic substances that

coat stable aggregates. Biological mechanisms include some microbial exudates that repel other organisms, transfer of SOC into biologically nonpreferred soil spaces (Ekschmitt et al. 2008; Kleber 2010; Kleber et al. 2011; Dungait et al. 2012), and substrate-driven biological rate limitations (Ekschmitt et al. 2005). Ecological mechanisms include coupled cycling of C with other soil constituents (H_2O, N, O, S and microelements), erosion cont rol, and deep translocation through biogeochemical processes (Fig. 2).

The SIC, a dominant form of C in soil of arid and semi-arid regions (rainfall <500 mm/year), consists of carbonates (CO_3^{-2}) and bicarbonates (HCO_3^-) of Ca^{+2}, Mg^{+2}, K^+ and Na^+. Further, SIC comprises of primary or lithogenic carbonates and secondary, or pedogenic carbonates. The SIC consists of elemental C; carbonate-bearing minerals (e.g., calcite, agronite, and gypsum); gaseous CO_2 as a by-product of heterotrophic respiration, and dissolved C as an equilibrium of H_2CO_3, HCO_3^-, and CO_3^{-2} (Jansson et al. 2010). Secondary carbonates are formed through reaction of HCO_3^- and CO_3^{-2} (in solution by dissolution of CO_2) with Ca^{+2} or Mg^{+2}. Formation of secondary carbonates leads to sequestration of atmospheric CO_2.

Secondary carbonates deposition can be of diverse morphological shapes such as rinds, coats, sheets, pendants, opal, etc. (Blank and Fosberg 1990). The rate of SIC sequestration through formation of secondary carbonates can be 0.12–0.38 MgC/ha.year to 160 cm depth by irrigation and fertilization (Bughio et al. 2015). However, leaching of bicarbonates into the subsoil or shallow water table and its reprecipitation can also be high (Lal 2008; Barta 2011; Ma et al. 2014; Monger et al. 2015). The pool of C in groundwater is 1404 Pg (petagram = 10^{15} = 1 billion metric ton = 1Gt) and the flux of bicarbonates can be as much as 2.1–7.4 g C/m^2.year, with a global flux of C into the groundwater at 0.2–0.36 PgC/year as bicarbonates. (Monger et al. 2015).

Dynamics of soil C pool can have a strong impact on atmospheric chemistry and the global C cycle (Lal 2004). For example, if it were possible to increase soil C pool globally by 4% to 3-m depth, it would cause a drawdown of atmospheric CO_2-C by 240 Pg, the amount equivalent to the reduction of >100 ppmv of CO_2. However, the logistics of achieving such an increase even over a decadal scale are insurmountable at the present level of scientific advances (Lal 2016). Nonetheless, the "4 per Thousand" program proposed at the COP21 in Paris in 2015 is a step in the right direction. Through improvements in soil health and the attendant pedospheric processes, sequestration of C in the soil solum has numerous ancillary benefits to human and nature. As Dyson (2008) stated, "if we control what the plants do with the carbon, the fate of the carbon in the atmosphere is in our hands."

Soil Functions and Ecosystem Services

Among numerous soil functions (Table 1) are those which form critical basis of all terrestrial life. The linkage between soil health and ecosystem services is depicted in Figure 3. Soil health impacts: (1) growing food through plants and animals by storage and availability of plant nutrients, cycling, and transformation of elements and delivering macro- and micronutrients when needed; (2) storing water

in the root zone to increase plant-available water capacity, denaturing, and filtering of pollutants, and appropriately using blue and gray/black water for mitigating drought stress and recycling nutrients; (3) moderating of climate through sequestration of C in the soil and biota, buffering against sudden/abrupt fluctuations in moisture and temperature regimes, and regulation of gaseous emissions (CO_2, CH_4, N_2O) into the atmosphere; (4) providing habitat and energy source to biota, especially the microbiota, and the storehouse of germplasm; and (5) providing industrial raw materials (e.g., clay, peat, minerals), and of antibiotics for human and animals and of other pharmaceuticals, and of organisms which create disease-suppressive soils (Fig. 2). Indeed, soil health is the engine of economic development. It impacts quality and magnitude of renewable water resources, adaptation/mitigation and stabilization of climate, production of biomass and net and ecosystem productivity, and the above and belowground productivity.

Soil Organic Carbon and Soil Health

Quantity, quality, and dynamics/turnover of SOC are critical to soil health (Lal 2014). Threshold level of SOC in the rootzone is 1.5–2.0%. Maintenance of SOC pool at above the threshold/critical level is essential to: (1) soil structure and aggregation which govern soil tilth and aeration; (2) water retention and use efficiency which control tolerance to drought, heat wave, and abrupt climate change; (3) nutrient retention and use efficiency which moderate nonpoint source pollution, water quality, and toxic algal blooms; (4) rhizospheric processes which influence elemental transformations and creation of disease-suppressive soils; and (5) gaseous emissions (e.g., CO_2, CH_4, N_2O) which moderate atmospheric chemistry and regulate climate change. Above all, numerous soil-related constraints to agronomic productivity (Fig. 3) can also be alleviated through enhancement and sustainable management of the SOC pool. Among these constraints are as follows: (1) inherent soil properties

Table 1. Soil functions related to soil quality.

Soil quality	Specific attributes	Soil functions
Physical	Texture, structure, depth, hydraulic conductivity, infiltration rate, aeration, surface area, bulk density	Water retention, transmission, filtration of pollutants, water cycling and renewability, foundation for civil structures, gaseous exchange
Chemical	Cation exchange capacity (CEC), pH, nutrient reserves, electrical conductivity	Cycling of elements, elemental transformation, buffering, leaching
Biological	SOC, microbial biomass C (MBC), species diversity, soil enzymes, respiration rate	Decomposition of waste, denaturing of pollutants, moderation of climate, carbon sequestration, habitat for biota, energy for soil organisms
Ecological	Soil depth, mineralogical composition, water storage and renewability	Production of biomass or NPP, moderation of climate, medium for plant growth, archive of planetary/human history and paleoclimate.

SOC, soil organic carbon; MBC, microbial biomass carbon; CEC, cation exchange capacity; NPP, net primary productivity.

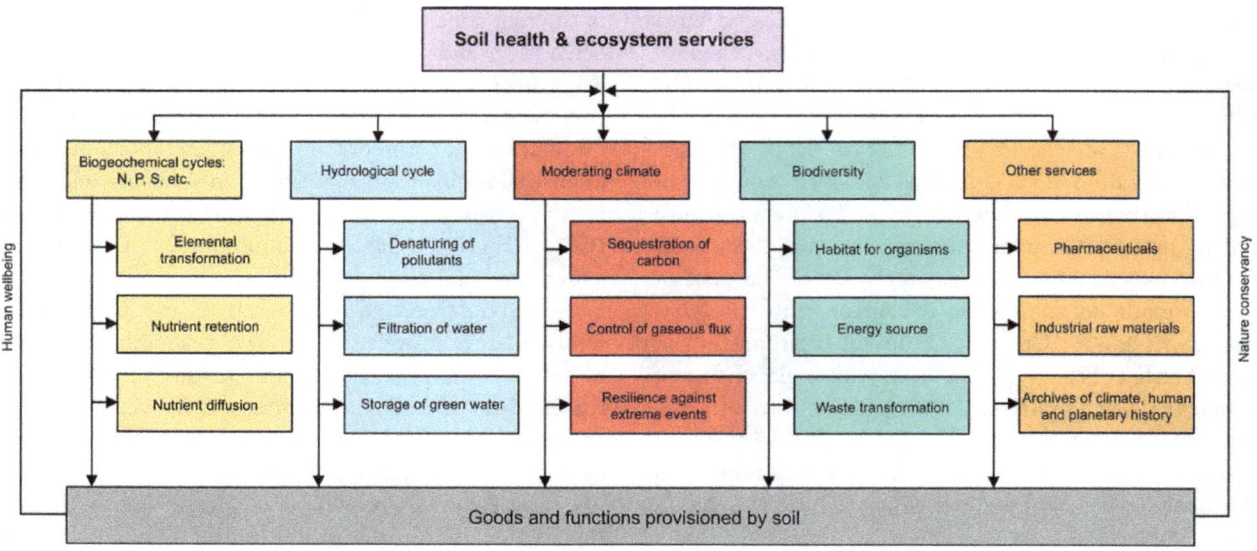

Figure 3. Soil health and ecosystem services for human well-being and nature conservancy.

(endogenous) related to the parent materials and soil formation; (2) environmental factors (exogenous) related to climate, landscape, and other biotic and abiotic factors; (3) water-related issues which influence the drought-flood syndrome; and (4) the human dimensions that affect availability and access to inputs, along with social, cultural, and environmental issues. Improvement in soil health through SOC enhancement can address these issues. Similar to soil quality, soil health also cannot be measured directly. Thus, there are key soil attributes which can be used as surrogates to develop soil health indicators (Fig. 4).

Soil Health and Sustainable Development Goals

Sustainable Development Goals (SDGs), launched by the U.N. in September 2015 as a continuation of the Agenda 21 and the Millennium Development Goals, are aimed at improving the environment, conserving nature, and enhancing human well-being. This is also called "Agenda 2030." Sustainable management of soil health is critical to advancing several SDGs (Table 2), especially those related to alleviating poverty (#1), ending hunger (#2), improving health (#3), clean water (#6), economic growth (#8), and climate action (#13). Implementation of "4 per Thousand" initiative proposed at the Paris Climate Summit in November/December 2015 highlights the importance of sequestering C in soil at the aspirational rate of 0.4%/year to 40-cm depth (Lal 2016). It is an important mechanism of advancing SDG #2, 6, 8, 13, and 15 (Table 2).

Technological Options to Manage Soil Health

Sustainable management of SOC is critical to enhancing and managing soil health. Thus, management of soil health involves management of SOC pool. The SOC pool can be enhanced by technological options that create a positive C budget (Fig. 5). Important among these are conservation agriculture (CA), integrated, and diverse cropping/farming systems, use of organic amendments and those options that restore soil/ecosystem functions. Among numerous soil properties (Fig. 4), it is pertinent to identify site-specific indicators of soil health. Thus, key soil properties (e.g., physical, chemical, biological, ecological) must be identified to develop an appropriate soil health index. Soil biological properties (MBC, enzymes) are among the most dynamic characteristics, which have a rapid response to landuse, landuse change, and soil/crop/animal management (Cardoso et al. 2013). Appropriate indicators of soil health may be those attributes that can enhance the following functions (Kibblewhite et al. 2008): (1) C transformations; (2) nutrient cycles; (3) soil structure maintenance; and (4) regulation of pests and disease. These functions are moderated by a range of biological processes moderated by diverse soil organisms under specific ambient environment.

Soil Health and Disease-Suppressive Attributes

Soil is a living system. Soil's capacity to perform these functions depends on: a range of biogeochemical processes

Figure 4. Global soil-related constraints to agronomic productivity.

Table 2. Advancing sustainable development goals through management of soil health.

Goal #	Objective	Impact of soil health
1	No poverty	Increase farm income
2	End hunger	Enhance quantity and quality of food
3	Good health	Produce nutritious food
5	Gender equality	Improve crop productivity of women farmers
6	Clean water and sanitation	Improve water quality
8	Economic growth	An engine of economic development
10	Reduce inequalities	Enhance and sustain farm productivity
12	Responsible consumption	Reduce input of water, nutrients and energy by decreasing losses
13	Climate action	Sequester C and mitigate climate change
15	Life on land	Increase activity and species diversity of soil biota

that occur in the soil, and the functionality of soil biodiversity (Smith et al. 2015). Increasing MBC and soil biodiversity (e.g., rhizobacteria, fungi) also leads to a greater suppression of crop pathogens and pests (Baker and Cook 1974; Larkin and VanAlfen 2015). Robust microbial communities can lead to either general suppression or specific suppression of diseases (Janvier et al. 2007). Thus, depending on microbial communities, soil may range from conducive to suppressive.

Mechanisms of disease-suppressive attributes include (Janvier et al. 2007): (1) slow establishment and persistence of pathogens; (2) lower severity of diseases; and (3) ineffectiveness of pathogens. General principles of enhancing soil health, and thus disease-suppressive attributes, include the following: (1) improve SOC pool; (2) adopt CA; (3) increase soil biodiversity; (4) diversify land use and maintain a live vegetable cover; (5) use organic amendments such as mulch, compost; and (6) adopt integrated nutrient management options. A judicious management of chemical fertilizers, in conjunction with the use of organic amendments, is also important to enhancing soil health (Singh and Ryan 2015). However, total disease control cannot be achieved by techniques which improve soil health (Fig. 5), but the incidence of soil-borne diseases can be reduced. Soil health improvement may not strongly influence the

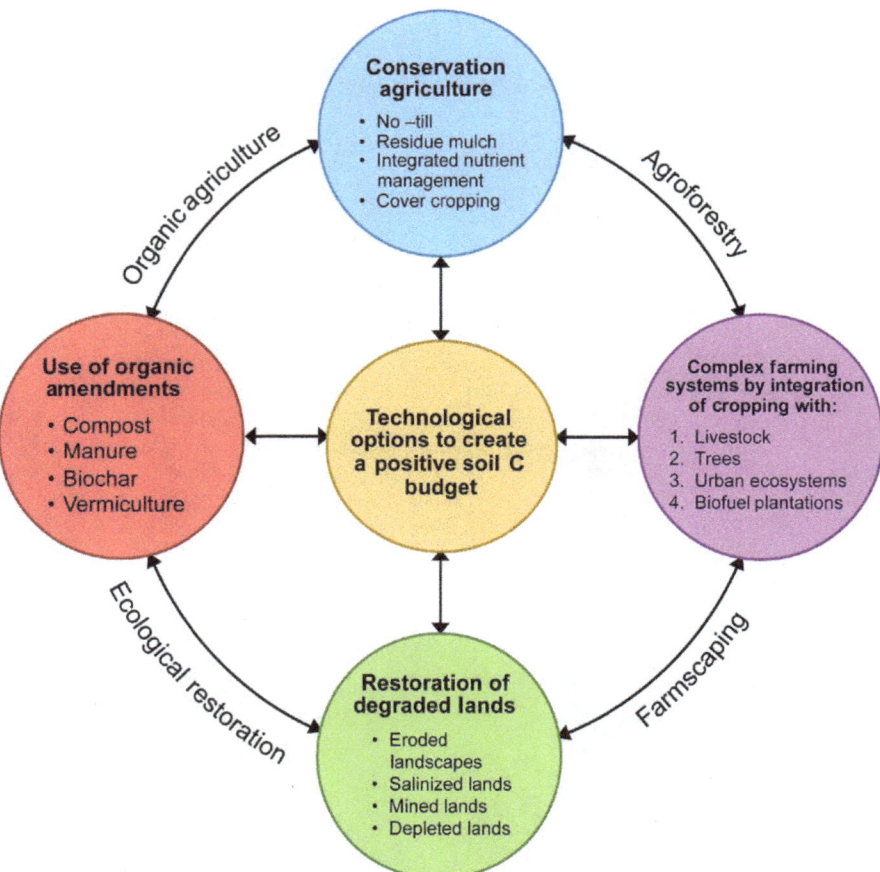

Figure 5. Technological options for soil carbon sequestration.

foliar-borne diseases, but healthy soils support population of beneficial microorganisms and can lead to induced resistance to both soil-borne and foliar disease. Further, plants grown on healthier soils are relatively more resilient and are less susceptible to pathogens (Larkin and VanAlfen 2015). Use of some organic amendments (e.g., compost) can impact disease-suppressive properties. Management of soil through these concepts would do more than just improving plant nutrients, it would also enhance the environment (Ehmke 2013). Healthy soils, with highly diverse and active microbial communities, also are a source of antibiotics for human and livestock (Ness 2015).

Managing Global Soil Carbon Pool

The importance of enhancing and sustaining global soil C (both SOC and SIC) pool is gaining momentum for provisioning of several ecosystem services, but especially to mitigate climate change and advance food security. For example, recommendation of the "4 per Thousand" program at COP21 in Paris in November/December 2015 is indicative of the political significance of this program

to address global issues. Therefore, management of soil health is crucial to understanding the dynamics and management of the SOC and SIC pools, and vice versa. In the context of climate change, it is also important to understanding the temperature-sensitivity of SOM decomposition and its effect on the global SOC pool in a warming earth (Zhang 2010). The projected increase in global temperature may also aggravate the risks of positive feedback between elevated CO_2 and SOM decomposition (Wolf et al. 2007). Better understanding of the SOC pool and dynamics in diverse ecoregions may be essential to their judicious management (Scharlemann et al. 2014). Yet, there is a huge potential of SOC sequestration in diverse regions, such as in Europe (Aertsens et al. 2013), USA (Lal et al. 2003), Brazil (Sa et al. 2016), etc. In addition to modeling (Campbell and Paustian 2015), understanding of the SOC pool at the regional-scale (e.g., the U.S. Corn Belt) is also important (Collins et al. 2000). Understanding of such basic processes is critical to the choice of site-specific management through the use of organic amendments (Cooperband 2002), tillage systems (Franzluebbers 2008; Overstreet and DeJong-Huges 2008), appropriate systems of intensification of agroecosystems (Liao et al. 2015), or management of irrigated systems (Cochran et al. 2010). Identification of researchable priorities should be based on a thorough analysis of the known and unknowns (Stockmann et al. 2013).

In general, soil health is affected more by the dynamics of SOC than SIC. Yet, formation of secondary carbonates, transport of bicarbonates into the shallow groundwater, and weathering of silicates can also affect atmospheric CO_2 and the global climate change. Management systems (e.g., composting, mulching, manuring, irrigation, limiting) can affect both SOC and SIC pools.

A judicious use of fertilizers implies that they supplement the supply of plant nutrients through natural systems of biological fixation, recycling of biomass and management of soil resources for efficient use in a manner compatible to environmental quality (Lal 2016).

Conclusions

Hayne (1940) stated that, "if we feed the soil, it will feed us," and that "only productive soil can support a prosperous people." Thus, maintaining soil health is essential to human health, ecosystem functions and nature conservancy. However, impact of soil health goes beyond human health, it also has a profound impact on atmosphere, biosphere, and the hydrosphere. The importance of soil health on mitigating climate change, improving water quality, enhancing biodiversity etc. cannot be overemphasized. The environmental consequence of soil health is also determined by the SOC pool, its dynamics and the turnover time, and that of climate change by both SOC and SIC dynamics.

Soil health is more pertinent to global issues now than ever before. Its management is essential to advancing food and nutritional security, critical to mitigating and adapting to changing and uncertain climate, important to reducing nonpoint source pollution, and eutrophication, pertinent to enhancing soil biodiversity, and needed to sustainable intensification of agroecosystems through enhancing use efficiency of inputs (water, nutrients) and reducing losses.

Conflict of Interest

The author is director of the Carbon Management and Sequestration Center, School of Environment and Natural Resources, The Ohio State University, Columbus, OH 43210 USA.

References

Aertsens, J., L. De Nocker, and A. Gobin. 2013. Valuing the carbon sequestration potential for European agriculture. Land Use Policy 31:584–594.

Albrecht, W. A. 1945. Soil fertility and its health implications. Am. J.Orthod. Oral Surg. Orthod. 31:279–286.

Albrecht, W. A. 1951. Pattern of caries in relation to the pattern of soil fertility in the United States. Dent. J. Australia 23:1–6.

Albrecht, W. A. 1957. Soil fertility and biotic geography. Geogr. Rev. 47:86–105.

Andrews, S., D. Karlen, and C. Cambardella. 2004. The soil management assessment framework: a quantitative soil quality evaluation method. Soil Sci. Soc. Am. J. 68:1945–1962.

Baker, K. F., and R. J. Cook. 1974. Biological control of plant pathogens. American Phytopathology Society, San Francisco, CA, 433 pp.

Balfour, E. B. 1943. The living soil. Faber and Faber Ltd., London, U.K.

Barta, G. 2011. Secondary carbonates in loess-paleosoil sequences: a general review. Cent. Eur. J. Geosci. 3:129–146.

Blank, R. R., and M. A. Fosberg. 1990. Micromorphology and classification of secondary calcium carbonate accumulations that surround or occur on the undersides of coarse fragments in Idaho (USA). Dev. Soil Sci. 19:341–346.

Brevik, E. C., and T. J. Sauer. 2015. The past, present, and future of soils and human health studies. Soil 1:35–46.

Bughio, M. A., P. Wang, F. Meng, C. Chen, Y. Kuzyakov, X. Wang, et al. 2015. Neoformation of pedogenic carbonates by irrigation and fertilization and their contribution to carbon sequestration in soil. Geoderma 262:12–19.

Campbell, E. E., and K. Paustian. 2015. Current developments in soil organic matter modeling and the expansion of model applications: a review. Environ. Res. Lett. 10:123004. doi:10.1088/1748-9326/10/12/123004.

Cardoso, E. J. B. N., R. L. F. Vasconcellos, D. Bini, M. Y. H. Miyauchi, C. A. dos Santos, P. R. L. Alves, et al. 2013. Soil health: looking for suitable indicators. What should be considered to assess the effects of use and management on soil health? Sci. Agricola 70:274–289.

Cochran, R. L., H. P. Collins, A. Kennedy, and D. F. Bezdicek. 2010. Soil carbon pools and fluxes following land conversion to irrigated agriculture in a semi-arid shrub-steppe ecosystem. Biol. Fertil. Soils 43:479–489.

Collins, H. P., E. T. Elliott, K. Paustian, L. C. Bundy, W. A. Dick, D. R. Huggins, et al. 2000. Soil carbon pools and fluxes in long-term corn belt agroecosystems. Soil Biol. Biochem. 32:157–168.

Cooperband, L. 2002. Building soil organic matter with organic amendments: a resource for urban and rural gardeners, small famers, turfgrass managers and large-scale producers. University of Wisconsin-Madison, Center for Integrated Agricultural Systems, Madison, WI.

Daily, G. C. 1997. The potential impacts of global warming on managed and natural ecosystem: Implications for human well-being. Abstracts of Papers of the American Chemical Society 213: 12-ENVR.

Demmings, E.. 1943. Statistical adjustment of data. Dover Publications, Oxford, U.K., 261 pp.

Doran, J. W., and M. R. Zeiss. 2000. Soil health and sustainability: managing the biotic component of soil quality. Appl. Soil Ecol. 15:3–11.

Doran, J. W., A. J. Jones, (Eds.), 1996. Methods for Assessing Soil Quality. Soil Science Society of America Special Publication 49: Soil Science Society of America, Madison, Wisconsin.

Doran, J. W., T. B. Parkin, J. W. Doran, D. C. Coleman, D. F. Bezdicek, and B. A. Stewart. 1994. Defining and assessing soil quality. In Doran, J. W., Coleman, D. C., Bezdicek, D. F., Stewart, B. A. (Eds.), Defining Soil Quality for a Sustainable Environment. Soil Science Society of America Special Publication, 35: Soil Science Society of America, Madison, Wisconsin, pp. 3–21.

Doran, J., M. Sarrantonio, M. Liebig, and D. Sparks. 1996. Soil health and sustainability. Adv. Agron. 56:1–54.

Dungait, J. A. J., D. W. Hopkins, A. S. Gregory, and A. P. Whitmore. 2012. Soil organic matter turnover is governed by accessibility not recalcitrance. Glob. Change Biol. 18:1781–1796.

Dyson, F. 2008. The question of global warming. New York Review of Books, 30 June 2010.

Ehmke, T. 2013. Soil health: feature article. Crop and Soils Magazine:4–9.

Ekschmitt, K., M. Q. Liu, S. Vetter, O. Fox, and V. Wolters. 2005. Strategies used by soil biota to overcome soil organic matter stability - why is dead organic matter left over in the soil? Geoderma 128:167–176.

Ekschmitt, K., E. Kandeler, C. Poll, A. Brune, F. Buscot, M. Friedrich, et al. 2008. Soil-carbon preservation through habitat constraints and biological limitations on decomposer activity. J. Plant Nutr. Soil Sci. 171:27–35.

Franzluebbers, A. J. 2008. Soil organic carbon sequestration with conservation agriculture in the southeastern USA: potential and limitations. Available at: http://www.fao.org/ag/ca/CarbonOffsetConsultation/Carbonme

Harris, R., D. Bezdicek. 1994. Descriptive aspects of soil quality health. In Doran, J. W., Coleman, D. C., Bezdicek, D. F., Stewart, B. A. (Eds.), Defining Soil Quality for a Sustainable Environment. Soil Science Society of America Special Publication, 35: Soil Science Society of America, Madison, Wisconsin, pp. 23–25.

Hayne, R. A. 1940. Make the soil productive: We can't grow crops on poor land, Education Series 2, Chicago, IL.

Howard, A. 1940. An agricultural testament. Oxford Univ. Press, London, U.K.

Howard, A. 1947. The soil and health: a study of organic agriculture. Devin-Adair Company, New York, NY.

Jansson, C., S. D. Wullschleger, U. C. Kalluri, and G. A. Tuskan. 2010. Phytosequestration: carbon Biosequestration by plants and the prospects of genetic engineering. Bioscience 60:685–696.

Janvier, C., F. Villeneuve, C. Alabouvette, V. Edel-Hermann, T. Mateille, and C. Steinberg. 2007. Soil health through soil disease suppression: which strategy from descriptors to indicators? Soil Biol. Biochem. 39:1–23.

Karlen, D., N. Wollenhaupt, D. Erbach, E. Berry, J. Swan, N. Eash, et al. 1994. Long-term tillage effects on soil quality. Soil Till. Res. 32:313–327.

Karlen, D., M. Mausbach, J. Doran, R. Cline, R. Harris, and G. Schuman. 1997. Soil quality: a concept, definition, and framework for evaluation. Soil Sci. Soc. Am. J. 61:4–10.

Kibblewhite, M. G., K. Ritz, and M. J. Swift. 2008. Soil health in agricultural systems. Philos. Trans. R. Soc. B Biol. Sci. 363:685–701.

Kleber, M. 2010. What is recalcitrant soil organic matter? Environ. Chem. 7:320–332.

Kleber, M., P. S. Nico, A. F. Plante, T. Filley, M. Kramer, C. Swanston, et al. 2011. Old and stable soil organic matter is not necessarily chemically recalcitrant: implications for modeling concepts and temperature sensitivity. Glob. Change Biol. 17:1097–1107.

Knight, H. G., C. E. Kellogg, C. P. Barnes, M. A. McCall, B. W. Allin, A. L. Patrick, et al. 1938. Soils and men. USDA Yearbook of Agriculture, Washington, DC.

Lal, R. 1993. Agronomic sustainability of different farming systems on Alfisols in Southwestern Nigeria. J. Sustain. Agric. 4:33–51.

Lal, R. 1994. Methods and Guidelines for Assessing Sustainable Use of Soil and Water Resources in the Tropics. USDA/SMSS Bull. 21, Washington, DC, 78 pp.

Lal, R. 1997. Degradation and resilience of soils. Philos. Trans. R. Soc. Lond. B. 352:997–1010.

Lal, R. 2004. Soil carbon sequestration impacts on global climate change and food security. Science 304:1623–1627.

Lal, R. 2008. Carbon sequestration. Philos. Trans. R. Soc. B Biol. Sci. 363:815–830.

Lal, R. 2009. Soil degradation as a reason for inadequate human nutrition. Food Sec. 1:45–57.

Lal, R. 2014. Societal value of soil carbon. J. Soil Water Conserv. 69:186A–192A.

Lal, R. 2016. Beyond COP21: potential and challenges of the "4 per Thousand" initiative. J. Soil Water Conserv. 71:20A–25A.

Lal, R., R. F. Follett, and J. M. Kimble. 2003. Achieving soil carbon sequestration in the US: a challenge to policy makers. Soil Sci. 168:1–19.

Larkin, R. P., and N. K. VanAlfen. 2015. Soil health paradigms and implications for disease management. Annu. Rev. Phytopathol. 53:199–221.

Larson, W. E., and F. J. Pierce. 1991. Conservation enhancement of soil quality. Int. Board Soil Res. Manage. Proc. 2:175–203.

Liao, Y., W. L. Wu, F. Q. Meng, P. Smith, and R. Lal. 2015. Increase in soil organic carbon by agricultural intensification in northern China. Biogeosciences 12:1403–1413.

Lovelock, J. 1979. Gaia: new look at life on earth. Oxford University Press, Oxford, U.K.

von Lützow, M., and I. Kögel-Knaber. 2009. Temperature sensitivity of soil matter decomposition- what do we know? Biol. Fertil. Soils 46:1–15.

Ma, J., R. Liu, L. S. Tang, Z. D. Lan, and Y. Li. 2014. A downward CO_2 flux seems to have nowhere to go. Biogeosciences 11:6251–6262.

McCarrison, R. 1921. Studies in deficiency disease. Hazell Watson and Viney Ltd, London, U.K.

Monger, H. C., R. A. Kraimer, S. Khresat, D. R. Cole, X. J. Wang, and J. P. Wang. 2015. Sequestration of inorganic carbon in soil and groundwater. Geology 43:375–378.

Ness, E. 2015. The hunt for antibiotics in soil. CSA News. doi:10.2136/sh2015-56-5-f.

Overstreet, L. F., and J. DeJong-Huges. 2008. The importance of soil organic matter in cropping systems of the Northern Great Plains. University of Minnesota Extension. Available at: http://www.extension.umn.edu/agriculture/tillage/importance-of-soil-organic-matter/

Parr, J. F., R. I. Papendick, S. B. Hornick, and R. E. Meyer. 1992. Soil quality: Attributes and relationship to alternative and sustainable agriculture. Amer. J. Alternative Agric. 7:5–11.

Rodale, J. I. 1945. Pay dirt: farming and gardening with composts. Devin-Adair Company, New York, NY.

Sa, J. C. M., R. Lal, C. C. Cerri, K. Lorenz, M. Hungria, P. C. F. Carvalho. 2016. Low-carbon agriculture in South America to mitigate global climate change and advance food security. Catena (accepted).

Scharlemann, J. P. W., E. V. J. Tanner, R. Hiederer, and V. Kapos. 2014. Global soil carbon: understanding and managing the largest terrestrial carbon pool. Carbon Manag. 5:81–91.

Singh, B., and J. Ryan. 2015. Managing fertilizers to enhance soil health. IFA, Paris, France.

Six, J., E. T. Elliott, and K. Paustian. 2000. Soil macroaggregate turnover and microaggregate formation: a mechanism for C sequestration under no-tillage agriculture. Soil Biol. Biochem. 32:2099–2103.

Six, J., R. T. Conant, E. A. Paul, and K. Paustian. 2002. Stabilization mechanisms of soil organic matter: implications for C-saturation of soils. Plant Soil 241:155–176.

Smith, J., J. Halvorson, and R. Papendick. 1993. Using multiple-variable indicator kriging for evaluating soil quality. Soil Sci. Soc. Am. J. 57:743–749.

Smith, P., M. F. Cotrufo, C. Rumpel, K. Paustin, P. J. Kuikman, J. A. Elliott, et al. 2015. Biogeochemical cycles and biodiversity as key drivers of ecosystem services provided by soils. Soil 1:665–685.

Stockmann, U., M. A. Adams, J. W. Crawford, D. J. Field, N. Henakaarchchi, M. Jenkins, et al. 2013. The knowns, known unknowns and unknowns of sequestration of soil organic carbon. Agric. Ecosyst. Environ. 164:80–99.

Sylvia, D. M., J. J. Fuhrmann, P. G. Hartel, and D. A. Zuberer. 1998. Principles of soil microbiology. Prentice Hall, Upper Saddle River, NJ, 550 pp.

Voisin, A. 1959. Soil, grass, and cancer. Philosophical Library, New York, NY.

Warkentin, B. 1995. The changing concept of soil quality. J. Soil Water Conserv. 50:226–228.

Wolf, A. A., B. G. Drake, J. E. Erickson, and J. P. Megonigal. 2007. An oxygen-mediated positive feedback between elevated carbon dioxide and soil organic matter decomposition in a simulated anaerobic wetland. Glob. Change Biol. 13:2036–2044.

Zhang, J. 2010. Temperature sensitivity of soil organic matter decomposition and the influence of soil carbon and attributes. [Graduate Theses and Dissertations], Iowa State University, Paper 11234.

4

Estimation of above-ground live biomass and carbon stocks in different plant formations and in the soil of dry forests of the Ecuadorian coast

Carlos A. Salas Macías[1] (iD), Julio C. Alegre Orihuela[2] & Sergio Iglesias Abad[3]

[1]Facultad de Ingeniería Agronómica, Universidad Técnica de Manabí, Lodana, Santa Ana, Ecuador
[2]Universidad Nacional Agraria La Molina, Lima, Perú
[3]Universidad Católica de Cuenca, Cuenca, Ecuador

Keywords
Above-ground biomass, allometric equations, carbon stocks, dry forest, global change.

Correspondence
Carlos A. Salas Macías, Department of Agronomic Sciences, Universidad Técnica de Manabí (UTM), Lodana, Santa Ana, Ecuador.

E-mail: csalas@utm.edu.ec

Funding Information
No funding information provided.

Abstract

Dry forests are very fragile ecosystems as they are easily used as a source of subsistence products. In this sense, quantifying the carbon stock in these forests is of relevant importance for their conservation and to be able to quantify their participation as mitigation of the effects of climate change. Five 250 m^2-sample plots were established to estimate carbon stored in two pools for each of the plant formations identified (Dry Scrubland, DS; Dry Deciduous Forest, DDF; Dry Semideciduous Forest, DSF). The amount of carbon stored in soils was determined by analyzing the organic carbon randomly taken in each plot. Allometric equations were used to estimate the amount of carbon in above-ground biomass, taking the total height (H) and the diameter at breast height (DBH) of trees whose DBH is equal to or greater than 5 cm. The total carbon stored in each plant formation was estimated by adding the amount of carbon in biomass and in soils, resulting in 60.30, 69.62, and 123.05 Mg of carbon per hectare for the DS, DDF, and DSF, respectively.

Introduction

One of the environmental problems arising from human development is the increase in greenhouse gas emissions. Specifically, CO_2 emissions have prompted research related to climate change so as to contribute with knowledge that will help mitigate such emissions (Pérez et al. 2009; Fonseca et al. 2011; Aguilar-Arias et al. 2012). According to the United Nations Framework Convention on Climate Change (UNFCCC), adopted in New York in May 1992, natural climate variability observed over comparable time periods are attributed directly or indirectly to human activity (land-use change, use of fossil fuels, use of agrochemicals, among others), which cause an increase in "greenhouse gas" concentrations in the atmosphere, affecting the increase in the average global temperature.

In view of this, the IPCC (2007) warns that, in the future, gases such as nitrous oxide (N_2O), carbon dioxide (CO_2), methane (CH_4), and ozone (O_3) would produce a global temperature increase between 3°C and 5°C, which would affect current precipitation patterns due to its impact on the land-ocean-atmosphere system. Of these, carbon dioxide (CO_2) is the most significant because of the large quantities produced as a result of human activity. In addition, about 20% of CO_2 emissions result from degradation or removal of natural ecosystems such as forests (Schimel et al. 2001; Schlegel et al. 2001). However, and in order to gather information regarding mitigation and adaptation options, the premise is that it is possible to capture carbon dioxide from the atmosphere and store it in the ecosystems themselves, preventing them from accumulating in the atmosphere.

In this context, several studies point out the potential of forests in terms of carbon storage, including studies carried out by Yatskov (2016), Jew et al. (2016), Fonseca et al. (2015, 2011, 2009), De Britez (2007), Mutuo et al. (2005), Oelbermann et al. (2004), Schimel et al. (2001), Ávila et al. (2001), among others. This is how forest ecosystems appear as large carbon sinks containing more than 80% of all above-ground carbon.

Nevertheless, carbon storage capacity can vary markedly depending on the structure and composition of a forest. It could be therefore assumed that rainforests, because of their diversity and the size of the individuals living in them, have greater carbon storage capacity than dry forests. The latter are some of the least known and most threatened terrestrial ecosystems (Murphy and Lugo 1986); they are characterized by seasonal ecological and production processes, and, when compared to rainforests, are of lower stature and basal area (Gentry et al. 1995; Linares-Palomino 2004a,b).

In general, Ecuador's dry forests are scarcely known, highly threatened, and economically important for large segments of the rural population, as they provide timber and nontimber products for subsistence and sometimes for sale. Several researches have been carried out on this type of forest, but it was not until only a few years ago that, after an intense and successful project, Ecuador managed to obtain important data at the country level, thanks to the so-called National Forest Assessment.

This type of (dry) forest can be found along the Ecuadorian coastline, the most diverse of which are located in the province of Loja (219 species), Guayas (169 species), and Manabí (143 species; Aguirre et al. 2006). In Manabí, dry forests are little known and highly threatened because of the economic importance they represent for certain rural sectors for whom they provide timber and consumer products.

In this sense, and as research progresses, a significant number of methodologies and guidelines have been established. Often, inventories use permanent plots of measurement to obtain statistically reliable data and reduce monitoring costs. In this regard, there are two commonly used methods to estimate biomass: the direct and the indirect methods.

The direct or destructive method consists in cutting down trees and determining their biomass by directly weighing each component. However, in this case, an indirect method was used given that this research was carried out in a reserve area.[1] First, the above-ground biomass was estimated followed by the carbon stored in said biomass.

This indirect method consists in the use of models based on mathematical equations that relate biomass to tree variables (DBH, total height, wood density, crown diameter, among others). Above-ground biomass may be calculated using allometric equations provided that a statistically representative sampling is designed to measure the independent variables of the selected allometric equation.

The purpose of this study is to estimate the carbon stocks of the above-ground biomass expressed in megagrams (Mg) of oven-dry weight/unit area, in addition to the carbon stored in the soils of three plant formations in dry forests along the Ecuadorian coastline.

Materials and Methods

The research was carried out in an area located along the center of the Ecuadorian coast, consisting of the eastern and western slopes of the Pacoche, Los Lugos, Agua Fría, and Monte Oscuro mountains, which form part of the discontinuous massif of the coastal mountain range in Manabí. Politically, the area is part of the parishes of San Lorenzo and Montecristi, belonging to the cantons of Manta and Montecristi, respectively, and within microregion 4 of the province of Manabí. Specifically, the area under study was the terrestrial part of the *Refugio de Vida Silvestre Marino Costero Pacoche* (Pacoche Marine and Coastal Wildlife Refuge; *RVSMC-Pacoche*), which occupies around 5096.41 ha (MAE, 2009).

The area stretches from sea level to 363 m above sea level. It is crossed longitudinally by the E15 or Marginal Way of the Coast, which connects Manabí with provinces Península de Santa Elena, to the South, and Esmeraldas, to the North. The southern boundary is located 30 km from Puerto Cayo, a small town with tourist importance in the area. To the north, and 25 km from the boundary, lies the city of Manta, the closest city with tourism potential.

Even though the Ecuadorian State, through the National Forest Assessment developed by the Ministry of the Environment, has established a methodology for carbon studies, such methodology considers many other parameters that are not included in the scope of this research. However, both the methodology developed by the Ministry of the Environment and that of this paper use the same allometric equation to estimate carbon in above-ground live biomass.

Determining plot type and number

Determining the type and number of plots was subject to the type of coverage, the precision required, the availability of resources for the development of field activities, and laboratory analyses (Rügnitz et al. 2009), and the objectives of the research (IDEAM, 2010; Yepes and Duque 2011). Given the exploratory purpose, the number of plots and sampling intensity is based on minimum sampling for carbon estimation investigations in areas with low tree density and small diameters (Rügnitz et al. 2009). However, the calculations to determine the number of plots were verified using the Winrock Sample Plot

Calculator Spreadsheet Tool, a tool developed within the framework of the Clean Development Mechanism (CDM; Walker et al. 2014).

Thus, temporary plots, generally used in rapid exploratory sampling, were determined. The information gathered result from specific data that do not require delimiting the unit or marking individuals for a periodic evaluation (Melo and Vargas 2003). Each plot had an area of 250 m^2 (10 m × 25 m), and five replications (MINAM, 2009) were installed in each of the previously determined plant formations. Sampling intensity was set at ±20% (Pearson et al. 2005) of the average carbon at a 95% confidence level.[2] This means, for example, that in 95% of the cases in which a 100 Mg C per ha carbon value is identified, the actual amount will be between 80 and 120 Mg C per ha.

Carbon pools

The carbon pools of the various forest formations were selected based on logistic factors (ease to transport samples to laboratories, and technical aspects for REDD[3] projects). In this context, two of the five pools that can be measured (Brown 2002; Rügnitz et al. 2009; IDEAM, 2010; Yepes and Duque 2011) were selected: above-ground live biomass and soil carbon.

Carbon stored in above-ground biomass

Compared to the destructive method, the indirect method is less expensive, and requires less time and less resources, which is why the latter was used to determine the carbon stores in above-ground biomass. In any case, destructive methods would not have been possible given that the area of study is a protected area. Thus, the allometric equation for mixed dry forests proposed by Chave et al. (2005) was used, requiring to measure variables in trees within the plots, to be entered into the following model:

$$AGB_{est} = \exp\left(-2.187 + 0.916 \times \ln\left(\rho D2H\right)\right)$$
$$\equiv 0.0112 \times (\rho D2H)\, 0.916,$$

where AGB$_{est}$ = Estimated above-ground biomass (kg DM./tree); ρ = Wood density (g/cm^3); D = Diameter at chest height (cm); H = Total height of the tree (m).

Measurement of dasometric variables

Once the sampling plots were installed, the required measurements were taken to apply to the allometric model. Such model includes measurements of the total height (m), DBH (cm), and wood density (g/cm^3). Only trees with diameter at chest height >5 cm were measured.

With regard to wood density, species were identified in each sampling plot and the "Global Wood Density

Database was used (Zanne et al. 2009). The objective set was to obtain the wood density for each species. However, in cases where there were no data in the database or in other bibliographic sources, the density of the genus or family was used. (Honorio and Baker 2010).

Calculation of carbon stock in above-ground biomass per hectare

After calculating the above-ground biomass (kg dry matter/tree), the total biomass is calculated in megagrams per hectare (Mg/ha), and this value is extrapolated to the hectare, as follows:

$$AGB = \left(\sum AU/1000\right) \times (10{,}000/\text{plot area}),$$

where AGB = Above-ground tree biomass (Mg DM/ha); $\sum$ AU = Sum of the tree biomass of all trees in the plot (kg DM/plot area); Factor 1000 = Conversion of sample units of kg DM/Mg; DM Factor 10,000 = Conversion of the area (m^2) to hectare.

Above-ground biomass to carbon conversions were performed pursuant to the guidelines established in the IPCC Good Practice Guidance for Land Use, Land-Use Change and Forestry (Penman et al. 2003), which assumes carbon content to be 50% of the above-ground biomass of each living tree (Barrett and Christensen 2011; Barrett 2014; Hetland et al. 2016; Jew et al. 2016; Rajput 2016; Tashi et al. 2016; Vijayakumar et al. 2016; Yatskov 2016).

$$\Delta AGB = (AGB \times CF),$$

where ΔAGB = Carbon amount in above-ground biomass (Mg C/ha); AGB = Above-ground tree biomass (Mg DM/ha); CF = Carbon Fraction (Mg C/t DM). The default value is 0.50.

Soil carbon

The total amount of carbon in soil (%) was measured in each layer of the profile at depths of 0–10 cm, 10–20 cm, and 20–30 cm. Bulk density (g/cc) was measured at each depth using the cylinder method in undisturbed soils. Wet and dry soils were measured to calculate the dry soil in the cylinder and percentage of soil moisture using the following formula:

$$D_{ap} = \frac{Pss}{Vol_c},$$

where BD = Bulk Density; DS$_w$ = Dry Soil weight; WS$_w$ = Wet Soil Weight; Vol$_c$ = Volume of the Sampling Cylinder.

In the meantime, three soil samples were taken to determine organic carbon by applying the Walkley–Black method (wet oxidation method; Bazan 1996). The estimation of soil carbon stocks per unit area on a plot is calculated using the following formula (Eggleston et al. 2006; Rügnitz et al. 2009):

$$COS_t = \sum_{horizon=i}^{horizon=n} \left(\left[\left(BD_i \times TH_i \times \left[1 - \frac{CR_i}{100} \right] \right) \times C_i \right] \times 100 \right),$$

where COS_t = Full profile organic carbon (Mg/ha); BD_i = Bulk density of horizon i (g/cm³); TH_i = Thickness of horizon i (m); CR_i = Volume of thick fragments of the horizon i (vol. %); C_i = % of organic carbon in i horizon (%).

Carbon stored in each plant formation

The total carbon stored in each plant formation will be given by the sum of the components as follows:

$$C_T = C_S + C_B,$$

where: C_T = Total Carbon; C_S = Carbon in soil; C_{BA} = Total carbon stored biomass (Mg C/ha).

Results and Discussion

The variability in the biophysical characteristics in the different forest formations in the study area (microclimate, land cover, use or conservation status) causes differences in the carbon stored inside each formation (IDEAM, 2010; Phillips et al. 2011; Yepes and Duque 2011).

Total carbon stored in the above-ground biomass was higher in the Dry Semideciduous Forest (DSF), decreasing in the dry deciduous forest and dry scrubland (DS; Table 1). This situation responds to the fact that in the DSF there are trees of larger size and diameter and more diversity. Moreover, MS forests are characterized by shrub vegetation and are more threatened by human activity.

The results obtained are in agreement with the research carried out on this type of plant formation where carbon storage in above-ground biomass in dry forests could be between 25 and 60 Mg C/ha (Brown et al. 1989; Brown and Lugo 1992; Sánchez and Méndez 2003). Similarly, the Ministry of Environment of Ecuador (MAE), in its publication "*Estadísticas de Patrimonio Natural*" (Natural

Heritage Statistics; MAE 2015) reports a mean dry forest carbon data of 37 Mg/ha which is in line with what was found in this study.

Furthermore, soil is an important carbon sink, containing more carbon than the sum in vegetation and the atmosphere (Swift 2001). This is why the IPCC recommends that it be considered as one of the compartments that should be evaluated in greenhouse gas inventories, for which the estimation is suggested to a depth of 30 cm (Eggleston et al. 2006; Solomon 2007). Accordingly, the results of the estimation of carbon stored in the study area are shown in Table 2.

In general terms, the results show low bulk densities at different depths and were not significantly different; they fluctuated between 0.95 and 1.15 g/cm³, which indicate that DDF and DSF organic soils are rich in humus, approaching the characteristics of loam soils; these being a little more clayey. These values tend to increase with depth, due to the greater biological activity in the horizon A and in DS.

The organic matter content for DS showed values ranging from 1.50% to 0.90%. Because many of these areas are intended for livestock, organic matter in the first horizon could increase. However, the values found do not show significant differences. Similarly, DDF results range from 1.00% to 3.30%, with the highest value corresponding to the first 10 cm of soil. In DSF, the highest organic matter value in the soil (2.00–5.40%) was found in the first 10 cm of the soil profile. This could be explained by the greater littering and biological activity in horizon A.

Therefore, based on the recommendation issued by the United States Department of Agriculture (USDA)'s Soil Survey Laboratory (SSL), the Van Bemmelen correction factor (1.724) was used to calculate the total carbon stored, assuming that the organic matter has 58% of organic carbon, yielding the following values: 26.83, 31.13, and 63.28 Mg C/ha for DS, DDF, and DSF, respectively.

Table 2. Carbon stored in soils of plant formations in the area of study.

| Plant formation | Organic matter (%) | Bulk density | | Carbon stored (Mg C/ha) | Total carbon (Mg C/ha) |
		Depth (cm)	BD (g/cm³)		
DS	1.50	0–10	1.08	15.66	26.83
	0.90	10–20	1.05	5.48	
	0.90	20–30	1.09	5.69	
DDF	3.30	0–10	0.95	18.18	31.13
	1.00	10–20	1.01	5.86	
	1.10	20–30	1.11	7.08	
DSF	5.40	0–10	1.11	34.77	63.28
	2.40	10–20	1.09	15.17	
	2.00	20–30	1.15	13.34	

DS, Dry Scrubland; DDF, Dry Deciduous Forest; DSF, Dry Semideciduous Forest; BD, Bulk density.

Table 1. Above-ground biomass carbon (Mg/ha) stored in the plant formations present in the study area.

Plant formations	Carbon stored (Mg C/ha)	SD
DS	33.47	13.26
DDF	38.49	24.43
DSF	59.77	25.93

DS, Dry Scrubland; DDF, Dry Deciduous Forest; DSF, Dry Semideciduous Forest; SD, standard deviation.

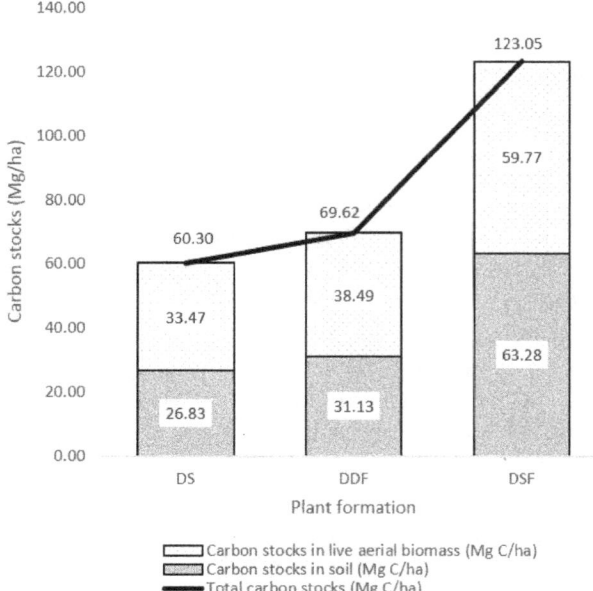

Figure 1. Carbon stocks (Mg/ha) for each of the plant formations under study. DS, Dry Scrubland; DDF, Dry Deciduous Forest; DSF, Dry Semideciduous Forest.

Moreover, the estimated values of soil carbon storage in the study area are in agreement with Balesdent and Arrouays (1999) and Trumbmore et al. (1995) who reported stocks between 60 and 70 Mg C/ha in forest soils. Along with evidence of soil carbon storage, it should also be considered that the change in soil carbon content due to land use does not exceed 20 Mg C/ha (IPCC, 2007).

Figure 1 shows the estimated data of carbon stored in biomass, carbon stored in soil, and total carbon for each of the plant formations under study. Thus, it can be observed that DSF contains more carbon in the above-ground biomass (59.77 Mg/ha) than DDF and DS (38.49 and 33.47 Mg/ha, respectively). Carbon stored in soil followed this trend with 63.28 Mg/ha of carbon stock in DSF, followed by DDF with 31.13 and 26.83 Mg/ha for DS. The total carbon stored in each plant formation was represented by the sum of carbon in above-ground live biomass and soil carbon, yielding values of 60.30, 69.62, and 123.05 Mg of Carbon per hectare for DS, DDF, and DSF, respectively.

Conclusions

The carbon stored in live above-ground biomass was higher in Dry Semideciduous Forest (59.77 Mg C/ha) followed by the formation of Dry Deciduous Forest (38.49 Mg C/ha) and Dry Scrubland (33.47 Mg C/ha).

The soils of the Dry Semideciduous Forest formation have more stored carbon (63.28 Mg C/ha) than the Dry Deciduous Forest (31.13 Mg C/ha) and Dry Scrubland (26.83 Mg C/ha).

The formation of dry semi-deciduous forest contains more total carbon stocks (123.05 Mg C/ha) than the formations of Dry Deciduous Forest (69.62 Mg C/ha) and Dry Scrubland (60.30 Mg C/ha).

For this case, the carbon stock was related to altitude; at higher altitude, higher carbon stocks.

Acknowledgments

Thanks are extended to the Secretaría de Educación Superior, Ciencia, Tecnología e Innovación (SENESCYT); Instituto de Fomento al Talento Humano del Ecuador; Universidad Nacional Agraria La Molina (UNALM); Universidad Técnica de Manabí (UTM); Ministerio de Ambiente del Ecuador (MAE); Dirección del Refugio de Vida Silvestre Marino Costero Pacoche; Ing. Juan Manuel Moreira Castro (Jardín Botánico UTM).

Conflict of Interest

None declared.

Notes

[1]Declared as a wildlife refuge by Ministerial Agreement No. 131, dated September 2, 2008.
[2]Usually, for forest projects, a precision level (sampling error) of ±10% of the average carbon value is used at a 95% confidence level. However, in the case of small-scale CDM projects, a level of accuracy of ±20% is used (Pearson et al. 2005; Emmer 2007).
[3]In carbon projects, it is essential to include above-ground biomass as a sink, as it is the pool that is most affected by deforestation/degradation of forests (BioCarbonFund, 2008).

References

Aguilar-Arias, H., E. Ortiz-Malavasi, B. Vílchez-Alvarado, and R. L. Chazdon. 2012. Biomasa sobre el suelo y carbono orgánico en el suelo en cuatro estadios de sucesión de bosques en la Península de Osa, Costa Rica. Rev. For. Mesoamericana Kurú 9:22–31.

Aguirre, Z., L. P. Kvist, and O. Sánchez. 2006. Bosques secos en Ecuador y su diversidad. Bot. Econ. Andes Centrales 2:162–187.

Ávila, G., F. Jiménez, J. Beer, M. Gómez, M. Ibrahim. 2001. Almacenamiento, fijación de carbono y valoración de servicios ambientales en sistemas agroforestales en Costa Rica. (CATIE) 8:32–35.

Balesdent, J., and D. Arrouays. 1999. Usage des terres et stockage de carbone dans les sols du territoire francais. Une estimation des flux nets annuels pour la periode 1990-1999. An estimate of the net annual carbon storage in French soils induced by land use change from 1900 to 1999 (note p). Comptes Rendus 85:265–277.

Barrett, T. 2014. Storage and flux of carbon in live trees, snags, and logs in the Chugach and Tongass National Forests.

Barrett, T. M., and G. A. Christensen. 2011. Forests of Southeast and South-Central Alaska, 2004–2008: Five-year forest inventory and analysis report.

Bazan, R. 1996. Manual para análisis químico de suelos, aguas y plantas. Universidad Nacional Agraria la Molina. Ed. F Perú, Lima, Perú.

BioCarbonFund. 2008. Methodology for estimating reductions in GHG emissions from mosaic deforestation. BioCarbon Fund, Washington, DC, USA.

Brown, S. 2002. Measuring, monitoring, and verification of carbon benefits for forest–based projects. Philos. Trans. Roy. Soc. Lond. A Math. Phys. Eng. Sci. 360:1669–1683.

Brown, S., and A. E. Lugo. 1992. Aboveground biomass estimates for tropical moist forests of the Brazilian Amazon. Interciencia 17:8–18.

Brown, S., A. J. R. Gillespie, and A. E. Lugo. 1989. Biomass estimation methods for tropical forests with applications to forest inventory data. For. Sci. 35:881–902.

Chave, J., C. Andalo, S. Brown, M. A. Cairns, J. Q. Chambers, D. Eamus, et al. 2005. Tree allometry and improved estimation of carbon stocks and balance in tropical forests. Oecologia 145:87–99.

De Britez, R. M.. 2007. Estoque e incremento de carbono em florestas e povoamentos de espécies arbóreas com ênfase na Floresta Atlântica do sul do Brasil. Embrapa Florestas; Sociedade de Pesquisa em Vida Selvagem e Educação Ambiental, Curitiba.

Eggleston, H. S., L. Buendia, K. Miwa, T. Ngara, and K. Tanabe. 2006. IPCC Guidelines for National Greenhouse Gas Inventories. Prepared by the National Greenhouse Gas Inventories Programme, vol. 4. IGES, Japan.

Emmer, I. 2007. Manual de contabilidad de carbono y diseño de proyectos. Proyecto Encofor. Quito, Ecuador., Ecofor, p. 22.

Fonseca, W., F. Alice, and J. M. Rey. 2009. Modelos para estimar la biomasa de especies nativas en plantaciones y bosques secundarios en la zona Caribe de Costa Rica. Bosque (Valdivia) 30:36–47.

Fonseca, W., J. M. R. Benayas, and F. E. Alice. 2011. Carbon accumulation in the biomass and soil of different aged secondary forests in the humid tropics of Costa Rica. For. Ecol. Manage. 262:1400–1408.

Fonseca, G., J. Galán, A. Enciso, R. Garduño, and E. Méndez. 2015. Cambio climático. s.l., CULCyT, 18.

Gentry, A. H., S. H. Bullock, H. A. Mooney, and E. Medina. 1995. Seasonally dry tropical forests. Pp. 146–194.

Hetland, J., P. Yowargana, S. Leduc, and F. Kraxner. 2016. Carbon-negative emissions: systemic impacts of biomass conversion: a case study on CO_2 capture and storage options. Int. J. Greenhouse Gas Control 49:330–342.

Honorio, E. N., and T. R. Baker. 2010. Manual para el monitoreo del ciclo del carbono en bosques amazónicos. Instituto de Investigaciones de la Amazonia Peruana, Lima, Perú.

IDEAM. 2010. Segunda comunicación nacional ante la Convención Marco de las Naciones Unidas sobre Cambio Climático. Instituto de Hidrología, Meteorología y Estudios Ambientales, Bogotá, Colombia.

IPCC. 2007. IPCC Guidelines for national greenhouse gas inventories: Reference Manual. 1996. 2007.

Jew, E. K. K., A. J. Dougill, S. M. Sallu, J. O'Connell, and T. G. Benton. 2016. Miombo woodland under threat: consequences for tree diversity and carbon storage. For. Ecol. Manage. 361:144–153.

Linares-Palomino, R. 2004a. Los bosques tropicales estacionalmente secos: I. El concepto de los bosques secos en el Perú. Arnoldia 11:85–102.

Linares-Palomino, R. 2004b. Los bosques tropicales estacionalmente secos: II. Fitogeografía y composición florística. Arnoldia 11:103–138.

MAE. 2009. Plan de manejo. Refugio de vida silvestre marina y costera Pacoche.

MAE. 2015. Estadísticas del Patrimonio Natural: Datos de bosques, ecosistemas, especies, carbono y deforestación del Ecuador continental. Pp. 1–20.

Melo, O., and R. Vargas. 2003. Evaluación Ecológica y Silvicultural de Ecosistemas Boscosos.

MINAM. 2009. Identificación de Metodologías existentes para determinar stock de carbono en ecosistemas forestales. Ministerio de Ambiente del Perú, Lima, Perú.

Murphy, P. G., and A. E. Lugo. 1986. Ecology of tropical dry forest. Annu. Rev. Ecol. Syst. 1986:67–88.

Mutuo, P. K., G. Cadisch, A. Albrecht, C. A. Palm, and L. Verchot. 2005. Potential of agroforestry for carbon sequestration and mitigation of greenhouse gas emissions from soils in the tropics. Nutr. Cycl. Agroecosyst. 71:43–54.

Oelbermann, M., R. P. Voroney, and A. M. Gordon. 2004. Carbon sequestration in tropical and temperate agroforestry systems: a review with examples from Costa Rica and southern Canada. Agr. Ecosyst. Environ. 104:359–377.

Pearson, T., S. Walker, and S. Brown. 2005. Sourcebook for Land use. Land-use change and forestry projects. Development 21:64.

Penman, J., M. Gytarsky, T. Hiraishi, T. Krug, D. Kruger, R. Pipatti, et al. 2003. Good practice guidance for land use, land-use change and forestry. Institute for Global Environmental Strategies, Hayama, Japan.

Pérez, A. R. Y., V. Pocomucha, and Y. Vargas. 2009. Carbono almacenado en diferentes sistemas de uso de la tierra del Distrito de José Crespo y Castillo, Huánuco, Perú. 48.

Phillips, J. F., A. J. Duque, K. R. Cabrera, A. P. Yepes, D. A. Navarrete, M. C. García, et al. 2011. Estimación de las reservas potenciales de carbono almacenadas en la biomasa aérea en bosques naturales de Colombia. Instituto de Hidrología, Meteorología, y Estudios Ambientales-IDEAM, Bogotá, CO.

Rajput, P. 2016. Carbon storage, soil enrichment potential and bio-economic appraisal of different land use systems in mid hill and sub-humid zone-II of Himachal Pradesh.

Rügnitz, M., M. Chacón, and R. Porro. 2009. Guía para la determinación de carbono en pequeñas propiedades rurales, 1st ed. Centro Mundial Agroflorestal (ICRAF)/ Consórcio Iniciativa Amazônica (IA), Lima, Perú.

Sánchez, M. D., and M. R. Méndez. 2003. Estudio FAO producción y sanidad animal; 155. Agroforestería para la producción animal en América Latina-II.

Schimel, D. S., J. I. House, K. A. Hibbard, P. Bousquet, P. Ciais, P. Peylin, et al. 2001. Recent patterns and mechanisms of carbon exchange by terrestrial ecosystems. Nature 414:169–172.

Schlegel, B., J. Gayoso, and J. Guerra. 2001. Manual de procedimientos para inventarios de carbono en ecosistemas forestales. Universidad Austral de Chile. Proyecto FONDEF D98I1076. Valdivia, Chile 18: 19–20.

Solomon, S.. 2007. Climate change 2007-the physical science basis: working group I contribution to the fourth assessment report of the IPCC, vol. 4. Cambridge University Press, New York, USA.

Swift, R. S. 2001. Sequestration of carbon by soil. Soil Sci. 166:858–871.

Tashi, S., B. Singh, C. Keitel, and M. Adams. 2016. Soil carbon and nitrogen stocks in forests along an altitudinal gradient in the eastern Himalayas and a meta-analysis of global data. Glob. Change Biol. 22:2255–2268.

Trumbore, S. E., E. A. Davidson, P. de Camargo, D. C. Nepstad, and L. A. Martinelli. 1995. Belowground cycling of carbon in forests and pastures of Eastern Amazonia. Global Biogeochem. Cycles 9:515–528.

Vijayakumar, D. B. I. P., F. Raulier, P. Bernier, S. Gauthier, Y. Bergeron, and D. Pothier. 2016. Cover density recovery after fire disturbance controls landscape aboveground biomass carbon in the boreal forest of eastern Canada. For. Ecol. Manage. 360:170–180.

Walker, S., T. Pearson, and S. Brown. 2014. Winrock Sample Plot Calculator Spreadsheet Tool.

Yatskov, M. A. 2016. The impact of disturbance on carbon stores and dynamics in forests of coastal Alaska.

Yepes, A. P., and Á. J. Duque. 2011. Protocolo para la estimación nacional y subnacional de biomasa-carbono en Colombia. Bogot DC, Colombia, 162.

Zanne, A., G. Lopez-Gonzalez, D. Coomes, J. Ilic, S. Jansen, S. Lewis, et al. 2009. Global wood density database. Dryad Identifier. http://hdl.handle.net/10255/dryad. 235.

5

The effect of aboveground biomass removal on soil macronutrient over time in Munesa Shashemane, Ethiopia

Aklilu Bajigo Madalcho

Department of Natural Resource Management, College of Dryland Agriculture, Jigjiga University, P.O. Box 1020, Jigjiga, Ethiopia

Keywords
Allometric, *Cupresus*, farmlands, *Saligna*

Correspondence
Aklilu Bajigo Madalcho, Department of Natural Resource Management, College of Dryland Agriculture, Jigjiga University, P.O. Box, 1020 Jigjiga, Ethiopia.

E-mail: bajigoaklilu05@gmail.com

Funding Information
No funding information provided.

Abstract

This study aims to compare the proportion of macronutrients in different parts of the aboveground biomass; and to estimate the amount of macronutrient loss from continuous farmlands (mechanized farm and traditional farm) and plantation stands (*Cupresus lusitanica* and *Eucalyptus saligna*) through harvesting. From 20×20 m^2 sized sample plots, a representative samples were collected from leaf, branch and stem of the plantation stands. Macronutrient analyses were carried out for the samples taken from plantations and maize crops on continuous farmlands. All data collected were subjected to ANOVA using the general linear model of SAS 9.0 at $P < 0.05$. The results of macronutrient in aboveground biomass of maize in the continuous farm lands shows that, the grain component contained as much as 50% of N and 60% of P compared to leaf and stem. However, a high proportion of K, Ca, and Mg were found in the leaf and stem components. In the case of *C. lusitanica*, 64.4% N, 70% P, 68.4% K, 65.3% Ca, and 68.75% of Mg was found in the stem compared to the leaf and branch. In the *E. saligna*, the stem contained 22% N, 30% P, 11.8% K, 26.6% Ca, and 12.5% Mg; with the remaining proportion of each nutrient found in the leaf and branch. Based on the results, much of the nutrients can be removed through the leaf and stem removal in the continuous farmlands. Complete removal of the biomass with little residue returned to the soil, for long periods under plantation land use, may also lead to macronutrient depletion. Therefore, to minimize this effect, the aboveground biomass harvest should focus on grain only on the farmlands, and stem only on the plantations.

Introduction

Tropical forest plantations are the most productive ecosystems in the world due to their fast growth (Binkley and Ryan 1998). Fast-growing trees have the ability to accumulate high amounts of nutrients in their biomass but may cause nutrient depletion in the soil. The studies conducted by Binkley and Ryan (1998) on the stands of *E. saligna* and *Albizia facaltaria* state that at stand age of 16 years, *Eucalyptus* plots had accumulated about 323 Mg/ha of aboveground biomass, and this was about 50% more than *Albezia* which had accumulated about 216 Mg/ha aboveground biomass. Some plant species shed their foliage in the dry season and release nutrients to the soil surface through mineralization while some evergreen species keep it in the leaf biomass for a longer period. The nutrient accumulation in the plant body also varies between different tree parts. The study conducted on *E. grandis* by Tiarks et al. (2004), shows that the fraction of macronutrients in utilizable stem wood is small as compared to total reserves at the site. The crown and bark accounts for the highest proportion of the nutrients that is, about 54% to 82% of the aboveground biomass macronutrient pools. The below-ground biomass remains in the soil and returns nutrients through decomposition of root debris whereas the aboveground biomass is almost always harvested. If not only logs, but also branches and twigs, are used as fuel wood, most of the nutrients will be lost and not returned to the soil nutrient store as when decomposed in situ. The forest floor contains large quantities of N, P, and Ca relative to the nutrients in the entire system (Tiarks et al. 2004). If left at site, they could also contribute to the formation of soil organic matter which is important for storing nutrients in the soil (Schroth and Sinclair 2003).

Among nutrient cycling processes, nutrient retranslocation becomes an important component of the cycle as the tree biomass increases over time (Bowen and Nambiar 1984). In order to understand nutrient cycling processes and evaluate site quality changes in a forest ecosystem, it is very important to know the nutrient distribution among different parts of the plant biomass.

Plant biomass contains different proportions of nutrients in different parts of the plant body. This implies that the soil nutrients can be affected differently depending on the parts of the plant biomass being harvested. In a tropical rain forest biomass, a high proportion of nutrients are found in the bark and foliage; with an equal proportion occurring in the litter. As much as 54% and 82% of the aboveground biomass macronutrient pools are found in the crown and bark fraction (Tiarks et al. 2004). The study on C, N, and P allocation in agro-ecosystems conducted between 1993 and 1997 in high Casamance, in southern Senegal reveals that the plant biomass in the crop and fallow plots contained 18 Mg C, 0.25 Mg N, and 0.02 Mg P/ha; of which more than 40–50% were stored in the aboveground woody biomass (Manlay et al. 2001).

Continuous land cultivation, without sufficient nutrient input, is the major reason for soil quality decline in most of sub-Saharan Africa (Buresh et al. 1997). This is related to two major factors; (1) the lack of financial capital in households, that hinders farmers' ability to purchase sufficient amounts of fertilizer; (2) shrinking land holdings, that leads to deforestation and increased reliance on crop residues for fuel and animal fodder, thus preventing input of organic residues to the soil. However, some of the nutrients removed for animal fodder and fuel wood can be returned back to the soil through intensive management of the manure and household waste (Lupwayi et al. 2000). In a study undertaken to investigate the contents of manures from six experimental stations and twenty small-scale farms in the Ethiopian highlands, the manures contained on average, 21.3 g K, 18.3 g N, 16.4 g Ca, 5.6 g Mg, 4.5 g P, 10.8 g Fe, 0.78 g Mn, 0.09 g Zn (Lupwayi et al. 2000).

A study conducted in West Africa on the differential N uptake by maize cultivars, and soil nitrate dynamics under N fertilization, indicated that N uptake was highest at the early stage of the plant stem elongation growth phase. The high aboveground dry-matter accumulation and low soil total N at crop maturity correlates with the aboveground biomass and total N uptake among tropical maize populations, with the soil N concentration (Lafitte and Edmeades 1994; Oikeh et al. 2003). Increased nutrient removal associated with harvesting crop residue for biofuel or other bioproducts has a short-term economic impact and a potential long-term impact sustainable land use system (Hoskinson et al. 2007). At the site where this study was made, the maize stover and the branches and twigs of the plantation trees are used for fuel wood and animal feed to some extent. Therefore, this study was aimed at comparing the proportion of macronutrients in different parts of the plant's aboveground biomass in four land use types; and to estimate the amount of macronutrient loss from these lands use types through harvesting.

Materials and Methods

The farmlands come from the conversation of natural forests. After some years of continuous cultivation, inorganic fertilizer is being added in the form of urea and Diamonium phosphate (DAP), at an approximate rate 50 kg/ha/year for each of the two fertilizers, in traditional farmlands. However, some farmers cannot afford to apply the recommended amount and do not use any fertilizers at all. The mechanized farming (MF) site has been plowed since 1968 and wheat, barley, and maize were cultivated as the major crops in rotation with the use of a fertilizer at a maximum rate of 100 kg/ha/year of urea and DAP, respectively, since clearance of the natural forest, and converted in to crop cultivation and tree plantations.

To determine the aboveground biomass, macronutrients under these farmlands, that is, mechanized (MF) and traditional farm (TF), a 20×20 m^2 plot was selected and subsampled to 1×1 m^2 at each corner and at the center. All of the maize (Zea mays, L.) crops in these five subsample plot areas were harvested, and the fresh weight was measured. Representative samples from the stem, leaf, and grain were then taken to give a total of 5×3 samples from both farmlands.

The fresh weights for each of the biomass components under each land use were recorded and allowed to oven dry at $65 \pm 5°C$ for 24 h. The dry mass of each biomass component was calculated by using the formula as stated by Chidumaya (2009) as follows:

Dry mass = (subsample dry weight/subsample fresh weight) × fresh weight of the whole sample. And then, the expansion factor = 10, 000/area of plot in m^2 was used to convert it to hectare basis.

A total of 30 samples were analyzed for macronutrient (N, P, K, Mg, and Ca) concentration in the aboveground biomass of the maize crop. Plant nutrient concentrations were determined on each plant component at each sampling site. The air-dried samples, from the leaf, grain, and stem, were ground to pass through a 2-mm sieve.

Dried and ground samples were wet digested. The N concentration was analyzed with a LECO TruSpec total combustion system. The P, Ca, Mg, and K were analyzed by using inductively coupled plasma atomic emission spectrometry "(ICP-AES)". The amount of macronutrient contained in the aboveground biomass of the maize crop was calculated with the help of measured nutrient concentrations, and the amount of biomass corrected for water content to a dry weight basis.

To determine the aboveground biomass macronutrients in plantation stands, developing biomass equations is a laborious work that requires considerable effort, time, and money for collecting the data. Once established, they can be used to estimate forest biomass. There are published equations for forest biomass assessment (Abate 2004; Aboal et al. 2005; Cole and Ewel 2006; Arevalo et al. 2007; Gil et al. 2010). The choice of equations depends on the applicability to the objective of the specific study, i.e., factors determining such as geographic location, land cover type/similarity in species, and forest management practices (DeGier 2003; Zewdie et al. 2009).

Hence, the aboveground biomass for the two plantations (C. lusitanica and E. saligna) was estimated by measuring DBH and height of all trees on a 20 × 20 m² sample plot from each of the plantation stands. For C. lusitanica, the aboveground biomass was calculated based on the already developed allometric functions taking DBH as an independent variable. For three different plant parts of C. lusitanica, the linear regression model developed at Munesa Shashemane was adopted according to Abate (2004). Hence, the three equations (dry weight (kg) of leaf is = 1.36 × DBH (cm) − 19.84) with $R^2 = 0.91$, dry weight (kg) of branch is = 3.61 × DBH (cm) − 61.23 with $R^2 = 0.96$, and dry weight (kg) of stem is = 22.32 × DBH (cm) − 299.05 with $R^2 = 0.91$) were used.

For E. saligna, the aboveground biomass of the trees was estimated by adopting the dry weight predictive allometric equation which was developed on the same study area based on the height (m) and DBH (cm) as reported by Teklu (1997). Thus, $\ln W = \ln b_0 + b_1 \ln DBH + b_2 \ln H$; Where W stands for dry weight of E. saligna (kg), ln stands for natural logarithm, b_0, b_1, and b_2 stands for the regression coefficients.

From each sample plot, one sample tree, having an approximately mean value of all the trees by height and DBH, was selected. After felling three selected trees in each plantation stand, it was separated in to stems, branches, and leaves as suggested by Abate (2004). Each stem was cut into three equal logs and 5-cm thick discs were cut from each log. The branches and leaves were subsampled as belonging to the upper, medium, and lower third of the crown. Then the samples that were taken from each

part of the trees were bulked together to produce three representative samples, from each part of the trees. The bulked samples taken from the stems, branches, and leaves, in three replications from both plantations, give a total of nine samples (3 replication × 3 treatments (branches, leaves, and stems)) from each plantation for macronutrient analysis.

Samples from all components were ground for macronutrient analysis. The elemental analysis was carried out by focusing on some of the macronutrients (N, P, K, Mg and Ca) in the aboveground biomass. The samples were analyzed for N concentration with a LECO truSpec analyzer. The P, Ca, Mg, and K were analyzed by using inductively coupled plasma atomic emission spectrometry "(ICP-AES)". To calculate the amount of macronutrients in the aboveground biomass of E. saligna and C. lusitanica, the mean element concentrations per tree component were multiplied by their respective mean dry biomass to determine the amounts of nutrients in the sampled tree components. Mean nutrient values of the whole tree were calculated on a per plot and per hectare basis.

The statistical analyses for the aboveground biomass macronutrient concentrations in the crop compartments (leaf, grain, and stem) and tree compartments (leaf, branch, and stem), and between land uses (MF, TF, E. saligna, and C. lusitanica) were carried out by using the software SAS 9.0 and the two-way ANOVA was computed at $P < 0.05$.

Results and Discussion

The total N concentration in grain, leaf, and stem did not vary significantly at $P < 0.05$ under MF, but under TF land use, the total N in grain, and leaf was significantly higher than that of stem. Under both MF and TF, the grain contained significantly higher concentrations of P than that of leaf and stem. Overall, higher concentrations of P were observed in the maize crop under MF compared to TF (Table 1). Under MF, the stem contained significantly higher and the grain significantly lower concentrations of K. Significantly higher concentrations of K were observed in the maize crop under MF compared to TF (Table 1). Although, the maize crop under MF contained a significantly higher concentration of Ca compared to TF (Table 1), the leaves had the highest concentrations followed by stem and grain under both farmlands. Regarding Mg concentration, the grain contained a significantly lower concentration under TF.

The variation in nutrients in different crop parts indicates the direction to concentrate on in managing soil properties, by managing residues from different crop components. According to the macronutrient concentrations observed in the aboveground biomass of the maize

Table 1. Mean macronutrient concentration (SD) in different plants parts of a maize crop under MF and TF. Means followed by different letters by column are significantly different at $P < 0.05$.

Plant parts		TN (g/kg)	P (g/kg)	K (g/kg)	Ca (g/kg)	Mg (g/kg)
MF	Grain	11.8 (4.50)[a]	2.67 (0.44)[a]	5.75 (0.66)[c]	2.95 (0.94)[c]	1.72 (0.31)[a]
	Leaf	11.0 (1.70)[a]	1.54 (0.44)[b]	12.2 (1.58)[b]	18.3 (2.88)[a]	2.51 (0.61)[a]
	Stem	7.10 (3.20)[a]	1.05 (0.57)[b]	21.1 (4.99)[a]	9.96 (0.89)[b]	2.39 (0.60)[a]
	Overall	9.97[a]	1.75[a]	13.02[a]	10.4[a]	2.21[a]
TF	Grain	12.7 (1.30)[a]	2.21 (0.34)[a]	5.70 (0.48)[a]	1.87 (0.88)[c]	1.15 (0.26)[c]
	Leaf	12.6 (2.40)[a]	1.16 (0.20)[b]	11.0 (2.87)[a]	15.9 (1.78)[a]	2.72 (0.48)[a]
	Stem	6.30 (3.00)[b]	0.58 (0.12)[c]	9.65 (5.14)[a]	9.17 (0.46)[b]	2.27 (0.28)[a]
	Overall	10.53[a]	1.32[b]	8.78[b]	8.98[b]	2.05[a]

Table 2. Estimated total amount of macronutrient in different plant parts of a maize crop under mechanized (MF) and traditional (TF) farm lands.

	Plant parts	N (Mg/ha)	P (Mg/ha)	K (Mg/ha)	Ca (Mg/ha)	Mg (Mg/ha)
MF	Leaf	0.4	0.06	0.42	0.67	0.09
	Grain	0.47	0.11	0.24	0.12	0.07
	Stem	0.08	0.01	0.24	0.12	0.03
	Total	0.95	0.18	0.9	0.91	0.19
TF	Leaf	0.28	0.03	0.24	0.35	0.06
	Grain	0.42	0.08	0.19	0.06	0.04
	Stem	0.15	0.01	0.25	0.24	0.06
	Total	0.85	0.12	0.68	0.65	0.16

Where, MF is for Mechanized, and TF is for Traditional farm respectively.

crop (Table 1), P concentration can be significantly affected by grain harvest under both MF and TF. The K concentration can be affected at a higher rate with stem harvest than that of grain and leaves under MF. A higher concentration of Ca can be removed from the soil with leaf export compared to grain and stem.

The macronutrient contents presented in Table 2 indicates that in comparison between MF and TF, the maize crop biomass under MF contained higher amounts of all the studied macronutrients. With a higher rate of mineral N fertilizer application, the P, Ca, and Mg uptake by maize aboveground biomass were significantly increased (Yu et al. 2009). The higher K content in the crop biomass could be attributed to the P and N fertilization under MF land, as NP treatment enhances a greater uptake of K by crops (Yu et al. 2009). Since there is no fertilization with K, the higher quantities of K may have been released from the reservoir pool with the help of NP fertilization. Fertilizer management practices through the addition of higher N and P in the fields resulted in a higher nutrient content in grain and in the aboveground biomass as a whole at harvest (Yu et al. 2009).

The macronutrient concentration in the crop biomass varies between the plant components. If a macronutrient is highly concentrated in a specific plant part, then the removal of that plant part has a pronounced effect on the soil nutrient balance. Under TF, 49.4% N and 66.7%

P; and under MF 49.5% N and 61.1% P was contained in the grain. During the grain harvest, a higher proportion of these nutrients can be removed (Table 2). However, for the other macronutrients investigated, the harvesting of only the grain may not bring a severe loss from the soil nutrient budget. For instance, under TF, the grain contained only 27.9% K, 9.2% Ca, and 25% Mg. Under MF, the grain contained only 26.5% K, 13.2% Ca, and 36.4% Mg. This indicates that only harvesting the grain part, whilst retaining the leaves and stems back on the field may return 72.1% K, 90.7% Ca, and 75% Mg under TF and 73.3% K, 86.8% Ca, and 63.2% Mg under MF. As shown by Mapfumo and Mtambanengwe (undated), from the 73 kg/ha of K stored within the plant in the study site, an average of 72% is stored within the stalks, leaves, and husks. The study by Blanco-Canqui and Lal (2008) indicated that stover removal resulted in a consistent decrease in K^+ from the soil with ≥75% of K content removed in crop residues. They also reported that stover removal leads to a decline in the total SOC pool, total N pool, and the available P concentration at a soil depth of 0–10 cm.

Crop management activities play an important role in the soil nutrient balance. Of the amount of macronutrients lost by the aboveground biomass, removal under continuous farmland uses, only N and P were higher in grain harvest. The K, Ca, and Mg were found in higher amounts in the leaf and stem and lower in grain. The results suggest that if at harvest only the grain part of the crop was removed, and the other crop residues returned to the soil, it would have a much less negative effect on the soil base cation budget than on N and P. Soil management through the management of the biomass harvest during each growing season has an important role in ensuring long-term soil fertility. A study conducted in Mexico which reported that highly declined soil fertility values are recorded in maize cultivation with 12 years of monoculture indicates that the age of land use for specific purposes, with low amounts of residue returning to the site, can cause soil fertility depletion (Marroquin et al. 2005).

The leaves of *C. lusitanica* and *E. saligna* contain significantly higher concentrations of total N, P, K, Ca, and Mg than other tree parts (Table 3). This is in line with the finding of Zewdie et al. (2009) on the *E. globulus* plantation trees where nutrient concentrations (N, P, K, Ca, and Mg) followed the order: leaves > stemwood > stembark > twigs > branches. Thus, the leaf of the plantation species is more important compared to other plant components for the conservation and recycling of macronutrients in the soil. Comparing the two plantation species, it is expected that under *C. lusitanica*, soil Mg may be more sensitive to the aboveground biomass removal than under *E. saligna* due to the differences in concentration.

The amount of macronutrients in plant aboveground biomass may be dependent on the age of the plantation. A high proportion of nonutilizable biomass will be produced if the trees are harvested at a young age or by harvesting compartments with high densities (Tiarks et al. 2004). Likewise with the trees of *E. saligna*, stand of one to three trees per stool contained higher aboveground biomass macronutrients in the nonutilizable components. This may be attributed to the relatively short rotation period of the trees leading them to have a higher vegetative part and less biomass accumulation in the stem part. A study by Zewdie et al. (2009) on *E. globules* indicated

that total aboveground biomass increased from 11 t/ha at a stand age of 1 year to 153 t/ha at 9 years of which the highest dry weight was allocated to stemwood. For these reasons, the frequent harvest of the nonutilizable part of *E. salinga* trees is more drastic to the soil nutrients.

The *C. lusitanica* aboveground biomass in the present study shows a greater share of macronutrients was contained in the stem (Table 4). This may be related to the higher biomass allocation to the stem in relation to the tree's age. As reported by Abate (2004) for about 34 year old *C. lusitanica* stand, 85% of the aboveground biomass was allocated to the stemwood, 11% to the branches, and the remaining 4% was allocated to the foliage. A high proportion of the aboveground biomass is in stemwood, particularly in trees of larger diameter, where the stemwood contained more than 50% of the aboveground biomass, followed by twigs and branches. A study made in an Amazonian forest for nutrient distribution in plant components also reported that the accumulation of biomass was higher in the stem than other plant components (Stark 1971). The stem biomass of *C. lusitanica* stand of 25 years old showed that the highest content for all macronutrients investigated were found in the stem. Since the greater part of the biomass is found in the stem (Table 6), even low concentrations result in export of high amounts of nutrients through stem harvest. A study undertaken in a tropical natural forest indicated that a large nutrient loss occurred through the stem harvest since the largest amount was found in the stem followed by branches, root, and leaves. Among the nutrients, loss of K was largest followed by N, Mg, Ca, and P (Shanmughavel et al. 2001).

The differences in the amount of nutrients in different plant components give important information about the soil nutrient loss from the sites, through uptake by the biomass and subsequent harvesting of that plant part. The nutrient loss from the soil under vegetation cover can mainly be attributed to the plant parts harvested or consumed on site. Harvesting the biomass of whole trees

Table 3. Mean macronutrient concentration (SD) in different plant parts of *Cupresus lusitanica* (CL) and *Eucalyptus saligna* (ES). Different letters are significantly different at *P* < 0.05.

Plant parts		TN (g/kg)	P (g/kg)	K (g/kg)	Ca (g/kg)	Mg (g/kg)
C. lusitanica	Branch	3.80 (1.10)[b]	0.38 (0.06)[b]	2.02 (0.87)[b]	4.26 (1.49)[b]	0.75 (0.07)[b]
	Leaf	19.6 (3.30)[a]	1.70 (0.50)[a]	9.50 (0.50)a	17.9 (4.95)[a]	2.70 (0.54)[a]
	Stem	2.47 (0.30)[b]	0.28 (0.03)[b]	1.50 (0.20)b	4.02 (0.63)[b]	0.48 (0.08)[b]
	Overall	8.62[a]	0.79[a]	4.34[a]	8.73[a]	1.31[a]
E. saligna	Branch	4.20 (2.30)[b]	0.29 (0.06)[b]	1.02 (0.03)[b]	3.97 (0.68)[b]	0.21 (0.03)[b]
	Leaf	16.6 (3.60)[a]	1.39 (0.11)[a]	11.8 (0.95)[a]	22.7 (1.55)[a]	1.49 (0.24)[a]
	Stem	2.40 (0.70)[b]	0.29 (0.10)[b]	0.66 (0.31)[b]	3.84 (0.83)[b]	0.11 (0.08)[b]
	Overall	7.73[a]	0.66[a]	4.49[a]	10.2[a]	0.6[b]

Table 4. Estimated total amount of macronutrient uptake by different plant parts in the *Cupresus lusitanica* and *Eucalyptus saligna* biomass.

	Plant parts	Dry biomass (kg/ha)	N (kg/ha)	P (kg/ha)	K (kg/ha)	Ca (kg/ha)	Mg (kg/ha)
C. lusitanica	Leaf	11,300	220	20	110	200	30
	Branch	26,000	100	10	50	110	20
	Stem	233,600	580	70	350	400	110
	Total amount (kg/ha)		900	100	510	710	160
E. saligna	Leaf	700	11.7	1	8.4	16.1	1.1
	Branch	120	5.3	0.4	1.3	4.9	0.3
	Stem	2000	4.8	0.6	1.3	7.6	0.2
	Total amount (kg/ha)		21.8	2	11	28.6	1.6

may have a considerable impact on the soil nutrient budget. From the total macronutrients investigated for *C. lusitanica* (Table 6), although harvesting the stem has the highest effect on the soil macronutrient nutrient budget, leaving the branches and leaves on the site has a positive impact on the soil nutrient balance. For instance, 35.6% N, 30% P, 31.6% K, 34.7% Ca, and 25% Mg were contained in the leaf and branch parts of *C. lusitanica* and contributed to the soil nutrient pool if left on the site. Results from an *E. globulus* plantation indicated that, 67.5% of the N and 33.6% of the P were contained in the biomass of stem bark, branches and leaves, and that 56.6% of the K and 59.2% of the Ca were contained in components other than the stem (Gil et al. 2010). From the total macronutrients investigated for *E. saligna* (Table 4), 78% N, 70% P, 88.2% K, 73.4% Ca, and 87.5% Mg were contained in plant components other than the stem. Hence, more can be returned to the soil under *E. saligna* by leaving the branch and leaf biomass on the site.

Harvest management should consider the export of nutrients and try to adopt a sustainable nutrient management approach on the site, since the soil nutrient pool, plant growth, and soil degradation are all related through nutrient uptake, and intensive export of biomass may deplete the soil nutrients. Crop residues like litter, bark, and branches are considered as contributors to the soil nutrient pool. The study by Gil et al. (2010) indicated that with the plantation management option to harvest only stem wood, the amount of K removal could be as low as 43.4%, of Ca as low as 12.7%; and of Mg as low as 35.5%. The study by Tiarks et al. (2004) on *E. grandis* also indicated that by leaving the stem bark and all fractions of the crown on the site, nutrient export can be minimized. Abate (2004) has recommended that in order to make the plantation forests sustainable, the silvicultural practice in the future should consider on-site conservation of foliage and bark.

The removal of leaves, twigs, and other understorey biomass for fuel wood may have contributed to the removal of some macronutrients in the study area. The harvest intensity under the *E. saligna* was high in the study site. The leaves, branches, and twigs were removed as fuel wood by the local people after the stem harvest in the area was sold in the local market for their livelihood (Poultouchidou 2012). As a result, high amounts of nutrients may have been removed through these plant parts. The intensive and regular removal of plantation forest floor litter may be used as an important reason for substantial loss of nutrients from the site (Gil et al. 2010).

Conclusion and Recommendation

The N and P were found to be higher in the grains of maize and there is a high probability of being lost through harvesting the grain. However, the Ca, Mg, and S can be returned to the land with the corn stover through decomposition on the farm land since a higher concentration of these macronutrients was observed in the stover. Although the proportion of N and P is lower in the stover of maize crops compared to Ca, Mg, and K, the residue management during harvesting can still minimize the mining of these critical macronutrients in farm lands.

In spite of the higher aboveground biomass for *C. lusitanica* in the stem during harvesting, returning other parts of the tree to decompose in the site can reduce the loss of the studied nutrients. Regarding *E. saligna*, a higher proportion of the nutrients were found in the leaf and branch parts, and therefore the removal of branches, leafs, and twigs may have a high impact on the studied soil macronutrients. In general, knowledge of biomass nutrient partitioning and the impact of export of macronutrients through the aboveground biomass harvest is very important for the long-term success of production. Since different plant parts contain different amounts of nutrients, complete removal of the biomass for a long period from crop land and plantation stand may lead to nutrient depletion. Therefore, it is advised to retain the corn stover on the farmlands; and twigs and litter on the tree plantations to minimize the loss of macronutrients and thus prevent further degradation of the site.

Acknowledgments

My deepest gratitude to Wondo Gent College of Forestry and Natural Resources for providing me field materials and laboratory facilities.

Conflict of Interest

None declared.

References

Abate, A., 2004. Sustainable Biomass and Nutrient Studies of Selected Tree Species of Natural and Plantation Forests: Implications Management of the Munessa-Shashemene Forest, Ethiopia. PhD Thesis Universität Bayreuth. 149 pp.

Aboal, J. R., J. R. Arevalo, and A. Fernandez. 2005. Allometric relationships of different tree species and stand above ground biomass in the Gomera laurel forest (Canary Islands). Flora-Morphol. Distrib. Funct. Ecol. Plants 200:264–274.

Arevalo, C. B. M., T. A. Volk, E. Bevilacqua, and L. Abrahamson. 2007. Development and validation of aboveground biomass estimations for four *Salix* clones in central New York. Biomass Bioenergy 31:1–12.

Binkley, D., and M. G. Ryan. 1998. Net primary production and nutrient cycling in replicated stands of *Eucalyptus saligna* and *Albizia facaltaria*. For. Ecol. Manage. 112:79–85.

Blanco-Canqui, H., and R. Lal. 2008. Corn Stover removal for expanded uses reduces soil fertility and structural stability. Soil Sci. Soc. Am. 73:418–426.

Bowen, G. D., and E. K. S. Nambiar. 1984. Pp. 53–78. Nutrition of plantation forests. Academic Press, London.

Buresh, R. J., P. A. Sanchez, and F. Calhoun. 1997. Replenishing soil fertility in Africa: soil fertility in Africa is at stake. Pp. 47–61. Soil science society of America. American Society of Agronomy, Madison, WI.

Chidumaya, E. N. 2009. Aboveground woody biomass structure and productivity in Zambezian woodland. For. Ecol. Manage. 36:33–46.

Cole, T. G., and J. J. Ewel. 2006. Allometric equations for four valuable tropical tree species. For. Ecol. Manage. 229:351–360.

DeGier A., 2003. A new approach to woody biomass assessment in woodlands and shrublands. In: P. Roy (Ed), India. Geoinformatics for tropical ecos. Pp. 161–198.

Gil, L., W. Tadesse, E. Tolosana, R. López (eds), 2010. *Eucalyptus* species management, history, status and trends in Ethiopia. Proceedings from the congress held in Addis Ababa. September 15–17th, 2010. pp 297-309.

Hoskinson, R. L., D. L. Karlen, S. J. Birrell, C. W. Radtke, and W. W. Wilhelm. 2007. Engineering, nutrient removal, and feedstock conversion evaluations of four corn Stover harvest scenarios. Biomass Bioenergy 31:126–136.

Lafitte, H. R., and G. O. Edmeades. 1994. Improvement for tolerance to low soil N in tropical maize I. Selection criteria. Field Crops Res. 39:1–14.

Lupwayi, N. Z., M. Girma, and I. Haque. 2000. Plant nutrient contents of cattle manures from small-scale farms and experimental stations in the Ethiopian highlands. Agric. Ecosyst. Environ. 78:57–63.

Manlay, R. J., M. Kairé, D. Masse, C. Jean-Luc, G. Ciornei, and C. Floret. 2001. C, N and P allocation in agroecosystems of a West African savanna: the plant component under semi-permanent cultivation. Agric. Ecosyst. Environ. 88:215–232.

Mapfumo, P., and F. Mtambanengwe, undated. Base Nutrient Dynamics and Productivity of sandy soils under maize-pigeonpea rotational systems in Zimbabwe. Mount Pleasant, Harare. 16, pp 226-238.

Marroquin, A. F., J. Pohlan, M. J. J. Janssens, and J. Borgman 2005. Effects of production systems with maize (*Zea mays L.*) on Soil Fertility and Biological Diversity in the Soconusco, Chiapas, Mexico. "The Global Food & Product Chain Dynamics, Innovations, Conflicts, Strategies". pp 1-6.

Oikeh, S. O., R. J. Carsky, J. G. Kling, V. O. Chude, and W. J. Horst. 2003. Differential N uptake by maize cultivars and soil nitrate dynamics under N fertilization in West Africa. Agric. Ecosyst. Environ. 100:181–191.

Poultouchidou, A., 2012. Effects of forest plantations on soil C sequestration and farmers' livelihoods–A case study in Ethiopia. Master's Thesis in Biology. Swedish University of Agricultural Science, Department of Soil and Environment, Upsala. pp 37.

Schroth, G., and F. L. Sinclair, 2003. Trees, crops and soil fertility: Impacts of trees on the fertility of agricultural soils. CABI International, UK, 438 pp.

Shanmughavel, P., L. Sha, Z. Zheng, and M. Cao. 2001. Nutrient cycling in tropical seasonal rain forest of Xishuangbanna, Southwest China. Part 1: tree species: nutrient distribution and uptake. Bioresour. Technol. 80:163–170.

Stark, N. 1971. Nutrient cycling I: nutrient distribution on some Amazonian soil. Int. J. Trop. Ecol. 12:177–201.

Teklu, G., 1997. Aboveground biomass of *Equalyptus lusitanica* (Smith) coppiece trees. A case study from Munessa Shashemane, Oromia, Ethiopia. Ethiopian MSc in Forestry Program thesis work. 75pp.

Tiarks, A., J. Ranjer, E. K. S. Nambiar, T. Toma (eds), 2004. Site management and productivity in tropical plantation forestry: Effects of Harvesting and Site Management on Nutrient Pools and Stand Growth in A South Africa Eucalypt Plantation. CIFOR. Precedings of workshops in Congo July, 2001 and China February, 2003. pp 185-193.

Yu, W.-T., Z.-S. Jiang, H. Zhou, and Q. Ma. 2009. Effects of nutrient cycling on grain yields and potassium balance. Nutr. Cycl. Agroecosyst. 84:203–213.

Zewdie, M., M. Olsson, and T. Verwijst. 2009. Aboveground biomass production and allometric relations of *Eucalyptus globulus* Labill. Coppice plantations along chrono-sequence in the central highlands of Ethiopia. Biomass Bioenergy 33:421–428.

6

Biomass and elemental concentrations of 22 rice cultivars grown under alternate wetting and drying conditions at three field sites in Bangladesh

Gareth J. Norton[1] [iD], Anthony J. Travis[1], John M. C. Danku[1,2], David E. Salt[1,2], Mahmud Hossain[3], Md. Rafiqul Islam[3] & Adam H. Price[1]

[1]Institute of Biological and Environmental Sciences, University of Aberdeen, Aberdeen AB24 3UU, UK
[2]Centre for Plant Integrative Biology, School of Biosciences, University of Nottingham, Sutton Bonington Campus, Loughborough LE12 5RD, UK
[3]Department of Soil Science, Bangladesh Agricultural University, Mymensingh, Bangladesh

Keywords
Alternate wetting and drying, arsenic, cadmium, rice, yield, zinc.

Correspondence
Gareth J. Norton, Institute of Biological and Environmental Sciences, University of Aberdeen, Aberdeen AB24 3UU, UK.

E-mail: g.norton@abdn.ac.uk

Funding Information
Biotechnology and Biological Sciences Research Council (Grant/Award Number: 'BB/J003336/1')

Abstract

As the global population grows, demand on food production will also rise. For rice, one limiting factor effecting production could be availability of fresh water, hence adoption of techniques that decrease water usage while maintaining or increasing crop yield are needed. Alternative wetting and drying (AWD) is one of these techniques. AWD is a method by which the level of water within a rice field cycles between being flooded and nonflooded during the growth period of the rice crop. The degree to which AWD affects cultivars differently has not been adequately addressed to date. In this study, 22 rice cultivars, mostly landraces of the *aus* subpopulation, plus some popular improved *indica* cultivars from Bangladesh, were tested for their response to AWD across three different field sites in Bangladesh. Grain and shoot elemental concentrations were determined at harvest. Overall, AWD slightly increased grain mass and harvest index compared to plants grown under continually flooded (CF) conditions. Plants grown under AWD had decreased concentrations of nitrogen in their straw compared to plants grown under CF. The concentration of elements in the grain were also affected when plants were grown under AWD compared to CF: Nickel, copper, cadmium and iron increased, but sodium, potassium, calcium, cobalt, phosphorus, molybdenum and arsenic decreased in the grains of plants grown under AWD. However, there was some variation in these patterns across different sites. Analysis of variance revealed no significant cultivar × treatment interaction, or site × cultivar × treatment interaction, for any of the plant mass traits. Of the elements analyzed, only grain cadmium concentrations were significantly affected by treatment × cultivar interactions. These data suggest that there is no genetic adaptation amongst the cultivars screened for response to AWD, except for grain cadmium concentration and imply that breeding specifically for AWD is not needed.

Introduction

With an ever-growing world population, producing sufficient food in the coming decades will be a major focus of crop science. Within in the next 40 years, the world's population is predicted to increase from 7 billion to 9 billion people (Godfray et al. 2010). Rice is expected to play a key role in feeding this increased population. At present, rice provides 20% or more of the daily calorie intake for half of the world's population (Kush 2013). In the future, global rice demand is expected to rise from 676 million tons in 2010 to 852 million tons by 2035 (Kush 2013). Currently, irrigated lowland rice systems represent about 75% of global rice production (Fageria

2007). To produce 1 kg of rice grain, an average of 2500 L of water is needed (Bouman 2009). Globally this equates to one-third of the World's developed freshwater being used for rice irrigation (Bouman 2009). Not only does rice require large quantities of water, but rice cultivation under flooded conditions also contributes to global methane production (Smith et al. 2008). Therefore, approaches to rice production that require less water, while still maintaining yields are needed. One of the strategies being adopted across parts of Asia is alternate wetting and drying (AWD) (Lampayan et al. 2015). AWD is a technique in which rice fields undergo a number of drying phases during the growing season (Zhang et al. 2009). Farms are encouraged to start with a technique called safe-AWD, where the water level is allowed to drop 15 cm below the soil surface a number of times during vegetative growth and then the fields are re-flooded prior to flowering (Lampayan et al. 2015).

If AWD water saving techniques are to be widely adopted, one of the important factors will be the effect that this technique has on grain yield. Some studies have shown AWD has no effect on yield compared to other water management practices (Yao et al. 2012; Howell et al. 2015; Linquist et al. 2015; Shaibu et al. 2015; Liang et al. 2016), some have shown decreases in yield (Sudhir-Yadav et al. 2012; Linquist et al. 2015; Shaibu et al. 2015) while some show an increase in yield (Yang et al. 2009; Zhang et al. 2009; Wang et al. 2014; Norton et al. 2017). The evidence suggests that "safe-AWD" has little impact on yield compared to AWD where the soil is allowed to go through more severe periods of drying. In addition, it has been demonstrated that AWD can also reduce the methane emissions from paddy fields (Linquist et al. 2015; Liang et al. 2016).

A key question to address is if there are any cultivar differences in response to AWD. If detected it would suggest breeding efforts will be required specifically targeting AWD rather than traditional flooded conditions. A study by Zhang et al. (2009), using two high yield rice cultivars found that while both AWD treatment and cultivar had significant effects on grain yield there was no interaction between these two factors indicating that both cultivars responded similarly to the AWD treatment. In a study by Howell et al. (2015), two rice cultivars were grown under AWD and continuous flooding (CF) and grain yield was not affected by the cultivar, the AWD treatment or the interaction between cultivar and treatment. These studies suggest cultivars do not differ in response to AWD but a wider survey of rice cultivars is required to be confident this is will hold across the diverse rice germplasm.

The impact of AWD on grain element composition has attracted attention in recent studies because of the potential nutritional value of some elements (e.g., iron and zinc) or the toxic nature of others (arsenic and cadmium) combined with the known effect that soil redox status has on their bioavailability. Several studies have investigated the consequences of AWD on the concentration of single grain elements, for example; grain arsenic decreased under AWD (Somenahally et al. 2011; Linquist et al. 2015; Chou et al. 2016), while zinc (Wang et al. 2014) and cadmium increased under AWD (Yang et al. 2009). The impact that AWD has on a range of grain elements was explored in a single cultivar (Norton et al. 2017). In that study, it was demonstrated that AWD caused an increase in grain manganese (18.5–27.5%), copper (36.7–80.8%), and cadmium (27.8–67.3%) and a decrease in the concentration of sulfur (4.2–15.4%), calcium (6.3–8.7%), iron (10.7–15.5%), and arsenic (13.7–25.7%) compared to plants grown under CF.

It has been demonstrated that there are cultivar differences for the concentration of a large number of elements within the straw and grain of rice (e.g., Jiang et al. 2008; Norton et al. 2010a). The impact that location (field site) has on the accumulation of different elements has previously been investigated for rice (Norton et al. 2010a). In Norton et al. (2010a), 18 different rice cultivars were compared across four different field sites and it was established that all 10 elements measured in the grain showed variation based on cultivar, site and cultivar × site interactions. The identification of variation in grain elements has been exploited for the genetic mapping of genomic regions responsible for grain element concentration (Stangoulis et al. 2007; Lu et al. 2008; Garcia-Oliveira et al. 2009; Norton et al. 2010b, 2012a,b, 2014; Zhang et al. 2014).

As water saving techniques for rice cultivation become widely adopted, evaluation of the adaptation of cultivars to the different cultivation techniques is needed. In this study, 22 cultivars were tested to determine if the genetic differences between cultivars affected plant mass when grown under AWD compared to CF. In addition, the elemental composition of both straw and grains was determined to identify any effect that AWD has, and to determine if genetics affects the response to AWD treatment. To further explore the effect of different environmental conditions on both the effect of AWD and cultivar differences between cultivars, the same experiment was conducted at three different field sites in Bangladesh.

Methods

Rice cultivars

At each site, 22 cultivars were tested (Table 1). The cultivars used in this study are a subset of the cultivars previously genotyped, using a 384 SNP array (Travis et al.

Table 1. Cultivars used in this study, including country of origin/collection and subpopulation allocation.

Cultivar name	Cultivar identifier	Country of origin/collection	Rice subpopulation[1]
Assam 4 (Boro)	IRGC ID 11482	India	Aus-1
ARC 5977	IRGC ID 12166	India	Aus-2
AUS 130	IRGC ID 28984	Bangladesh	Aus-2
AUS 154	IRGC ID 28997	Bangladesh	Aus-1
AUS 362	IRGC ID 29149	Bangladesh	Aus-2
AUS Kushi	IRGC ID 66688	Bangladesh	Aus-2
Pura Nukna	IRGC ID 26413	Bangladesh	Aus-admix
Shada Boro	IRGC ID 34752	Bangladesh	Aus-1
Nai Dumur	IRGC ID 35057	India	Aus-admix
Dubhi Gora	IRGC ID 74567	India	Aus-admix
DJ 29	IRGC ID 76316	India	Aus-1
Jabahul	IRGC ID 86978	Bangladesh	Aus-admix
Black Gora	GSOR 301017	India	Aus-admix
Dhala Shaitta	GSOR 301041	Bangladesh	Aus-2
Kasalath	GSOR 301077	India	Aus-1
DD 62	GSOR 301306	Bangladesh	Aus-2
DJ 123	GSOR 301307	Bangladesh	Aus-admix
DM 59	GSOR 301312	Bangladesh	Aus-2
ARC 10376	GSOR 301341	India	Aus-admix
BR 6	–	Bangladesh	Indica
BRRI Dhan 28	–	Bangladesh	Indica
BRRI Dhan 47	–	Bangladesh	Indica

[1]Based on SNP analysis (Travis et al. 2015).

2015). The cultivars were either from the *aus* subpopulation originating from Bangladesh or India, or were improved Bangladeshi cultivars (BR 6, BRRI Dhan 28, and BRRI Dhan 47).

Field experiment

Three field experiments were conducted during the 2014 boro (dry) season in Bangladesh. The field sites were at Mymensingh (a noncalcareous floodplain soil; 24°42'58''; 90°25'26''), Madhupur (a Pleistocene terrace soil; 24°35'19''; 90°02'22'') and Rajshahi (a calcareous floodplain soil; 24°23'41''; 88°31'41''). Basic soil properties can be found in Table S1. Two different irrigation techniques were tested, and for each treatment four replicate blocks in a randomized block design were used. The water irrigation techniques used were CF and AWD.

For all three field experiments, the rice seeds were sown in a nursery bed at Mymensingh on the 17th December 2013. Prior to transplanting the seedlings at the three field sites, each site was ploughed, and then leveled. The day before transplanting the seedlings into the experimental plots started, the plots were fertilized with 40 kg/ha nitrogen, 15 kg/ha phosphorus, 50 kg/ha potassium, 15 kg/ha sulfur and 3 kg/ha zinc (see Table 2 for dates). A further 40 kg/ha nitrogen (as urea) was supplied during the tiller stage (see Table 2 for dates) and another 40 kg/ha nitrogen at the flowering stage (see Table 2 for dates). The seedlings were transplanted (see Table 2 for dates) into the

eight plots at Mymensingh each plot was 22.7 m × 11.8 m, at Rajhashi each plot was 12.4 m × 10 m, and at Madhupur each plot was 24 m × 10 m. Plants were planted in 2 m long rows as two plants per hill with a distance of 20 cm between each hill in a row, there was a 20 cm gap per row. The position of each cultivar in each replicate was randomized. Between each row of test cultivar, a row of a check variety (BRRI Dhan 28) was transplanted.

After the plants were transplanted, the plots were flooded. For the four CF plots, the surface water was kept at a depth of between 2 and 5 cm above the soil surface during the vegetative stage and reproductive stage. For the four AWD plots, plastic perforated tubes (pani pipe) were placed across the plots to monitor the depth at which the soil was saturated with water. The objective was to allow water to drain/percolate naturally from the AWD plots until the average depth of the water was 15 cm below the soil surface, at which point the plots were irrigated to bring the water depth to between 2 and 5 cm above the soil surface. At each site, the AWD plots went through four cycles of soil drying (Table 2). After the final cycle, the AWD plots were kept flooded and maintained the same as the CF plots.

After the cultivars had flowered and the grain matured, the grains and straw from each cultivar was harvested by hand from the six central hills of each row. The grain was then threshed by hand and the grain weighed to determine grain mass. Grain mass is expressed as the mass of grains harvested from the six central hills. The straw was harvested

Table 2. Date of sowing, transplanting, start and end of AWD cycles at the three field sites.

Date	Rajshahi	Mymensingh	Madhupur
Sowing in seed bed	17.12.2013[1]	17.12.2013[1]	17.12.2013[1]
Transplanted into the field	29.01.2014	05.02.2014–06.02.2014	08.02.2014–09.02.2014
Start of first AWD cycle	20.02.2014	21.02.2014	25.02.2014
Start of second AWD cycle	05.03.2014	05.03.2014	07.03.2014
Start of third AWD cycle	19.03.2014	15.03.2014	16.03.2014
Start of fourth AWD cycle	31.03.2014	27.03.2014	27.03.2014
End of AWD cycles	13.04.2014	15.04.2014	12.04.2014
Harvest	07.05.2014–18.05.2014	08.05.2014–28.05.2014	07.05.2014–31.05.2014
Initial fertilizer application	28.01.2014	04.02.2014	07.02.2014
First split nitrogen split	21.02.2014	27.02.2014	01.03.2014
Second split nitrogen split	22.03.2014	27.03.2014	30.03.2014

AWD, Alternative wetting and drying.
[1]All seeds were sown in seed beds in Mymensingh and transported to the other two sites.

approximately 5 cm above the soil, dried and then weighed to determine the straw weight. Straw mass is expressed as the mass of straw harvested from the six central hills. When dry, the straw was cut into small pieces ~1–2 cm long. A subsample of grains and straw was then sent to the University of Aberdeen, UK for elemental analysis.

Rice straw and grain analysis

Elemental analysis of rice straw and grains were conducted as described in Norton et al. (2017). Briefly rice grains were dehusked and oven dried, followed by microwave digestion with concentrated HNO_3 and H_2O_2 as described in Norton et al. (2012a). Straw was oven dried, powdered, and digested using nitric acid and hydrogen peroxide on a block digester (Norton et al. 2017). Total elemental analysis (sodium, magnesium, phosphorus, potassium, calcium, chromium, manganese, iron, cobalt, nickel, copper, zinc, arsenic, molybdenum, and cadmium) was performed by inductively coupled plasma – mass spectroscopy. Trace element grade reagents were used for all digests, and for quality control replicates of certified reference material (Oriental basma tobacco leaves [INCT-OBTL-5]) and rice flour [NIST 1568b]) were used; blanks were also included. All samples and standards contained 10 μg/L indium as the internal standard. In addition to elemental analysis on digested material described above, the concentration of nitrogen and carbon were determined on the powdered samples using an NCS analyser (NA2500 Elemental Analyser; Carlo Erba Instruments Wigan, UK).

Statistical analysis

All statistical analyses were performed, using the statistical software Minitab v.17 (State College, PA) and SigmaPlot v.13 (Systat Software Inc., San Jose, CA, USA). For the plant mass traits and the plant elemental concentration traits, a three-way ANOVA was conducted with treatment (AWD and CF), site and, cultivar as the explanatory variables. For the three-way ANOVA, the presence of an interaction between the three explanatory variables was also determined. For correlation analysis, a Spearman's rank correlation was used.

Results

The mean data for each of the cultivars grown under the different water treatments at each site are presented in Table S2. Graphs presenting the effect of treatment and site for all traits measured are provided in Figures 1, 3, 4, 5 and Figures S1, S2.

Straw biomass

There was a significant difference between the three field sites for straw biomass and a site by treatment interaction ($P < 0.001$; Table 3). The highest straw biomass was at Rajshahi (mean 142.3 g), followed by Madhupur (89.0 g) and the lowest average at Mymensingh (74.3 g) (Fig. 1a). Overall, there was no significant difference in straw biomass for plants grown under AWD compared to CF. However, there was a significant difference at the Rajshahi site, where plants grown under CF had a higher straw biomass compared to those grown under AWD. There were significant differences in straw biomass between cultivars and significant two-way interactions for site by treatment interaction and cultivar by site interaction (Table 3). There was a significant positive correlation between the straw biomass of cultivars grown under AWD when compared to CF for each of the three field sites (Fig. 2A).

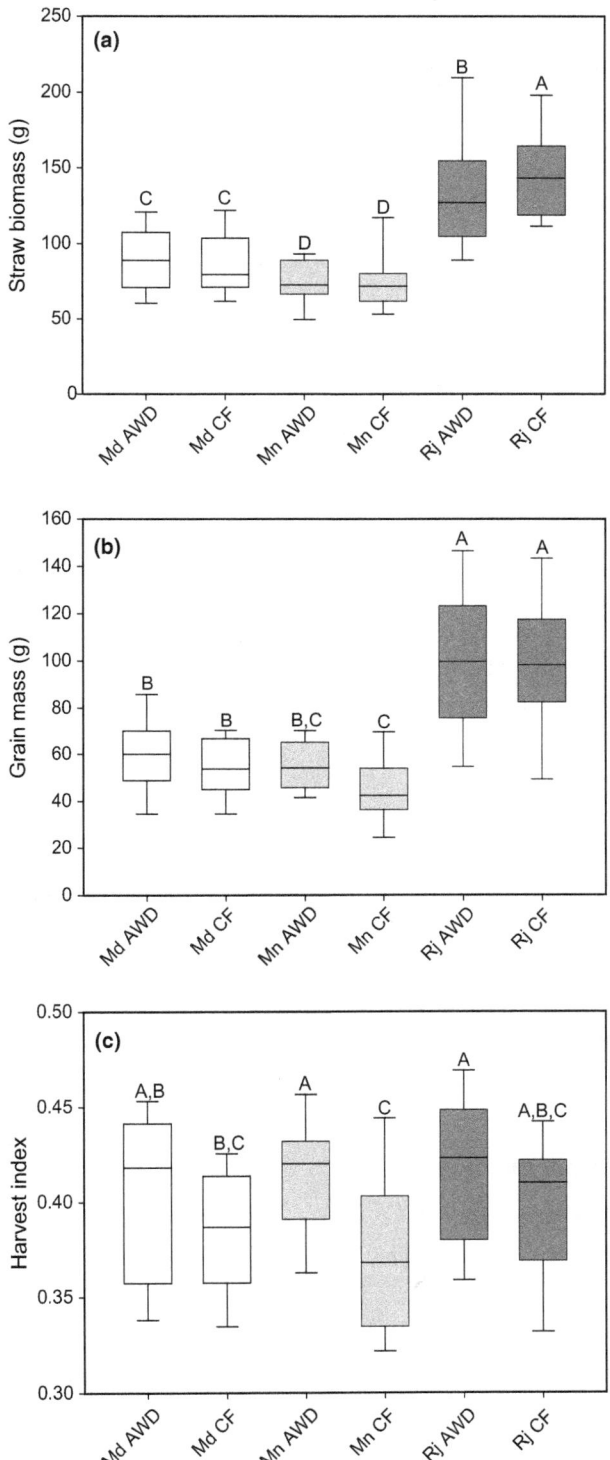

Figure 1. Yield traits for the 22 cultivars grown at the three different sites under AWD and CF. (a) straw biomass, (b) grain mass, (c) harvest index. Md = Madhupur, Mn = Mymensingh, Rj = Rajshahi. Data which had the same letter are not significantly different. Grain and straw mass are the sum of the mass from the six hills. AWD, Alternative wetting and drying; CF, continuous flooding.

Grain mass

There was a significant difference ($P < 0.001$) for grain mass between the three field sites, with the highest grain mass being at Rajshahi (mean 99.6 g), followed by Madhupur (58.3 g) and lowest at the Mymensingh (50.6 g) (Fig. 1b). Overall, there was a significant difference ($P = 0.02$) in grain mass for plants grown under AWD compared to CF, with the AWD plants on average having a grain mass of 71.7 g compared to 67.3 g for the CF plants. There was no significant site by treatment interaction. There was, however, a significant difference ($P < 0.001$) in grain mass between the cultivars and a significant ($P < 0.001$) cultivar × site interaction (Table 3). There was a significant positive correlation between the grain mass of cultivars grown under AWD when compared to CF for each of the three field sites (Fig. 2B).

The data reported is the grain mass per row of six cultivars. This can be converted into an approximation of grain yield, by scaling up the value based on the planting density, however this must be used with caution. The yields at the three sites under the two different treatments are as follows: 3.8 t/ha for Mymensingh CF, 4.5 t/ha for Mymensingh AWD, 4.6 t/ha for Madhupur CF, 5.0 t/ha for Madhupur AWD, 8.3 t/ha for Rajshahi CF, and 8.3 t/ha for Rajshahi AWD.

Harvest index

There was a significant difference in harvest index for the three sites, with plants grown at the site in Rajshahi having the highest harvest index and plants grown at the Madhupur site having the lowest harvest index (Fig. 1c). Overall the plants grown under AWD had a greater harvest index than those grown under CF conditions. However, this was only a small effect, with harvest index increasing on average by 6.9% under AWD. There was a site × treatment interaction revealing that the positive effect of AWD on harvest index was much stronger in Mymeningh than the other sites (Fig. 1c). There was also a significant cultivar difference for harvest index and site × cultivar interaction (Table 3). There were significant positive correlations between the harvest indexes of the cultivars grown under AWD when compared to CF at Madhurpur and Rajhashi, but not at Mymensingh (Fig. 2C).

Straw carbon and nitrogen concentration

The concentration of carbon in the rice straw was not affected by treatment, but it was significantly affected by the site where the rice plants were grown (Table 4), although the effect was small. The straw biomass of rice

Table 3. Statistical analysis of yield traits across the three different site for the 22 cultivars grown under AWD and CF (treatment). Values reported are the f-values from the ANOVA with the asterisk indicating the level of significance.

Trait	Site (S) df = 2	Treatment (T) df = 1	Cultivar (C) df = 21	S × T df = 2	S × C df = 42	T × C df = 21	S × T × C df = 42
Straw biomass (g)	315.66***	NS	22.61***	4.78**	4.95***	NS	NS
Grain mass (g)	263.85***	6.23*	17.62***	NS	3.81***	NS	NS
Harvest index	3.56*	35.26***	7.20***	3.15*	2.09***	NS	NS

AWD, Alternative wetting and drying; CF, continuous flooding; NS, not significant.
$*P < 0.05$, $**P < 0.01$, $***P < 0.001$.

plants grown at Madhupur and Mymensingh were approximately 43% carbon while the plants grown at Rajshahi were approximately 41.5% carbon on average. There was also a significant cultivar difference in the concentration of carbon measured in the cultivars across all sites and treatments (Table 4).

The concentration of straw nitrogen was significantly affected by the site where the rice plants were grown, if they were grown under AWD or CF and there was a significant difference between the cultivars (Table 4; Fig. 3). A higher average nitrogen concentration in the straw was observed at the Mymensingh site (0.8%), while the average nitrogen concentration in the straws of rice plants was the same at both the Madhupur and Rajshahi sites (0.6%). On average, AWD decreased straw nitrogen (~0.6%) compared to CF (~0.7%) across all sites and cultivars. There were also significant effects on straw nitrogen concentration for site × treatment and site × cultivar interactions (Table 4). AWD caused a significant decrease in straw nitrogen compared to the CF treatment when the Madhupur and Mymensingh sites were examined individually (Fig. 3).

Straw macro and micro elemental concentration

Straw arsenic

The concentration of straw arsenic was significantly affected by the site where the plants were grown (Table 4). The average straw arsenic concentration was higher at Rajshahi and Mymensingh (1.27 and 1.25 mg/kg, respectively) compared to Madhupur (0.76 mg/kg). Overall, AWD significantly decreased straw arsenic by 16.7% compared to CF. However, when the AWD treatment effect was explored at each site individually, there was only a significant difference between the AWD and CF treatments at the Mymensingh site, with a 35% decrease in grain arsenic in the AWD treatment (Fig. 4a). There was also a significant cultivar difference in the concentration of straw arsenic measured in cultivars across all sites and treatments (Table 4).

Straw iron

The concentration of straw iron was not significantly affected by the site where the plants were grown (Table 4). There was an overall treatment effect across all sites, with AWD on average causing a 12.9% decrease in straw iron compared to CF. However, when each site was examined separately, there was only a significant difference in straw iron at the Mymensingh site, with plants grown under AWD having on average 25% less iron in the straw (Fig. 4b). There was a significant cultivar difference in the concentration of straw iron measured in the cultivars across all sites and treatments (Table 4), as well as a site × cultivar interaction for straw iron.

Straw zinc

The only significant factor that affected zinc concentration in the straw was the site at which the plants were grown (Table 4). The highest concentrations of zinc in the straw were found in plants grown at the Madhupur site (61.04 mg/kg) followed by Mymensingh (30.19 mg/kg), with the lowest average concentration at the Rajshahi site (18.54 mg/kg) (Fig. 4c).

Straw cadmium

The concentration of straw cadmium was significantly affected by the site where the plants were grown (Table 4), with the average straw cadmium concentration being the highest at Madhupur (0.07 mg/kg) followed by Mymensingh (0.05 mg/kg) and Rajshahi having, on average, the lowest (0.02 mg/kg). Overall, across all sites, the AWD treatment did not affect cadmium concentration in the straw. However, there was a significant site × treatment interaction and when each site was examined separately there was a significant difference in straw cadmium at the Mymensingh site, with the plants grown under AWD having on average 70% more cadmium in straw (Fig. 4d). There was no significant cultivar difference in the concentration of straw cadmium measured in cultivars across all sites and treatments (Table 4).

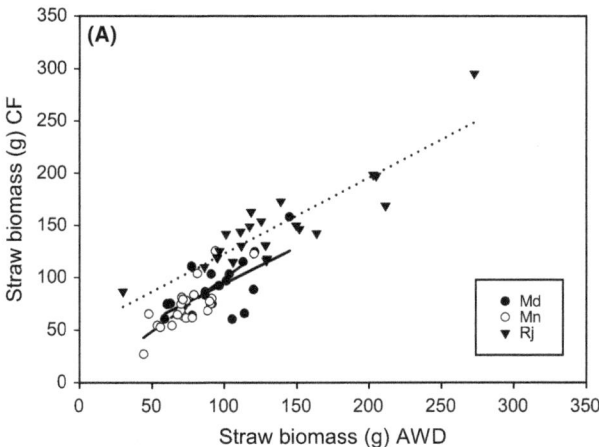

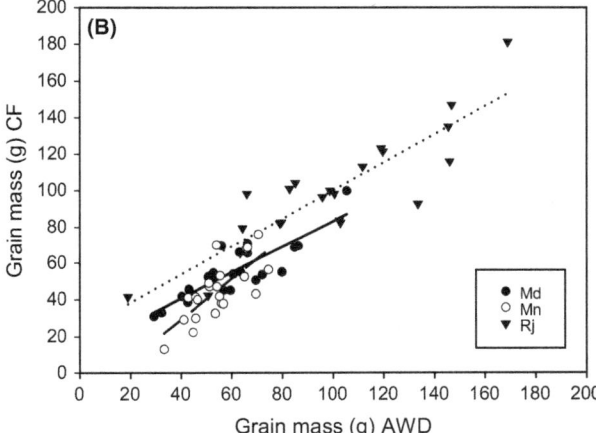

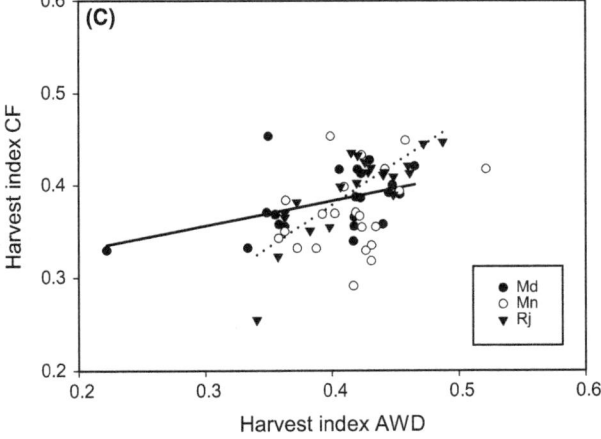

Figure 2. Correlation between yield components of the cultivars grown under AWD and CF conditions at the three field sites. (A) straw biomass, (B) grain mass, (C) harvest index. Filled circles (and solid line) are for the cultivars grown at Md, open circles (and dashed line) are for cultivars grown at Mn, and upside down filled triangles (and dotted line) are for cultivars grown at Rj. Only significant correlations ($P < 0.05$) are indicated with lines. For straw biomass at each site, there was a significant relationship between AWD and CF, Mn $r^2 = 55.2\%$, Md $r^2 = 71.7\%$, Rj $r^2 = 77.2\%$. For grain mass at each site, there was a significant relationship between AWD and CF ($P < 0.001$), Mn $r^2 = 70.3\%$, Md $r^2 = 52.9\%$, Rj $r^2 = 76.2\%$. For harvest index, there were a significant relationship between AWD and CF at two of the three sites Md $r^2 = 20.8\%$, Rj $r^2 = 67.4\%$. Grain and straw mass are the sum of the mass from the six hills. AWD, Alternative wetting and drying; CF, continuous flooding.

mg/kg). The AWD treatment caused a significant decrease in straw phosphorus across all sites (AWD average concentration 459 mg/kg and CF average concentration 590 mg/kg). There was no significant effect of AWD treatment at the field site in Rajshahi, but AWD caused a significant decrease in straw phosphorus compared to CF at the Madhupur (26.3%) and Mymensingh (29.9%) sites (Fig. 4e).

Straw molybdenum

Two significant factors affected the concentration of molybdenum in the straw. There was an overall AWD treatment effect across sites and a significant interaction between site × treatment (Table 4). AWD caused a 17.5% decrease in straw molybdenum compared to the plants grown under CF across all sites. AWD treatment only caused a significant effect at the Mymensingh site, with an average straw molybdenum concentration for the plants grown under AWD being 0.71 mg/kg compared to 1.09 mg/kg in plants grown under CF (Fig. 4f).

Other elements in the straw

The site where the rice was grown had a significant effect on elemental concentration in the straw for all other elements measured (Table 4). Across all sites and cultivars, the AWD treatment caused a significant change in the concentration of magnesium, chromium, cobalt, nickel, and copper (Table 4, Figure S1). Across the sites and treatments, cultivar caused a significant differences in straw concentration of sodium, magnesium, potassium, calcium, manganese, cobalt, and copper (Table 4). A number of two way interactions between site × treatment and site × cultivar were identified for some elements, however, no treatment × cultivar interactions or three-way site × treatment × cultivar interactions were detected.

Straw phosphorus

The concentration of straw phosphorus was significantly affected by site, treatment, and cultivar as well as interactions between site × treatment and site × cultivar (Table 4). The site with the highest phosphorus concentration across treatments was Madhupur (631.8 mg/kg), followed by Mymensingh (550.3 mg/kg), and Rajshahi having on average the lowest straw phosphorus concentration (390.5

Table 4. Statistical analysis of straw element traits across the three different site for the 22 cultivars grown under AWD and CF (treatment). Values reported are the *f*-values from the ANOVA with the asterisk indicating the level of significance.

Trait	Site (S) df = 2	Treatment (T) df = 1	Cultivar (C) df = 21	S × T df = 2	S × C df = 42	T × C df = 21	S × T × C df = 42
N	48.64***	46.34***	12.39***	10.00***	1.65**	NS	NS
C	28.96***	NS	1.70*	NS	NS	NS	NS
Na	18.32***	NS	7.03***	5.26**	NS	NS	NS
Mg	66.30***	6.77*	7.64***	NS	NS	NS	NS
P	25.46***	22.05***	4.22**	4.70*	1.44*	NS	NS
K	118.02***	NS	4.94***	3.18*	NS	NS	NS
Ca	6.04**	NS	3.56***	NS	1.47*	NS	NS
Cr	8.84***	13.12***	NS	4.22*	1.74**	NS	NS
Mn	154.10***	NS	2.80***	4.49*	NS	NS	NS
Fe	NS	17.59***	2.64***	7.03**	1.60*	NS	NS
Co	54.87***	7.47**	5.88***	NS	NS	NS	NS
Ni	NS	7.52**	NS	3.58*	1.72**	NS	NS
Cu	24.45***	7.71**	8.95***	NS	1.44*	NS	NS
Zn	123.80***	NS	NS	NS	NS	NS	NS
As	75.01***	27.56***	2.79***	23.74***	1.82**	NS	NS
Mo	NS	16.52***	NS	7.98***	NS	NS	NS
Cd	50.94***	NS	NS	9.96***	NS	NS	NS

AWD, Alternative wetting and drying; CF, continuous flooding; NS, not significant.
*P < 0.05, **P < 0.01, ***P < 0.001.

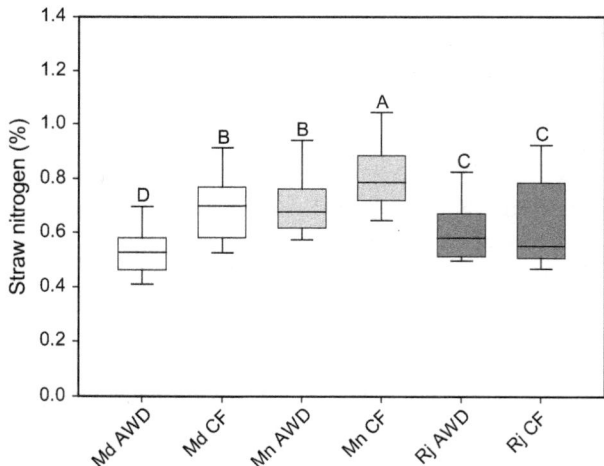

Figure 3. Straw nitrogen for the 22 cultivars grown at the three different sites under AWD and CF. Md = Madhupur, Mn = Mymensingh, Rj = Rajshahi. Data which has the same letter are not significantly different. AWD, Alternative wetting and drying; CF, continuous flooding.

Macro and micro elemental concentrations in the grains

Grain arsenic

There were significant effects of site, treatment, cultivar, and a site × treatment effect for grain arsenic (Table 5) with AWD on average decreasing grain arsenic by 3.2% across all sites. There was only a significant difference between

AWD and CF at the Mymensingh site, with a 16.0% decrease in grain arsenic in the AWD treatment (Fig. 5a).

Grain iron

The site where the plants were grown, the AWD treatment they received and the cultivar all had a significant effect on grain iron concentration (Table 5). In addition, significant effects on grain iron concentration were observed for both site × treatment and site × cultivar interactions. Overall, the plants with the lowest grain iron concentration were grown at Mymensingh (11.4 mg/kg) while the other two sites had a higher concentration of grain iron (12.1 and 12.2 mg/kg, for Rajshahi and Madhupur, respectively). Overall, AWD decreased grain iron by 5.8% across site and cultivars. There was only a significant difference between AWD and CF treatments at the Mymensingh site, where AWD caused a 12.4% decrease in grain iron (Fig. 5b).

Grain zinc

The site where plants were grown and the cultivar had a significant effect on grain zinc concentration, but there was no overall AWD treatment effect (Table 5). The site with the lowest average grain zinc concentration across both AWD and CF treatments was Rajshahi (25.8 mg/kg), followed by Mymensingh (30.2 mg/kg) and the site with the highest grain zinc concentrations was Madhupur (average 37.8 mg/kg). There was no significant effect of

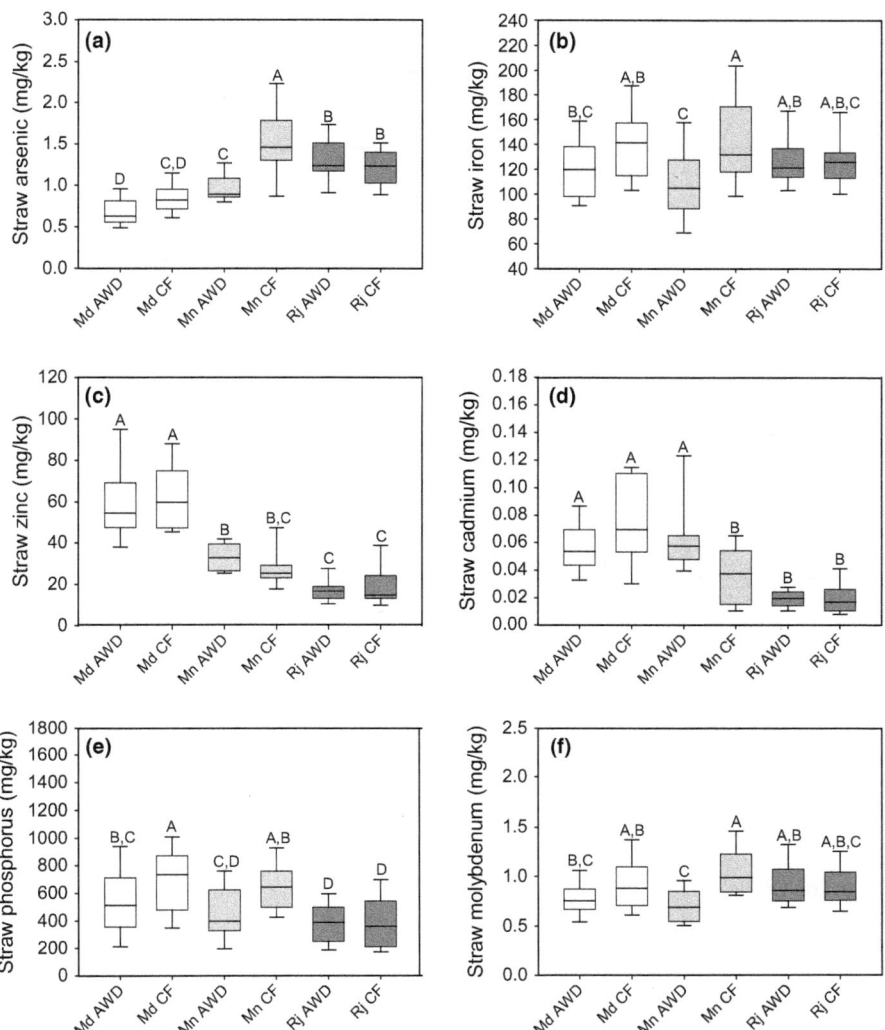

Figure 4. Straw element traits for the 22 cultivars grown at the three different sites under AWD and CF. (a) grain arsenic, (b) grain iron, (c) grain zinc, (d) grain cadmium, (e) grain phosphorus, (f) grain molybdenum. Md = Madhupur, Mn = Mymensingh, Rj = Rajshahi. Data which has the same letter are not significantly different. AWD, Alternative wetting and drying; CF, continuous flooding.

AWD treatment on grain zinc compared to CF at any of the three sites across all of the cultivars (Fig. 5c).

Grain cadmium

The site where plants were grown, the AWD treatment they received and the cultivar all had a significant effect on grain cadmium concentration (Table 5). In addition, significant two-way interactions affecting grain cadmium concentration were observed for all combinations of factors (site, treatment and genotype) as well as a significant three-way interaction between all of the factors. AWD caused a 43% increase in grain cadmium compared to the CF treatment on average across the three sites. A significant difference in grain cadmium between the AWD

and CF-treated plants was only observed at the Mymensingh site (Fig. 5d).

Grain phosphorus

The site where plants were grown, the AWD treatment they received and the cultivar all had significant effects on grain phosphorus concentration (Table 5). The plants grown at Madhupur had significantly higher concentrations of grain phosphorus (4390 mg/kg) compared to Mymensingh (4197 mg/kg) and Rajshahi (4191 mg/kg). AWD reduced grain phosphorus by 5.9% compared to plants grown under CF across all three sites. There was a significant decrease in grain phosphorus concentration between AWD treatments

Table 5. Statistical analysis of grain element traits across the three different site for the 22 cultivars grown under AWD and CF (treatment). Values reported are the *f*-values from the ANOVA with the asterisk indicating the level of significance.

Trait (mg/kg)	Site (S) df = 2	Treatment (T) df = 1	Cultivar (C) df = 21	S × T df = 2	S × C df = 42	T × C df = 21	S × T × C df = 42
Na	35.89***	17.7***	14.78***	12.84***	1.47*	NS	NS
Mg	NS	33.83***	6.03***	3.67*	NS	NS	NS
P	6.73**	27.64***	6.10***	5.67**	NS	NS	NS
K	23.95***	5.05*	3.27***	8.85***	NS	NS	NS
Ca	13.66***	6.78*	16.32***	3.34*	NS	NS	NS
Cr	NS	NS	NS	NS	NS	NS	NS
Mn	749.00***	NS	9.56***	30.68***	1.84**	NS	NS
Fe	10.12***	22.99***	12.27***	7.77**	1.95**	NS	NS
Co	146.96***	3.95*	23.21***	25.54***	1.57*	NS	NS
Ni	132.45***	5.68*	2.37*	NS	NS	NS	NS
Cu	282.03***	63.78***	11.73***	48.36***	1.64*	NS	NS
Zn	301.60***	NS	19.39***	4.28*	1.69*	NS	NS
As	30.46***	4.22*	12.70***	16.53***	NS	NS	NS
Mo	110.09***	56.53***	6.24***	11.02***	1.77*	NS	NS
Cd	150.71***	74.95***	13.19***	91.81***	2.31***	2.59***	5.12***

AWD, Alternative wetting and drying; CF, continuous flooding; NS, not significant.
*$P < 0.05$, **$P < 0.01$, ***$P < 0.001$.

at the Madhupur (9.8%) and Mymensingh (6.6%) site (Fig. 5e).

Grain molybdenum

The site where plants were grown, the AWD treatment they received and the cultivar all had significant effects on grain molybdenum concentration (Table 5). In addition, significant effects on grain molybdenum concentration were observed for site × treatment and site × cultivar interactions. AWD significantly decreased grain molybdenum concentration by 16.9% across all sites. Plants with the lowest grain molybdenum were grown at the Madhupur site (0.50 mg/kg). While those with the highest grain molybdenum concentration were grown at the Mymensingh site (0.79 mg/kg). However, AWD also significantly decreased grain molybdenum at Mymensingh (22.8%) and at Madhupur (25.6%) (Fig. 5f).

Other elements in the grain

For all the other elements measured except magnesium and chromium, the site where the rice was grown had a significant effect on elemental concentration in the grains (Table 5, Fig. 2). Across all the sites and cultivars, the treatment AWD caused a significant change in the concentration of sodium, magnesium, potassium, calcium, cobalt, nickel, and copper. There were significant cultivar effects between cultivars for the concentration of all elements measured in the grain except chromium across all sites and treatments. A number of two-way site ×

treatment and site × genotype interactions were also detected (Table 5).

Impact of flowering time on grain traits

At the Mymensingh field site, the flowering time of the accessions was monitored. This was not done at the other two sites. At this site, all the 22 accessions initiated flowering prior to the last AWD cycle finishing. There was variation in flowering time with the earliest flowering cultivars initiating flowering 17 days before the AWD cycles finished and latest flowering cultivar initiating flowering 3 days prior to the last AWD cycle finishing. Under both AWD and CF flowering time did not significantly correlate with grain mass, straw biomass, or harvest index. For the plant grown under AWD conditions, a number of grain elements concentration correlated with flowering time; there was a positive correlation for grain nickel ($r = 0.600$, $P = 0.004$) and negatively correlations with grain phosphorous ($r = -0.452$, $P = 0.040$), grain zinc ($r = -0.499$, $P = 0.021$) and grain arsenic ($r = -0.629$, $P = 0.002$) (Figure S3). For the plants grown under CF, flowering time also correlated with a number of grain elements; there were negative correlations for grain magnesium ($r = -0.515$, $P = 0.017$), grain phosphorus ($r = -0.577$, $P = 0.006$), and grain zinc ($r = -0.450$, $P = 0.041$). To compare if the impact of flowering time differed across treatments, for each cultivar, the ratio of the AWD concentration for each element was compared to the concentration of the element in the cultivar grown under CF. This ratio was then tested for its relationship with flowering time (Figure S3). Only the ratio for arsenic significantly correlated with flowering time ($r = -0.510$, $P = 0.018$).

Discussion

One of the most important aspects of the adoption of a rice cultivation practice will be the impact that cultivation practice has on yield. Previous studies on the effect of AWD on yield compared to other rice cultivation practices have reported varying impacts from a reduction in yield to increases in yield (Yang et al. 2009; Zhang et al. 2009; Sudhir-Yadav et al. 2012; Yao et al. 2012; Wang et al. 2014; Howell et al. 2015; Linquist et al. 2015; Shaibu et al. 2015; Norton et al. 2017). In this study, plants grown under AWD across all three sites did show an increase in grain mass. However, at each individual field site, different magnitudes of treatment effects were observed. The AWD treatment effects were greatest in Mymensingh but no significant effects were observed in Rajshahi. Straw biomass was not consistently affected by AWD. The largest and most consistent effect on yield traits was for the harvest index. A significant difference was observed at the Mymensingh site with plants grown under AWD having an increased harvest index, and at the other two sites a similar trend was also observed. This may indicate that there is a change in the allocation of resources for plants grown under AWD. It has been shown in a number of studies that either the number of tillers or productive tillers increases in plants grown under AWD (Yang and Zhang 2010; Howell et al. 2015; Norton et al. 2017). As part of this field experiment, the number of productive tillers was determined for a single cultivar (these results are presented in Norton et al. (2017)). It was observed that plants grown under AWD had a greater number of productive tillers. A similar process may contribute to effects on harvest index in other cultivars observed in this study.

A few previous studies have tested a small number of rice cultivars to identify variation in yield trait responses to AWD (e.g., Zhang et al. 2009; Howell et al. 2015). In this study, 22 cultivars were tested for their yield response under AWD compared to CF. Significant cultivar differences were identified for yield as well as a significant interaction between cultivar and the site at which plants were grown, but importantly no significant interactions between cultivar and treatment were identified for the yield traits (Table 3). This suggests that while there is genetic variation among the 22 cultivars tested for yield traits (grain mass, straw yield and harvest index), the variation does not impact their response AWD. It can be concluded that most traits or genes that maximize yield under AWD are the same as those that maximize yield in conventional CF paddy irrigation, which means

that current high yielding varieties should perform well (relative to other cultivars) under AWD cultivation as they already do under CF paddy cultivation.

Straw nitrogen concentration was reduced by AWD when compared to CF at two of the three sites. It has been suggested that AWD can affect the fate of nitrogen in paddy fields (Tan et al. 2015). Those authors demonstrated that nitrogen losses due to volatilization and denitrification would increase under AWD compared to CF, using a simulation model of water movement, transport, and transformation. These nitrogen losses are due to the intensified nitrification-denitrification processes caused by a high concentration of ammonium ions and the cycling between anaerobic and aerobic condition in the soil (Tan et al. 2015). In our study, it could be proposed that more nitrogen was lost in the AWD treatment due to volatilization and denitrification, which directly caused a decrease in plant nitrogen accumulation.

In both the grain and straws of the rice plants grown under AWD at the Madhupur and Mymensingh, the concentration of phosphorus was lower than in the plants grown under CF. This would be expected under anaerobic conditions, because redox-sensitive mineral constituents (e.g., iron and manganese) release associated (adsorbed or co-precipitated) phosphorus anions making P more available in the CF treatment. In addition, anaerobic conditions could also result in a release of P from the organic fraction. The difference between soil chemistry under the CF (submerged anaerobic soil) and AWD (soil which is fluctuating between submerged and aerobic soil), could explain a number of the observed effects of AWD on straw and grain element concentrations. The concentration of iron, arsenic, and manganese would be expected to decrease in the soil solution under aerobic conditions, whereas the concentrations of cadmium and copper would be expected to increase under oxidizing conditions (Rinklebe et al. 2016).

For some of the elements at some of the sites, these soil effects can be seen in the plants but this is not the case for all elements at all sites (Tables 4, 5; Figs. 4, 5). While all sites underwent AWD treatment, it appears that the effect of AWD compared to CF was almost negligible at the Rajshahi site for all traits measured. There was no significant difference between the AWD and CF treatments for any of the straw and grain elements measured. In addition, the only yield trait significantly affected by AWD treatment in Rajshahi was straw biomass (Fig. 1a). Because the same AWD treatment was applied across all sites the major difference between sites was, therefore, the soil properties and environmental conditions at each field site. In addition to noting the soil properties of the Rajshahi site, it may be significant that the plants grew much more

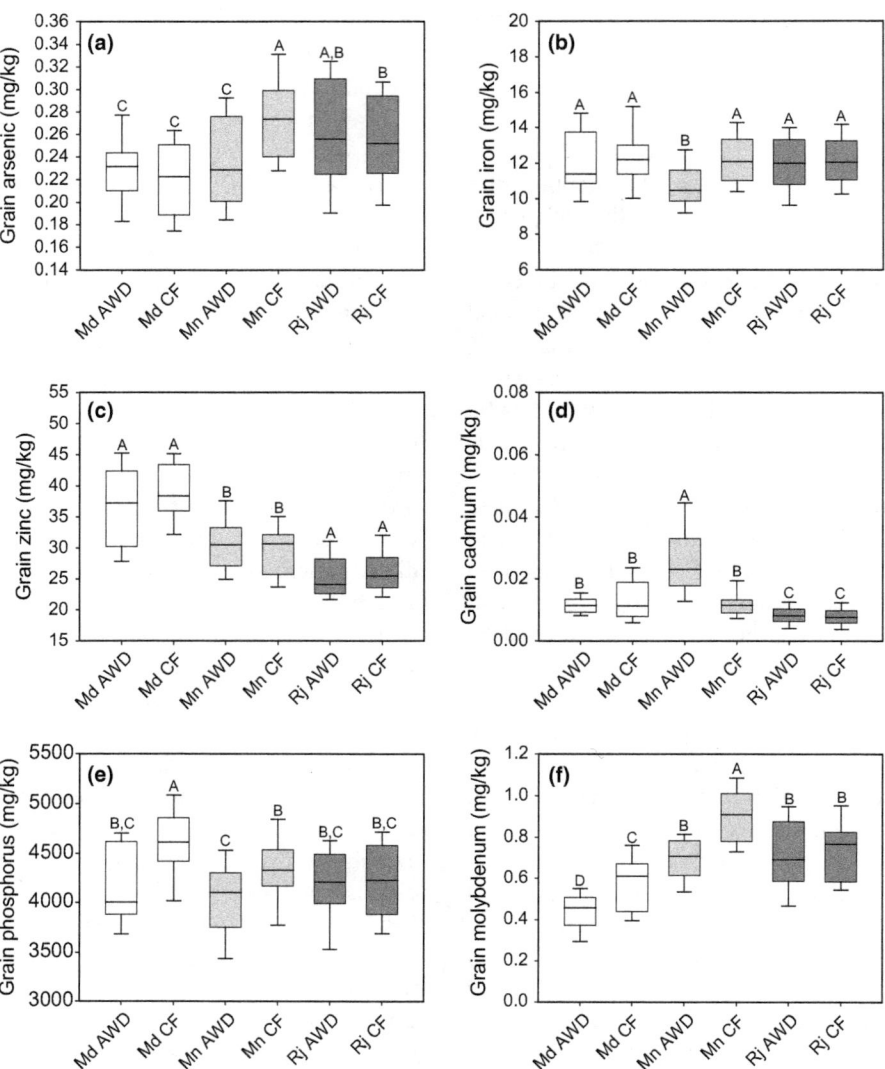

Figure 5. Grain element traits for the 22 cultivars grown at the three different sites under AWD and CF. (a) grain arsenic, (b) grain iron, (c) grain zinc, (d) grain cadmium, (e) grain phosphorus, (f) grain molybdenum. Md = Madhupur, Mn = Mymensingh, Rj = Rajshahi. Data which has the same letter are not significantly different. AWD, Alternative wetting and drying; CF, continuous flooding.

in this site (higher grain and straw). Both of these factors should be investigated further.

As flowering time was only measured in a single replicate at Mymensingh and not at the other sites, it is not possible to determine the impact that AWD has on flowering time. As there was a range in flowering time for the cultivars and flowering occurred prior to the last AWD cycle finishing, the data can be used to explore if this had an impact on the grain concentration of elements. During the later parts of the AWD cycle, the soil chemistry should change as it moves from an anaerobic soil to an aerobic soil, and this will affect the availability of elements to the plants (Rinklebe et al. 2016). For example, under anaerobic conditions, arsenic availability will be greater than under aerobic conditions, and it has been

demonstrated that this impacts the accumulation of arsenic within rice grains, with plants grown under anaerobic conditions have an order of magnitude higher grain arsenic (Xu et al. 2008; Norton et al. 2012a,b, 2013). It is also proposed that the arsenic grain is filled directly from pools of arsenic accumulated by the plant rather than remobilization of arsenic from leaves (Carey et al. 2011). Therefore, it could be hypothesized that plants flowering earlier (when the AWD cycle had not finished) would be exposed to periods of dry (aerobic soil) with lower concentrations of mobile arsenic and therefore would accumulate lower concentrations of arsenic in their grains. However, this was not the case and in fact for arsenic it was the opposite (Figure S3). It is already known that there are strong cultivar differences in elemental concentration in grains (e.g., Jiang et al. 2008; Norton et al. 2010a), it makes the unraveling of the observation of the relationships between flowering time and grain element concentration difficult. Further work on the timing of AWD and how this effects elements is essential, as not only was arsenic effected, but the key nutritional element zinc. However, grain zinc concentration was correlated to flowering time in both AWD and CF grown plants indicating that this is directly related to flowering time rather than the impact that AWD has.

Despite the lack of response to AWD treatment at Rajshahi, the AWD treatment affected the overall concentration of a large number of elements within both the straw and the grains of the rice plants. There were also cultivar differences for a majority of the elements measured in straw and all the elements except for chromium measured in the grains. However, only grain cadmium concentration was affected by the interaction between treatment × cultivar and site × treatment × cultivar. This suggests that for grain cadmium, the genetic mechanism responsible for grain concentration is different when plants are grown under AWD and CF. This means that the genes and quantitative trait loci that have been shown to regulate cadmium accumulation under paddy conditions (Isikawa et al. 2010; Norton et al. 2010b; Zhang et al. 2014; Huang et al. 2015) may not be applicable under AWD. The accumulation of cadmium in grains has been highlighted as a potential risk to human health in rice crops (Meharg et al. 2013). Understanding the mechanisms that regulate the accumulation of cadmium in rice will be beneficial, especially an understanding of the impact that AWD has on the uptake and accumulation mechanisms. For all the other grain elements, the results suggest that progress in regulating the accumulation of these elements (to reduce toxic arsenic or increase beneficial Fe and Zn, for example) would still be equally applicable under AWD cultivation.

In conclusion, AWD in our study has been shown to affect some components of yield and element concentration in plant tissues when rice was grown at three different field sites, but the response differed between sites as indicated by site × treatment interactions. These interactions did not reflect conflicting direction of responses for different sites, rather differences in the magnitude of response in that some sites (especially Rajshahi) did not respond to AWD for some measured traits. The impact that AWD had on the concentration of elements in the grain indicates that AWD may be useful for reducing grain arsenic but care should be taken for areas with high soil available cadmium. Importantly, only grain cadmium concentration was significantly affected by the interaction between treatment and cultivar. This suggests that breeding rice for CF conditions will be equally effective for producing cultivars suited to AWD. Probably it is only if cadmium accumulation is potentially problematic for a specific region that breeding efforts will need to be tailored to AWD.

Acknowledgments

This work was supported by the Biotechnology and Biological Sciences Research Council [BB/J003336/1].

Conflict of Interest

None declared.

References

Bouman, B. 2009. How much water does rice use. Rice Today 8:28–29.

Carey, A. M., G. J. Norton, C. Deacon, K. G. Scheckel, E. Lombi, T. Punshon, et al. 2011. Phloem transport of arsenic species from flag leaf to grain during grain filling. New Phytol. 192:87–98.

Chou, M. L., J. S. Jean, G. X. Sun, C. M. Yang, Z. Y. Hseu, S. F. Kuo, et al. 2016. Irrigation practices on rice crop production in arsenic-rich paddy soil. Crop Sci. 56:422–431.

Fageria, N. K. 2007. Yield of rice. J Plant Nut. 30:843–879.

Garcia-Oliveira, A. L., L. Tan, Y. Fu, and C. Sun. 2009. Genetic identification of quantitative trait loci for contents of mineral nutrients in rice grain. J. Integr. Plant Biol. 51:84–92.

Godfray, H. C. J., J. R. Beddington, I. R. Crute, L. Haddad, D. Lawrence, J. F. Muir, et al. 2010. Food security: the challenge of feeding 9 billion people. Science 327:812–818.

Howell, K. R., P. Shrestha, and I. C. Dodd. 2015. Alternate wetting and drying irrigation maintained rice yields despite half the irrigation volume, but is currently unlikely to be adopted by smallholder lowland farmers in Nepal. Food Energy Secur. 4:144–157.

Huang, Y., C. Sun, J. Min, Y. Chen, C. Tong, and J. Bao. 2015. Association mapping of quantitative trait loci for mineral element contents in whole grain rice (*Oryrza sativa* L.). J. Agric. Food Chem. 63:10885–10892.

Isikawa, S., T. Abe, M. Kuramata, M. Yamaguchi, T. Ando, T. Yamamoto, et al. 2010. A major quantitative trait loci for increasing cadmium-specific concentration in rice grain is located on the short arm of chromosome 7. J. Expt. Bot. 61:923–934.

Jiang, S. L., J. G. Wu, N. B. Thang, Y. Feng, X. E. Yang, C. H. Shi. 2008. Genotypic variation of mineral elements contents in rice (*Oryza sativa* L.). Eur. Food Res. Technol. 228:115–122.

Kush, G. S. 2013. Strategies for increasing the yield potential of cereals: case of rice as an example. Plant Breed. 132:433–436.

Lampayan, R. M., R. M. Rejesus, G. R. Singleton, and B. A. M. Bouman. 2015. Adoption and economics of alternate wetting and drying water management for irrigated lowland rice. Field. Crop. Res. 170:95–108.

Liang, K., X. Zhong, N. Huang, R. M. Lampayan, J. Pan, K. Tian, et al. 2016. Grain yield, water productivity and CH4 emission of irrigated rice in response to water management in south China. Agric. Water Manage. 163:319–331.

Linquist, B. A., M. M. Anders, M. A. A. Adviento-Borbe, R. L. Chaney, L. L. Nalley, E. F. F. da Rosa, et al. 2015. Reducing greenhouse gas emissions, water use, and grain arsenic levels in rice systems. Glob. Change Biol. 21:407–417.

Lu, K., L. Li, X. Zheng, Z. Zhang, T. Mou, Z. Hu. 2008. Quantitative trait loci controlling Cu, Ca, Zn, Mn and Fe content in rice grains. J. Genet. 87:305–310.

Meharg, A. A., G. Norton, C. Deacon, P. Williams, E. E. Adomako, A. Price, et al. 2013. Variation in rice cadmium related to human exposure. Environ. Sci. Technol. 47:5613–5618.

Norton, G. J., M. R. Islam, G. Duan, M. Lei, Y. Zhu, C. M. Deacon, et al. 2010a. Arsenic shoot-grain relationships in field grown rice cultivars. Environ. Sci. Technol. 44:1471–1477.

Norton, G. J., C. M. Deacon, L. Xiong, S. Huang, A. A. Meharg, and A. H. Price. 2010b. Genetic mapping of the rice ionome in leaves and grain: identification of QTLs for 17 elements including arsenic, cadmium, iron and selenium. Plant Soil 329:139–153.

Norton, G. J., G. L. Duan, M. Lei, Y. G. Zhu, A. A. Meharg, and A. H. Price. 2012a. Identification of quantitative trait loci for rice grain element composition on an arsenic impacted soil: influence of flowering time on genetic loci. Ann. Appl. Biol. 161:46–56.

Norton, G. J., S. R. M. Pinson, J. Alexander, S. Mckay, H. Hansen, G. L. Duan, et al. 2012b. Variation in grain arsenic assessed in a diverse panel of rice (Oryza sativa) grown in multiple sites. New Phytol. 193:650–664.

Norton, G. J., E. E. Adomako, C. M. Deacon, A. M. Carey, A. H. Price, and A. A. Meharg. 2013. Effect of organic matter amendment, arsenic amendment and water management regime on rice grain arsenic species. Environ. Pol. 177:38–47.

Norton, G. J., A. Douglas, B. Lahner, E. Yakubova, M. L. Guerinot, S. R. M. Pinson, et al. 2014. Genome wide association mapping of grain arsenic, copper, molybdenum and zinc in rice (Oryza sativa L.) grown at four international field sites. PLoS One 9:e89685.

Norton, G. J., M. Shafaei, A. J. Travis, C. M. Deacon, J. Danku, D. Pond, et al. 2017. Impact of alternate wetting and drying on rice physiology, grain production, and grain quality. Field Crop Res. 205:1–13.

Rinklebe, J., S. M. Shaheen, and K. Yu. 2016. Release of As, Ba, Cd, Cu, Pb, and Sr under pre-definite redox conditions in different rice paddy soils originating from the U.S.A. and Asia. Geoderma 270:21–32.

Shaibu, Y. A., H. R. Mloza Banda, C. N. Makwiza, and J. Chidanti Malunga. 2015. Grain yield performance of upland and lowland rice varieties under water saving irrigation through alternate wetting and drying in sandy clay loams of Southern Malawi. Exp. Agric. 51:313–326.

Smith, P., D. Martino, Z. Cai, D. Gwary, H. Janzen, P. Kumar, et al. 2008. Greenhouse gas mitigation in agriculture. Philos. Trans. R Soc. Lond. B Biol. Sci. Sust. Agric. II 363:789–813.

Somenahally, A. C., E. B. Hollister, W. Yan, T. J. Gentry, and R. H. Leoppert. 2011. Water management impacts on arsenic speciation and iron-reducing bacteria in contrasting rice-rhizosphere compartments. Environ. Sci. Technol. 45:8328–8335.

Stangoulis, J. C. R., B. L. Huynh, R. M. Welch, E. Y. Choi, and R. D. Graham. 2007. Quantitative trait loci for phytate in rice grain and their relationship with grain micronutrient content. Euphytica 154:289–294.

Sudhir-Yadav, E. Humphreys, T. Li, G. Gill, and S. S. Kukal. 2012. Evaluation of tradeoffs in land and water productivity of dry seeded rice as affected by irrigation schedule. Field Crops Res. 128:180–190.

Tan, X., D. Shao, W. Gu, and H. Liu. 2015. Field analysis of water and nitrogen fate in lowland paddy fields under different water managements using HYDRUS-1D. Agric. Water Manage. 150:67–80.

Travis, A. J., G. J. Norton, S. Datta, R. Sarma, T. Dasgupta, F. L. Savio, et al. 2015. Assessing the genetic diversity of rice originating from Bangladesh, Assam and West Bengal. Rice 8:35.

Wang, Y., Y. Wei, L. Dong, L. Lu, Y. Feng, J. Zhang, et al. 2014. Improved yield and Zn accumulation for rice grain by Zn fertilisation and optimized water management. J. Zhejiang Univ. Sci. B 15:365–374.

Xu, X. Y., S. P. McGrath, A. A. Meharg, and F. J. Zhao. 2008. Growing rice aerobically markedly decreases arsenic accumulation. Environ. Sci. Technol. 42:5574–5579.

Yang, J. C., and J. H. Zhang. 2010. Crop management techniques to enhance harvest index in rice. J. Exp. Bot. 61:3177–3189.

Yang, J. C., D. Huang, H. Duan, G. Tan, and J. Zhang. 2009. Alternate wetting and moderate drying increase grain yield and reduces cadmium accumulation in rice grains. J. Sci. Food Agric. 89:1728–1736.

Yao, F. X., J. L. Huang, K. H. Cui, L. X. Nie, J. Xiang, X. J. Liu, et al. 2012. Agronomic performance of high-yielding rice variety grown under alternate wetting and drying irrigation. Field. Crop. Res. 126:16–22.

Zhang, H., Y. Xue, Z. Wang, J. Yang, and J. Zhang. 2009. Alternate wetting and moderate soil drying improves root and shoot growth in rice. Crop Sci. 49:2246–2260.

7

Bioenergy from agroforestry can lead to improved food security, climate change, soil quality, and rural development

Navin Sharma[1], Babita Bohra[2], Namita Pragya[2], Rodrigo Ciannella[1], Phil Dobie[1] & Sarah Lehmann[3]

[1]World Agroforestry Centre, UN Gigiri, Nairobi, Kenya
[2]World Agroforestry Centre, NASC Complex, New Delhi, India
[3]Intern at GIZ, Untergasse 15, 65510 Idstein, Germany

Keywords
Agroforestry, bioenergy, biofuels, food security, livelihood.

Correspondence
Babita Bohra, World Agroforestry Centre, NASC Complex, New Delhi, India.

E-mail: b.bohra@cgiar.org

Funding Information
No funding information provided.

Abstract

Well-designed bioenergy systems can contribute to several objectives, such as mitigating climate change, increasing energy access, and alleviating rural poverty. With adequate technical assistance and land management, farm yields and income can be increased, food security strengthened, carbon sequestration improved, and pressure for land clearing reduced. There are, nonetheless, risks involved on bioenergy production and several initiatives worldwide have failed to achieve proposed positive outcomes. Overreliance on monoculture plantations, negative land-use change impacts, and use of cereal crops as feedstocks are among the main causes. Agroforestry systems and practices can address most of these risks and thus play an important role in sustainable production of several bioenergy outputs, including efficient solid biomass, biogas, liquid biofuels, and dendro power. This article assesses the potential of such integrated approaches to provide multiple benefits, including the coproduction of food, animal feed, and organic fertilizers, while respecting economic, social, and environmental sustainability indicators. Building on experiences from sub-Saharan Africa, developing Asia, and Latin America, promising perennial species, production models, and value chains are analyzed. Finally, key challenges and potential solutions for larger scale adoption of integrated food-energy approaches are also identified and discussed.

Introduction

Globally, 1.2 billion people lack access to electricity, while 2.7 billion depend on traditional fuels (e.g., firewood, charcoal, and dung) for cooking (IEA, 2011); a figure linked to approximately 4.3 million deaths a year from exposure to household air pollution (WHO). Most of these people can be found in rural areas of developing Asia or sub-Saharan Africa. They constitute a large part of the one billion global workforces whose livelihoods depend on agriculture (FAO), as well as the 500 million smallholder farmers who produce 80 % of the food in these regions (Nwanze 2011). Not by chance, many of them are also part of the 980 million people living in extreme poverty in rural areas worldwide (IFAD, 2011).

Improving access to affordable and reliable modern forms of energy services is essential to reduce poverty and promote economic growth in the developing world (Malla 2013). Energy is required not only for basic human needs, such as for lighting, cooking and heating, but also for improving agricultural productivity, allowing the use of advanced machinery from sowing to postharvest processing. Decreasing costs and scaling-up of several technologies for decentralized renewable energy (RE) production is undoubtedly essential to address this issue. Although enhancing energy access, efficiency and the use of renewables might have been missed by the Millennium Development Goals; this target is nowadays largely acknowledged as crucial to the global development agenda. In particular, the Sustainable Development Goals (SDGs) launched in 2015 has adopted a specific goal to "ensure access to affordable, reliable, sustainable and modern energy for all" (UNDP). The Sustainable Energy for All initiative pioneered by the United Nations and the World Bank is also dedicated to achieving similar objectives (The Secretary-General's High-Level Group, 2012).

On the other hand, what is often overlooked by policy makers and governments worldwide is the potential to reduce poverty by encouraging smallholder farmers to become, in addition to consumers, also producers of renewable energy. Traditional agricultural markets (for food, feed, and fiber mainly) account for only 3 % of global GDP (FAO). They are thus too small to sustain so many people whose livelihoods depend on this sector. Rural development requires not only encouraging energy provision to meet local needs, be it to provide cleaner cooking fuel or to intensify agricultural production. Moreover, smallholder farmers need alternatives to enhance their incomes. Energy markets, through sustainable bioenergy production, provide such opportunities. Out of the 7.7 million people estimated to be employed (directly or indirectly) in the renewable energy sector, approximately 40 % are in the bioenergy segment – producing liquid biofuels, modern biomass, or biogas (Ferroukhi et al. 2015). No other renewable energy source has the potential to generate so many jobs for farmers in developing countries. Solar technologies, for instance, have created a significant number of jobs (approximately 3.2 million globally), but these are highly concentrated in a few industrialized countries, in addition to focusing on urban areas.

Well-designed bioenergy production systems can contribute to several objectives, such as mitigating climate change, increasing energy access, and alleviating rural poverty (Casillas and Kammen 2010). With adequate technical assistance and land management, farm yields and income can be increased, food security strengthened, carbon sequestration improved, and pressure for land clearing reduced (Dale et al. 2011). There are, nonetheless, risks involved with bioenergy production and several initiatives have failed to achieve these positive outcomes. The benefits of *Jatropha curcas*, for instance, were clearly oversold in developing countries such as Tanzania (Sosovele 2010). Particular features of this species, such as the toxicity of its seedcake, require further investment in research before larger scale field deployment. Many governments have also encouraged large-scale monoculture of bioenergy feedstocks in poor regions. Ethiopia, for instance, set aside 17.2 million ha of land for biofuel production in 2010 (Sosovele 2010). Land-use changes had negative impacts in some cases. In Indonesia, for instance, conversion of peat swamp forests to oil palm led to significant losses in terms of biodiversity and aboveground biomass carbon, in addition to annual emissions of about 4.6 million Mg of belowground carbon from peat oxidation (Koh et al. 2011). In other instances, the diversion of food crops toward fuel production played a relevant role in food

price increases, jeopardizing food security in poor food-importing countries (HLPE, 2013).

Agroforestry systems and practices can address most of these risks and thus play an important role in sustainable bioenergy production. It is well-known that models that integrate woody perennials (trees, shrubs, palms, bamboos, etc.) with crops and/or animals can contribute to climate change adaptation (Lin et al. 2008) and mitigation (Budiadi and Ishii 2010), improve soils (Glover et al. 2012) and natural resource management (Ong and Kho 2015), increase yields (Bayalaa et al. 2012), income (Sood 2006), and thus strengthen food security (Sanchez et al. 1997). Furthermore, such systems allow for food, feed, and several types of bioenergy to be produced respecting economic, social, and environmental sustainability indicators. In an integrated strategy together with improved kilns and stoves, agroforestry can sustainably provide woodfuel, reducing wood harvest pressures in forests in sub-Saharan Africa (Iiyama et al. 2014). It can also provide modern alternatives to cooking fuel, such as biogas produced from waste biomass materials. This approach has been largely adopted in countries such as Vietnam and Nepal (Bogdanski 2012). As Sri Lanka's experience demonstrates, agroforestry can also supply biomass for economical electricity (dendro power) generation (Kulatunga 2012). Regarding liquid biofuels, it is worth noting that global investment in this sector has plummeted in the last 5 years, and most of it is now concentrated on advanced technologies (IREA, 2016), such as second-generation ethanol produced from lignocellulosic biomass or third-generation fuels derived from algae, which are coming to age (UNCTD, 2016). However, as section Programme for the Development of Alternative Biofuel Crops – An ongoing project of the World Agroforestry Centre (ICRAF) discusses with examples in India, Kenya, and Brazil, biofuels can still be sustainably and competitively produced through agroforestry systems that tap into the potential of native (and sometimes less popular) oilseed tree species. These systems have low production cost, not only because the technologies deployed are relatively simple and well-established, or due to the sustainable supply of cheap raw materials, but also because all components of the resulting biomass, including by-products and waste, are used in a bio refinery system (Gupta et al. 2014).

The objective of this article is to evaluate the potential of bioenergy from agroforestry systems to improve livelihoods and food security, to assess progress made in this area and to identify knowledge gaps. It explores the role of trees in providing energy and suggests ways of making bioenergy a mainstream energy source. It cites examples existing of national and regional programs that require farmers to incorporate trees into agricultural systems and describes various activities undertaken in these programs.

The paper also answers specific questions relating to constraints to and motivations for, including trees in farming systems.

Agroforestry, Gender, Health, and Energy Production – Increasing the Quality of Life

The World Agroforestry Centre (ICRAF) defines agroforestry as "a collective name for land-use systems and practices where woody perennials (trees, shrubs, palms, bamboos, etc.) are deliberately integrated with crops and/or animals on the same land management unit in some form of spatial arrangement or temporal sequence. There are three broad types of agroforestry: (i) agrosilviculture which integrates annual crops and trees, (ii) silvopastoral which defines the integration of livestock and trees and (iii) agrosilvopastoral which integrates annual crops, plus livestock, and trees. Furthermore, there are different forms of agroforestry systems, such as alley cropping systems, intercropping, or hedgerow systems.

There are normally both ecological and economic interactions between the woody and non-woody components in agroforestry" (Lundgren and Raintree 1982). For example, agroforestry integrates perennials on farmlands. Trees and crops provide mutual benefits. The roots of crops force the tree roots to colonize a deeper soil horizon, protecting the tree from droughts and affording it access to otherwise unused nutrients. The same roots allow for a much better percolation of rainwater into the deep soil, reducing waterlogging and storing much needed water

that becomes available through capillary action to crops during dry spells. The tree roots on the other hand bind the soil, reduce soil erosion, especially on sloping grounds. The trees slow down winds and reduce soil desiccation. They increase soil moisture and reduce nutrient leaching. Their roots act as a pump, bringing nutrients to the surface from deep soil horizons. These become available to crops through leaf, twig and tree-root litter, all of which also contributes to soil organic matter.

Smart agroforestry systems can improve livelihoods, provide food and nutrition security, and clean energy in the form of bioenergy (refer Fig. 1). In this way they contribute to sustainable rural development. Trees provide goods, including timber for fuel, fodder for animals, and food for people (fruits, vegetables, nuts, and oils), which are essential to meeting nutrition requirements, and trees also provide ecosystem services as they help improve the water management and soil fertility, for example, through maintenance of the nitrogen cycle. They support further environmental services, such as carbon storage, water purification, and biodiversity conservation.

Wood-based fuels have been associated with poverty and are believed to pollute the environment. However, when trees are replanted and woodfuels applied in modernized technologies, they can be clean and truly renewable. Trees have the potential to meet the urgent need for energy among poor people. The inclusion of tree-based energy in policy discussions is therefore required.

Agroforestry reduces pressure on forests as trees grown on farms provide fuelwood, helping to meet local energy needs. Forests have been exploited for centuries to provide

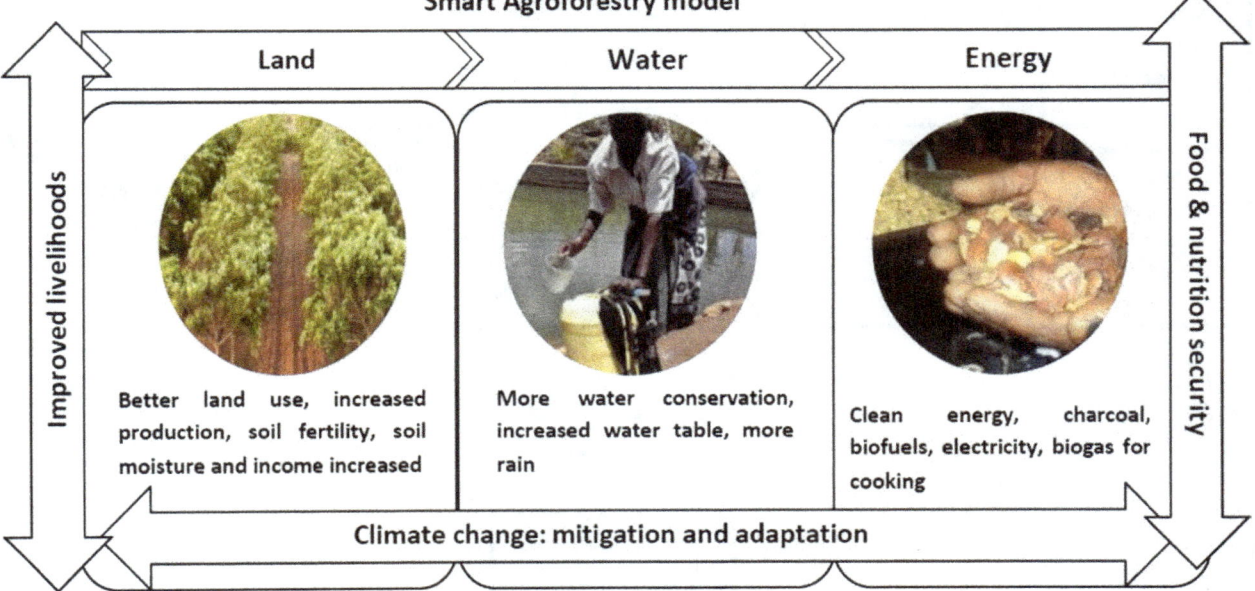

Figure 1. Integration of energy and food security through smart Agroforestry model. Source: Babita Bohra, ICRAF

the basic energy needs of human beings. This remains unabated. In particular, charcoal production for urban energy is one of the major causes of the systematic degradation of dryland forests worldwide (Ouya 2013). To fight forest degradation, it is imperative to foster more efficient production and use of firewood and charcoal, while also promoting alternative sustainable bioenergy options.

In order to address issues related to agriculture, forestry and livelihoods simultaneously, adopting a landscape approach is essential. Landscapes are complex systems with sets of social, biophysical, human, ecological, and economic dimensions that interact with each other. Such interactions happen at multiple levels: the plot, field, farm, and beyond. Understanding and building on interactions and feedback loops is thus important for success (Selvarajah 2013). Furthermore, the landscape approach places forest-dependent communities at the center.

Agroforestry holds potential in climate change mitigation and adaptation. Interestingly, global climate change initiatives tend to target reforestation and forest protection and largely do not mention agroforestry. However, agroforestry provides a win-win solution to a seemingly difficult choice between reforestation and agricultural land-use changes and is thus a powerful way to address mitigation. According to estimates by the International Panel on Climate Change (IPCC), land-use change activities generate 1.6 billion tons of carbon annually (Denman 2007). Modern, well-planned, wisely-implemented, and locally adapted bioenergy systems hold promise in addressing increased carbon emissions.

Agroforestry can further help farmers and communities to adapt to changes in climate through maintaining and enhancing environmental services. It can provide opportunities for diversification that build resilience and generate additional income. There are innumerable examples of traditional land-use practices involving the growing of trees with agricultural crops to augment food production and buffer farmers against the risks of crop failure by providing alternate sources of income. The incorporation of livestock and bioenergy also provide diversification opportunities to farmers. Agroforestry interventions are uniquely placed to achieve multiple objectives of food and environmental security (refer Fig. 2).

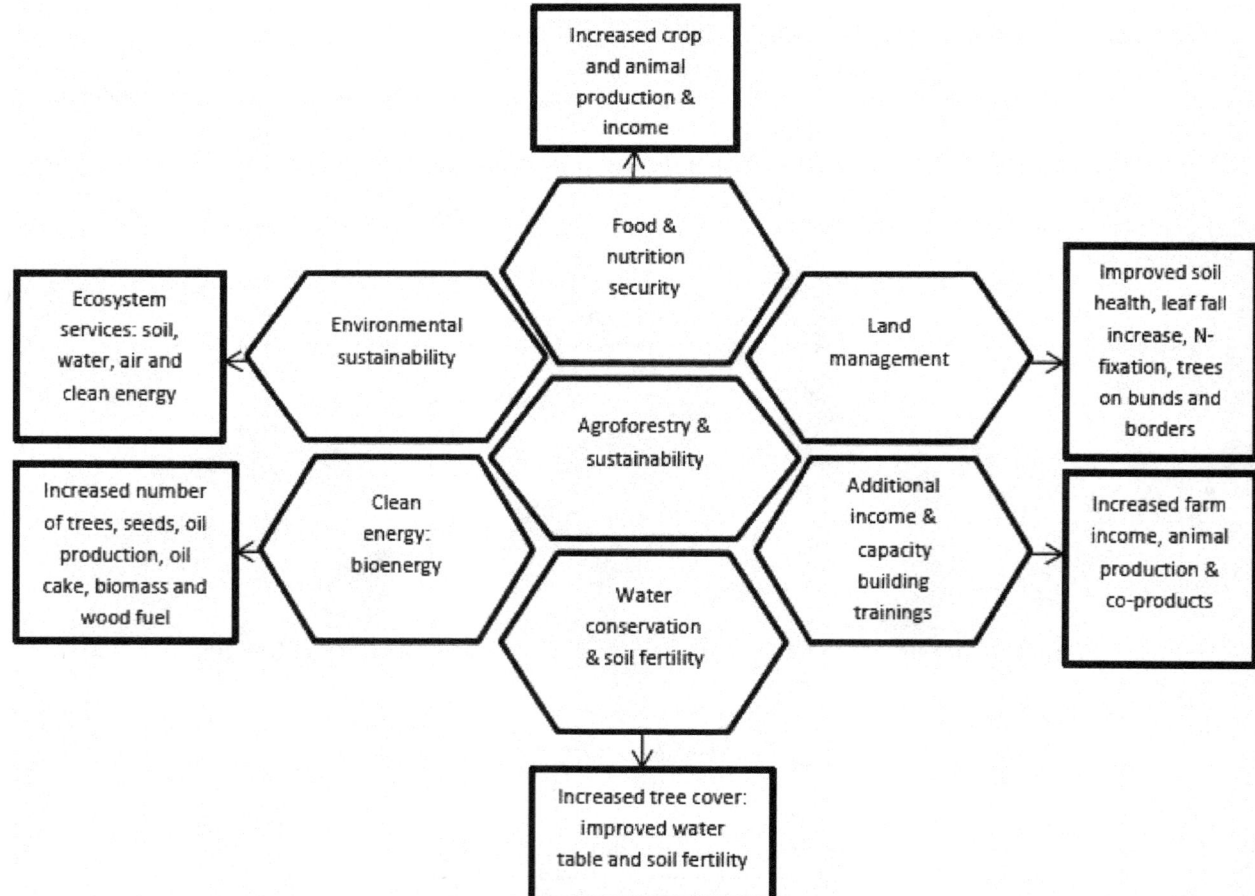

Figure 2. Multiple benefits from Agroforestry system. Source: Babita Bohra, ICRAF

The FAO projects that global food production will need to increase by 60% by 2050 (from 2005/07 levels) to meet the world's demand under a business-as-usual scenario (Millennium Institute, 2013). When the landholding size is small, such as in Africa and the Indian subcontinent, combining traditional agriculture with appropriate trees might be a good solution to optimize farm productivity. Thus, agroforestry interventions might act as potent instrument in achieving the required growth in agriculture.

Agroforestry is a "social forestry" that helps people to meet their economic, environmental, and social needs on their own land (refer Fig. 2). Through providing tree product-based economic opportunities, agroforestry can also help to create rural employment.

Agroforestry as a source of energy

Trees can provide feedstocks for production of various forms of bioenergy, for eg. solid biomass can be used as firewood, as charcoal and for electricity generation. Oilseeds can be used for production of liquid biofuels like biodiesel, while lignocellulosic biomass can be used to produce ethanol. Residues like leaves and oilseed cake can be used for biogas production (refer table 1 for various tree based technologies). Selecting the appropriate tree species to provide the desired benefits and integrating them into traditional agriculture without compromising and, if possible, improving agricultural productivity, remains a challenge. Agroforestry approaches will increasingly find a central place in these bioenergy initiatives.

Forms of bioenergy that can be derived from agroforestry systems

Fuelwood and charcoal

Fuelwood plays an important role in the lives of the poor and rural families, by providing a primary source of energy. Fuelwood has both domestic and industrial uses and is used in rural and urban regions of most economies of the developing world (Dovie et al. 2004). Seventy-nine percent (79%) of the total traditional energy (fuelwood, cow dung, biomass etc.) consumed in developing countries is fuelwood and between 60% and 69% of this is in sub-Saharan Africa (Adebimpe 2013). About 70% of the energy consumed in India is met by fuelwood collected from forests and marginal lands.

Despite such widespread reliance on fuelwood, there are frequently no adequate policies in place that could make fuelwood use more sustainable and efficient. Unsustainable harvesting of fuelwood from forests may lead to negative impacts like habitat fragmentation and loss of environmental services, the encroachment of unprotected natural forests, and finally to deforestation. If fuelwood species are grown on fallow land, such negative impacts can be avoided. Trees can then provide both fuel resources and help to address energy crises within a country (Jain 1994).

Agroforestry can make a significant contribution toward the provision of fuelwood. Although accurate estimates are difficult to obtain, it is assumed that agroforestry contributes to fuelwood production anywhere between 20% in Africa to 70% in Asia.

Charcoal is a vital source of energy for cooking for a large part of the rural population in developing countries (Mwampamba et al. 2013). Charcoal industry provides lucrative business opportunities worth millions of dollars (Mwampamba et al. 2013). Compared to untreated biomass, charcoal has a high energy density and a low water content which makes it easier to store and transport. According to a study done by Mwampamba et al. (2013), charcoal is used by people in wide range of income category and since it reduces indoor air pollution in comparison to firewood, it is more preferred by urban and people in higher income group. To reduce deforestation related to charcoal production, agroforestry systems can be enhanced to support local production of the source material wood. This may also provide the opportunity for employment in local charcoal production.

Liquid biofuel

Liquid biofuels have received increasing attention worldwide as renewable fuels with the potential to help improve energy security, mitigate climate change, and revitalize agricultural economies (Scovronick and Wilkinson 2014). Bioethanol and biodiesel are the two global liquid biofuels that hold promise for replacing gasoline and diesel fuel, respectively (Demirbas 2011).

Demand for ethanol is expected to more than double in the next 10 years. To fill the demand-supply gap, new feedstock and technologies must be developed and achieve commercial scale (Demirbas 2011). Trees and agroforestry systems will have an increasing role to play in the global production of ethanol. Although most of the current ethanol production comes from annual crops such as sugarcane and maize, many trees provide sugar-rich sap or fruit pulps which can be fermented and processed into ethanol. Examples include simarouba (*Simarouba glauca*), which is a promising raw material both for biodiesel (produced from its seed) and ethanol (produced from its fruit pulp) (Joshi and Joshi), as well as sugar palm (*Arenga pinnata*) and nypa palm (*Nypa fruticans*). Trees also provide lignocellulosic biomass that can be used for the production of second-generation ethanol, although these advanced technologies are only slowly reaching commercial scale.

Table 1. Summary of various tree-based energy technologies.

	Technology	Features	Uses & benefits	Successful examples	Remarks
Fuelwood	Direct combustion of fuelwood in fuel-efficient cookstoves Smokeless ovens	The Fuel Value Index (FVI) is used for screening suitable fuelwood species. High moisture content and ash reduce the FVI values and heat of combustion	Cooking Provides livelihood activity for poor rural communities	Fuel-efficient wood burning stoves by TIDE (Technology Informatics Design Endeavour), India Smokeless ovens of Vivekanand trust, Karnataka state, India Project Surya for improved cookstove intervention, India	There is a strong link between fuelwood uses and livelihoods, especially in low- income communities. To avoid forest decline, fuelwood should be harvested responsibly and under proper monitoring. Fast-growing trees should be planted and used trees should be replaced. Tree species like *Ixora parviflora*, *Schleichera trijuga*, *Parkinsonia aculeata*, and the shrubs like *Capparis aphylla*, *Dodonaea viscosa*, *Combretum ovalifolium*, *Vitex negundo*, *Rubus ellipticus*, *Coriaria nepalensis*, *Mimosa rubicaulis*, and *Antidesma ghaesembilla* make excellent firewood.
Charcoal	Slow pyrolysis of biomass in kiln	Roughly one-third of dry feed (based on mass) is converted to charcoal and the remaining into syngas and bio-oil/tar. Compared to untreated biomass, charcoal has high energy density and low water content and thus is easier than fuelwood to store and transport (Larsson et al. 2013).	Increases the efficiency of fuelwood Produces less smoke and requires little or no preparation before use Has higher energy content per unit mass and can be reused when left over after cooking. Can be easily transported and stored	Mafia Pilot project in Tanzania, initiated in 1985 and funded by the German development agency, GTZ Malawi Charcoal Project, funded by the World Bank and the Malawi Government was initiated in 1986 Dandora Charcoal Plant, established by the Kenya Planters Co-operative Union	The Mafia Pilot project uses coconut industry waste, the Malawi Charcoal project uses wood waste, and Dandora Charcoal Plant uses coffee husk as feedstock. Examples show that fuelwood supply limitations have now led to the use of other agricultural and forestry residues for charcoal preparation. Efficient kilns are required to improve production efficiency.
Liquid biofuels	*Bioethanol:* Cellulosic ethanol From fermentation of sugar	Ethanol is produced from bio-matter such as sugar cane, corn, or other grains, and is therefore also referred to as bioethanol.	Saves gasoline Reduces GHG emissions Reduces food-fuel competition	Indian Railways Project LIBERTY, US	Project LIBERTY converts baled corn cobs, leaves, husk, and stalk into renewable fuel; therefore, this new technology will help to reduce pressure on food crops like sugarcane, and soybean etc.
	Biodiesel: Transesterification	Biodiesel is usually produced from vegetable oils or animal fats. It can either be used directly as fuel with some engine modification, or in blends with petroleum diesel with few or no modifications in the diesel engine (Demirbas 2011)	Saves petro-diesel Cleaner than petro-diesel and thus leads to GHG emission reductions By-products like oil cake can either be used as fertilizer or as animal feed and the husk can be burned to release energy.	Indian Railways Bio-fuel Park, Hassan, India	Biofuel-park of Hassan is agroforestry model which grows seven species; Pongamia, Madhuka, Azadirachta, Simarouba, Jatropha, Calophyllum, and Aphanamixis. This system does not hinder the normal agricultural process and also provides extra income for farmers from the production of biofuel and its by-products.

(Continued)

Table 1. (Continued)

	Technology	Features	Uses & benefits	Successful examples	Remarks
Bio-electricity	Biomass gasifier plant	Biomass includes dedicated energy crops, agricultural residues, and forest residues as well as other biomass sources such as landfill gas, animal and human waste (Evans et al. 2010). There are two types: *Dedicated biomass-based power plant:* Generally smaller in size due to limited feedstock availability and high transportation costs (Bertrand et al. 2014). Have high investment cost and a low conversion efficiency of 25% (Bertrand et al. 2014). *Cofiring of biomass in coal-based power plant:* Requires little or no investment. The conversion efficiency is on average 36% in OECD countries.	Saves fossil fuel Reduces GHG emissions Promotes sustainable approach in meeting rural energy needs Electricity for irrigation, lighting, and small industries Enhances green cover, carbon sequestration, and can reclaim waste land.	BERI (Biomass Energy for Rural India) project, Tumkur district of Karnataka, India AVANI project, Pithoragarh, Uttarakhand, India Biomass gasifier by Ankur Scientific in Sankheda Taluka of Vadodara district, Gujarat, India SRE (Saran Renewable Energy), Garkha Village, Bihar, India	Has great potential in meeting the electricity needs of rural villages. Continuous supply of raw material is the first prerequisite of all such electrification projects. To ensure the same, BERI uses fast-growing species like *Prosopis juliflora, Lantana camara,* Epil-epil (Subabool), glyciridia, and bamboo. AVANI uses waste pine needles, while Ankur and SRE use waste from the agricultural fields. Further research is required to increase the conversion efficiencies of these projects.
Biogas	Anaerobic digestion/ bio-methanation of biomass in biogas plants Biogas stove	The main components are methane (50–80%) and carbon dioxide. Types of biogas plants: *KVIC (Floating Dome):* Consists of a pit on which a mild steel or plastic dome rests. Suitable for most areas except very cold climates. *Deenbandhu Model:* Has an underground masonry structure, which guards it against cold temperatures. Excavation is difficult in hard bedrock and needs high investment. *Plug Flow model:* Comprised of a cylindrical HDPE film. Suitable for warm climates. Has low cost with good gas production	Saves LPG or firewood as cooking and heating fuel Power generation and energy production for vehicles. Reduces exposure to thick smoke and chances of lung diseases (Bi and Haight 2007). The solid residue or sludge can be used as organic compost.	Biotech plants, Pathanapuram, Kerala, India The Biogas Support Programme (BSP), Nepal Arti biogas plant, India Promotion of Sustainable Agricultural Activities through Demonstration of Bio-gas Plants and Other Allied Activities, by Sarvangeen Vikas Samiti, Gagha Villages, India	Biogas digesters use food and agricultural waste and thus have particular potential, in areas where there are severe health hazards due to the production of large volumes of organic waste and pollution of water supplies. To realize this potential in the Panchayats, there is a need to work on creating awareness and secure the required finances.

Biodiesel is usually produced from animal fat or from several types of vegetable oils. There are many trees around the world with large potential for straight vegetable oil (SVO) and biodiesel production. Oil palm (*Elaeis guineensis*) and jatropha (*J. curcas*) are common examples. Oil palm has proven potential for transformation into a number of products and fuels, including biodiesel (Chew and Bhatia 2008). The original expected potential of jatropha has not been realized; so far, its yield has been poor and gives poor economic returns under smallholder management despite its ability to grow on marginal land (Ilyama et al. 2013). However, a few authors claim that for good and profitable yield from jatropha, its basic agronomic properties need to be properly understood and it needs to be cultivated with proper and careful agricultural practices (Pandey et al. 2011; Behera et al. 2010; Openshaw 2000).

Moreover, trees that could provide liquid biofuels will become more important as techniques for converting the lignin and cellulose of trees directly into ethanol improve. Trees can address many bioenergy forms which are summarized in Table 1. Clearly, our current knowledge suggests that traditional first-generation biofuel crops, especially those grown in monocultures, may not be suitable vehicles for realizing biofuel production at scale. Not only because they have contributed to negative food versus fuel land-use changes and impacts, but also because they have marginal to negative GHG emission benefits. By incorporating biofuel production in agroforestry systems, a positive effect on the overall carbon footprint may be achieved.

Electricity generation from woody biomass

Woody biomass can easily be used in gasification systems that provide fuel to drive machinery and generate electricity. These systems are cost-effective and already widely used in both developed and developing countries. These systems can be effective at both large scale (the electricity generating woody biomass plants) and small scale (community scale installations that are already in use in the developing countries).

The use of biomass to generate electricity not only saves fossil fuel, it also reduces the amount of harmful emissions entering the atmosphere (Ryabov et al. 2006). Dedicated energy species, such as fast-growing trees for wood pellet production, agricultural residues, forest residues, and other biomass sources like animal and human waste, are a few common sources of biomass used for the generation of electricity (Evans et al. 2010).

There has been a steady increase in global biomass-based electricity generation –, on average 13 TWh per year between 2000 and 2008. The market share of biomass-based electricity in total global generation is approximately 2% and this has been maintained over the last 20 years.

Agroforestry systems can be designed for such places to ensure sustainable and efficient production of biomass for local power generation without affecting food production, and in some cases even improving it. Ideal tree species would usually be fast growing and drought tolerant, have coppice capability, be suitable for short gestation periods of 2–3 years, and have high calorific value and less ash content, among other features. Several species meet these requirements, such as *Casuarina junghuhniana*, *Casuarina equisetifolia*, *Eucalyptus*, *Melia dubia*, *Leucaena leucocephala*, *Prosopis juliflora*, and bamboo (Shankar 2012).

Biogas for cooking or electricity generation

Biogas can be produced by anaerobic digestion of human and animal manure, as well as biomass such as agroforestry residues (Food and Agricultural Organisation of United Nations). The main components of biogas are methane and carbon dioxide, with a composition of about 50–80% methane, the balance being mainly carbon dioxide. Biogas can be used for heating, cooking, power generation, and energy production for vehicles (Coimbra-Araújo et al.).

Oilseed cakes left over from SVO and biodiesel production are among the residues which can significantly enhance biogas production. Anaerobic digestion has the potential to drastically reduce GHG emissions when replacing fossil fuels (Weiland 2010).

Under optimal conditions, about 90% of feedstock energy is converted into biogas which can readily be used for cooking and lighting purposes. The sludge produced during the digestion process is nontoxic and odorless in nature. A major portion of its nitrogen and other nutrients remain intact during the process, making it a good source of fertilizer. When compared to cattle manure left in the fields to dry under the sun, the slurry from the digester has higher nitrogen content. Some of this nitrogen also gets converted to urea, which is more readily accessible by plants than many nitrogen compounds.

The dual yield of anaerobic digestion of biomass in the form of energy and fertilizer makes biogas plants an ideal candidate for rural villages (Food and Agricultural Organisation of United Nations). Biogas also offers improved health benefits due to its cleaner combustion compared to fossil fuels and other biomass used in the domestic environment (Food and Agricultural Organisation of United Nations). The chances of exposure to thick smoke and lung diseases are reduced by shift from traditional fuels to biogas (Bi and Haight 2007).

Value chain development of agroforestry products

The major difficulty in demonstrating the value of a biomass-based energy system has been a lack of proper value chains. There have been many efforts to date to provide woodfuel in a sustainable way, but few projects have succeeded in creating value chains that simultaneously address energy needs and forest degradation. Where value chains have been addressed, attention is often only paid to one part of the chain, such as cookstoves or reforestation. In other attempts, livelihood aspects have been overlooked. This way, significant structural changes have not been induced. The entire wood energy value chain, right from production to conversion, transport and marketing to consumption, should be considered.

Factors which affect the formulation of an effective and inclusive wood energy value chain include: lack of reliable baseline data; lack of intersectoral cooperation and coordination; and a lack of adequate consideration in sector policies, mainly due to capacity deficits. Other factors include weak law enforcement capacities, lack of commitment and ownership from respective governments, corruption and oligopolistic structures of the value chain (Sepp et al. 2014).

Support to inclusive value chains should take place at two levels: institutional and technical development. Institutional development will help formalize and organize the value chain and also lead to capacity development of stakeholders, creation of networks of rural and urban wood energy markets, and support to wood energy depots. Technology improvement would help promote further development of kilns, stoves, fuel types etc. (Sepp et al. 2014).

The complex supply chain of biomass for energy generation, unpredicted quality of feedstocks, variability and uncertainty in the supply chain, economic conditions, and market fluctuations affect the amount and the cost of the energy that is produced. While many efforts have been made to make these initiatives economically viable, only a few have succeeded in creating value chains that simultaneously address energy needs and improvements in the livelihoods of smallholders.

The issue of rural energy is generally dependent on national programs which are often technology-centric or end-use-based, and in the majority of cases, without any interlinkages with the rural ecosystem. Such a system is likely to experience challenges in sustainability. In bioenergy production, it is essential that upfront forward business linkages are established, for example, direct market linkages, as this is one of the main drivers for growers to adopt agroforestry.

Practical Examples of Agroforestry Production Models

Benefits of leguminous tree species

Leguminous trees have an important role in agroforestry systems as they restore nutrient cycling and improve soil fertility. The rhizobium bacteria in their roots have the ability to extract nitrogen from air and accumulate it biologically. These nitrogen-fixing trees are often deep rooted and can thus easily access nutrients in subsoil layers. Their extensive root systems also stabilize soil. Leguminous trees provide many other benefits, including food, animal fodder, fuelwood, timber, shade, and can serve as living fence (Elevitch 2008). There are many such species, including *Acacia auriculiformis*, *Cajanus cajan*, *Faidherbia albida*, *Gliricidia sepium*, *L. leucocephala* which have successfully been integrated into various agroforestry systems.

Faidherbia albida

Faidherbia albida, which is indigenous to Africa, has a unique characteristic which makes it suitable for intercropping systems. The tree, also known as winterthorn, has a number of properties which make it ideal for agroforestry systems. As a legume *F. albida* provides a habitat for symbiotic bacteria, able to fixate nitrogen and therefore providing the tree with valuable nutrients, which are passed on to the surrounding soil when the tree sheds its leaves. *F. albida*'s phenology is such that it loses its leaves during the rainy season, hence not shading adjacent crops during their prime growing phase. Its deep reaching taproots make it very drought resistant and protect the landscape against soil erosion. The tree's pods and foliage can be used as animal fodder or even for human nutrition. The flower of *F. albida* is widely known to be a preferred pasture for honey bees. Furthermore, the tree's bark and leaves are used for traditional remedies against a number of diseases (Tijani et al. 2008).

Some 500,000 farmers in Malawi, Tanzania, and Zambia cultivate their crops alongside *F. albida* in agroforestry systems. In some cases, maize yields are reported to have doubled or tripled (World Agroforestry Centre).

In Niger, farmer-managed natural regeneration (FMNR) of degraded landscapes with agroforestry systems based on *F. albida* intercropped with sorghum and millet already covers more than 5 million hectares (World Agroforestry Centre). The success of this system is due to the multiple benefits it provides, including fuelwood and fodder production as well as improved soil fertility which generates higher crop yields.

Gliricidia sepium

A long-term trial conducted by the World Agroforestry Centre in Malawi which intercropped maize with *Gliricidia sepium* [a tree that can provide several bioenergy products (World Agroforestry Centre, 2009)] achieved yields three times higher than when maize was planted alone and left unfertilized (Akinnifesi et al. 2006). Other experiments in India demonstrate that the use of *Millettia pinnata* (Pongamia) (World Agroforestry Centre, 2009) seedcake (a coproduct from SVO or biodiesel production) as organic fertilizer can lead to yield improvements of between 49% and 87% for soybean and maize (Wani and Sreedevi).

In Sri Lanka, a novel agroforestry system involving *Gliricidia* and coconut cultivation is being trialed for electricity generation (Kulatunga 2012). This is an integrated approach involving tree-based food (feed)-energy systems and has been demonstrated at the commercial level. The approach overcomes concerns that growing crops for bioenergy might compete for resources with food production. The system enables energy, fertilizer-fodder-fuelwood trees to be incorporated into agricultural fields to provide feedstock for power generation. The project started with a capacity of 1 MW and only when the system was fully functional, was the capacity enhanced to 10 MW. Next to quality fodder, additional income stems from lopping (wood). The system enabled growers to increase their milk production and generate additional income from wood sales without losing the income from coconuts.

Leucaena leucocephala

Leucaena leucocephala is one of the most widely used species in alley cropping, where it is planted in hedges along contours at intervals of 3–10 m with crops in between. It is a valuable fodder tree in the tropics. The quality of the leaves compares with alfalfa or Lucerne in feed value (Orwa et al. 2009). It is planted, for example, in an agroforestry system with sorghum. Sorghum yields can be high even in poor rainfall conditions.

Leucaena is also being considered as an option for electricity generation due to its fast growth rates and higher biomass yield. *Leucaena* can yield around 2–20 MT per ha per annum. There has been some pilot project to evaluate its potential; however, there are few power plants yet using *Leucaena*.

Cajanus cajan

The GIZ Integrated Food Security Programme (IFSP) implemented in Mulanje district, Malawi, between 1996 and 2004 (FAO) promoted pigeon pea *Cajanus cajan* (pigeon pea) intercropped with maize, the local staple food. This was complemented with the introduction of energy-saving clay stoves. Through the program, smallholders were able to harvest about 5 tons of biomass per ha, helping to satisfy cooking energy needs through their own means, as they used the woody pigeon pea stalks as their main source of fuel. The farmers also have an additional protein-rich food crop and have decreased the need for chemical fertilizer inputs while increasing agricultural yields naturally; *Cajanus cajan* improves soil fertility through nitrogen fixation, mulching of leaves (improving water retention), and decrease soil erosion (prolonged ground cover).

Acacia auriculiformis

About 140 km east of Kinshasa on the Batéké plateau, 8000 hectares of *Acacia auriculiformis* were initially planted in an EU-funded project implemented by the Dutch company HAI. From 1994, the activities were taken over by the German Hans Seidel Foundation, supported by the NGO CADIM. Later CIRAD implemented a research development project about the fuelwood sector around Kinshasa.

Early during the project's life span, the so called Mampu plantation was divided into plots of 25 hectares and allocated to 320 farming families. Since 2009, farmers own official land titles and a farmers' union manages the plots independently. Corn and cassava are intercropped with *Acacia auriculiformis*. *Acacia auriculiformis* fixes nitrogen from the atmosphere and restores it in the soil. This way, higher crop yields can be achieved. Three years after planting, the acacias reach a height of three meters. After around 10 years, a veritable acacia forest has developed. Every 4 meters, a one meter wide strip of soil is left unfarmed, so that acacia seeds can germinate and reconstitute the future forest stand. Until the crop rotation can start (after 10 years) farmers generate income mainly from beekeeping (FAO, 2010).

Acacia auriculiformis is being used for charcoal production. The total charcoal production from the plantation varies from 8000 to 12000 tons per year, corresponding to 5 t/ha of wood and a charcoal yields after carbonization of 20%. In addition, the production system yields about 10 000 tons a year of cassava, 1200 tons of maize, and 6000 tons of honey per year. Some of the farmers can earn up to US$ 4000 a year. The charcoal production satisfies around 2 per cent of the demand of Kinshasa (FAO, 2010).

Benefits of nonleguminous tree species

Eucalyptus species

In Australia, a system to integrate eucalyptus with wheat crop is being implemented. Growing the Mallee variety is profitable and easy to integrate into farming systems in semi-arid regions of southern Australia (Wildy et al. 2003). The trees are integrated into wheat belts to: (i) reduce soil salinity; (ii) give shade and shelter for animals; (iii) to reduce erosion by acting as windbreaks; and (iv) to store carbon. The Commonwealth Scientific and Industrial Research Organization (CSIRO) has evaluated eucalyptus grown in this system for producing liquid biofuels and electricity. Electricity produced by eucalyptus biomass from such systems was found to be economically viable compared to liquid biofuels.

An example of a successful approach is the GIZ Village-based Individual Reforestation Project in Madagascar (Sepp et al. 2014). *Eucalyptus* is used in plantations to reforest degraded land patches. Communes allocate individual land titles to farmers for reforestation purposes, this way the wood produced on the land is the property of the farmers involved. The project addresses the entire value chain, from individual afforestation schemes at village level to harvesting, processing, conversion, distribution, and marketing, all the way to end-consumers and related technology. As a result, it is estimated that the project increases the income of the local population by about 40% compared to the average income in rural areas. It will also provide greater opportunities for women and strengthen their economic position in society as they play a very important role in the wood energy sector as producers, users, and collectors and sometimes also deal in trading of woodfuels. This offers many prospects for promoting gender equality. Moreover, it offsets the unregulated exploitation of more than 72,000 ha of natural forests, which would otherwise have been cleared for charcoal production (Sepp et al. 2014).

Dendrocalamus strictus

Dendrocalamus strictus (Bamboo), in particular, is proving successful as a feedstock for local power generation due to its fast-growing nature; it can grow up to 18 inches in 1 day and can be harvested in 2–4 year cycles. Bamboo also has the ability to regreen degraded lands while releasing 35% more oxygen than the equivalent volume of other trees. A bamboo plantation can absorb 12 tons of carbon dioxide per hectare annually. Bamboo groves also serve as windbreaks and acoustic barriers, and bamboo is said to be superior to regular wood in terms of certain physical and mechanical properties (Siraj 2014).

Bamboo, intercropped with sesame or chickpea as an intercrop was studied by the Indian Agricultural Research Institute in the semi-arid region of central India. Bamboo was planted at 10 × 10 m and 12 × 10 m spacing. The total bamboo culms were higher when bamboo was grown without intercrops. The grain yield of sesame and chickpea were reduced. Yields decreased especially close to the bamboo clumps. However, at more than 3 m distance from the bamboo clump, there was no reduction in crop yields. The income generated through bamboo can compensate the monetary losses of the intercrop. There are multiple benefits related to the planning of bamboo: the soil pH is raised, organic carbon and available phosphorus increases. In drought prone areas, bamboo-based agroforestry can increase the resilience of farmers, conserve and enrich the soil (Ahlawat 2014).

Programme for the Development of Alternative Biofuel Crops – An ongoing project of the World Agroforestry Centre (ICRAF)

The World Agroforestry Centre (ICRAF) is implementing the 4-year *Programme for the Development of Alternative Biofuel Crops* in partnership with centers of excellence in Asia, Africa, and Latin America. Launched in 2013 with seed capital from the International Fund for Agricultural Development (IFAD), this initiative was designed to conduct research in development along several biofuel value chains, while at the same time strengthening food security and improving the livelihoods of smallholders. Pursuing a landscape approach, the program has already started operations in India, Brazil, and Kenya, where it is developing sustainable agroforestry systems with integrated food and energy production. Some of the initiatives under the program are presented in the following boxed text for India, Brazil, and Kenya (World Agroforestry Centre).

India

In India, the Programme is being implemented mainly in the states of Karnataka and Maharashtra. It is supporting the improvement and scaling-up of a promising biofuel initiative from Karnataka (refer Picture 1). Essentially, smallholders in energy-deprived villages are provided with quality planting material and technical assistance for growing native or locally adapted bioenergy trees, such as *Millettia pinnata* (pongamia), *Simarouba glauca* (simarouba), *Madhuca longifolia* (mahua), *Azadirachta indica* (neem) and other species in the borders and bunds of their plots, in addition to communal marginal lands. These agroforestry systems improve food security and alleviate energy poverty, since all their products are consumed locally; the SVO, biodiesel, and biogas are used for running tractors, irrigation pumps and for cooking via adapted stoves, while the biofertilizer derived from the seedcake improves

Picture 1. Integrated food and energy model in Hassan district of Karnataka, India: a pilot site of the ICRAF's program (ICRAF, Biofuels team)

the productivity of traditional crops. This pilot builds on existing partnerships between ICRAF, the University of Agricultural Sciences - Bangalore (UASB), the Karnataka State Biofuel Development Board (KSBDB), and the PDKV Akola University in Maharashtra. In a nutshell, this initiative has been spearheaded by UASB and KSBDB in Karnataka, and has already attracted significant attention nationally and internationally, with other states and countries such as Maharashtra and Nepal starting to develop similar models.

ICRAF has been conducting studies to evaluate this model and its potential for improvement. Preliminary data suggest that the efforts undertaken in Karnataka have a positive impact on the livelihoods of the poorest, and that the model implemented in the state can be scaled-up, subject to some adjustments. It was found, for instance, that smallholders adopting such agroforestry systems can achieve an increase in annual income of up to 36% in the long term under the current scenario. The extra income tends to come during the nonharvesting periods of the year when it is most needed. With further investments, it is estimated that the potential annual income increase in the long term could reach 60%. However, to fully realize this potential, several gaps still need to be addressed, such as through further investments in germplasm collection, nursery techniques, technologies for oil expelling and processing, and value addition at the village level. The model also needs to transition into financially sustainable and bankable proposals, allowing for reduction in public subsidies. IFAD and ICRAF are working with the Government of Karnataka and other partners toward achieving these objectives, while facilitating the scaling-up of the model into Maharashtra and other states.

Brazil

In Brazil, the Programme is addressing research gaps that limit the development and scaling-up of pro-poor *Acrocomia aculeata* (macauba) value chains in Brazil's northeastern region. *Acrocomia aculeata*, a palm species native to the tropical regions of the Americas, is a highly productive tree with oil-bearing fruits (refer Picture 2). It can be integrated into existing pastures without reducing grass yields. The tree often grows on poor soils and has high tolerance to drought. Its fruits are nontoxic; so, *Acrocomia aculeata* coproducts can be used for many different purposes along the value chain, including the production of animal feed, fuel briquettes, activated charcoal, and cosmetics.

ICRAF is partnering with the Brazilian Agricultural Research Corporation (Embrapa) to develop silvopastoral and agroforestry systems that integrate food crops and/ or livestock with *Acrocomia aculeate*, benefiting smallholder producers (World Agroforestry Centre). In addition to improving livestock production, these systems will facilitate the integration of smallholder producers from several states into the National Biodiesel Production and Use Programme (PNPB), an initiative that has benefitted more than 100,000 family farmers in 10 years (Ubrabio).

Kenya

In Kenya, the Programme is partnering with Eco Fuels Kenya Ltd. (EFK), a social enterprise that produces liquid biofuel (straight vegetable oil), organic fertilizers, briquettes and poultry feed through a sustainable, no-waste manufacturing process, based entirely on the nut from *Croton megalocarpus* (croton). This is an abundant, indigenous tree found across central and west Kenya, as well as in

Picture 2. *Acrocomia aculeata*-based biofuels model in Brazil: a pilot site of the ICRAF's program (ICRAF, Biofuels team)

other countries in East Africa (refer Picture 3). Until recently, its nut had no commercial value (since it is not edible) and it is still a largely wasted natural renewable resource. *Croton's* potential for multiple uses has long been researched by ICRAF, and its value for energy production has been recognized by the Government of Kenya several years ago (Endelevu Energy & Energy for Sustainable Development, 2008). Nonetheless, it was only more recently that the species started attracting international interest, private capital, and professional management toward developing its value chain.

Founded in 2012, EFK sources the croton nut from rural communities around its factory, which harvest them from local forests and farmlands. Thus, EFK provides an assured market to the farmers. The oil is used mainly to run water pumps and off-grid generators, while the briquettes are targeted at rural and urban dwellers for cooking and heating purposes. Based on the EFK's business model, the Programme is assessing the sustainability of the croton value chain in Kenya and addressing research gaps to support its upscaling in East Africa, while maximizing its positive impact on livelihoods of the rural poor. In particular, it will contribute to increased additional income for approximately 5000 collectors by the end of 2016, in addition to improving the supply of affordable and cleaner energy options in the central region of the country.

Some of the gaps already identified by ICRAF and EFK, which are being tackled by the Programme, include the need to support the selection and growing of superior

Picture 3. *Croton megalocarpus*-based biofuels model in Nairobi, Kenya: a pilot site of the ICRAF's program (ICRAF, Biofuel team)

croton trees through establishment of germplasm, including natural seed collection, as well as to promote capacity development activities, such as training sessions on croton planting and management, targeted at smallholder producers. Although a previous analysis from ICRAF (Ndegwa et al.) indicates that the cultivation of croton can be profitable even when the tree is planted for hedging (as is customary in central Kenya) the project is conducting trials of different agroforestry systems that incorporate the species, so as to identify options and combinations that maximize its value and potential for energy poverty alleviation.

Gender Role in Agroforestry-Based Bioenergy

Women and men are affected in different ways by poverty. In many developing countries, women experience severe levels of deprivation due to several reasons, among which is inadequate access to clean energy options (Karekezi et al. 2002). Women and children are exposed to indoor smoke through the use of woodfuels in traditional cookstoves. Most rural households use biomass energy and women in poor households spend more time searching for firewood than those in households with higher incomes (Energia, 2008). In areas where there is decreased vegetation cover, women and children walk long distances in search of firewood. A study (Thorlakson and Neufeldt 2012) shows that, for example, in the Lower Nyando District of Kenya, households spend on average 9 hours per week on fuelwood collection, with some women reporting that they walked more than 20 km/day to purchase fuelwood in neighboring districts. After agroforestry interventions, the weekly average came down to 6 hours per week. The mature trees on their land made them feel safe and gave access to more stable fuelwood supply. Women with mature trees on their land felt that they now have access to a safer and more stable supply of fuel wood. These women found that once fuelwood was available nearby, they could devote more time to other income-generating activities (Thorlakson and Neufeldt 2012). Moreover, increasing the efficiency of cookstoves may easily reduce the health risks of women and children. Improved cookstoves are available for domestic use that can reduce fuel use and particulate matter emissions when compared to traditional cookstoves (World Bank, 2011).

In another case, during World Congress on Agroforestry in February 2014, held in New Delhi in India, a lady named Prabhavati, from Bijapur taluk of Karnataka, narrated her story on neem seed collection activity. Each woman used to collect about 4–5 tons of neem seeds every season and sell it for only Rs 2–3/kg (Rs 8000–15,000 per year). After the implementation of biofuel program by KSBDB (Karnataka State Biofuel Development Board) and intervention by an NGO called Biofuel Lead, they could access more information on marketing and prices. This helped them to directly sell their seeds to oil mills and also gave them a bargaining power to establish better rates for their seeds. They formed a self-help group of about 150 women, who collect and sell neem seeds and each woman makes, on average, about Rs 30,000/year. This additional income has helped them to improve their standard of living and quality of life.

Key Constraints for Setting up Agroforestry Systems

The adoption of agroforestry still faces major constraints. The domination in agriculture of monocultures and short-term benefits is a challenge and trees in agricultural fields can hamper mechanized operations. A key bottleneck hindering the development of agroforestry is poor access to the resources needed to establish such systems, such as capital, labor, farming inputs, land, extension services or markets. Inadequate legislation, regulations, and policies can further jeopardize agroforestry development. For instance, in many countries and regions, agroforestry has no clear status and falls between agricultural and forestry sectors, leaving its regulation in a gray area.

Lack of suitable policy

Although agroforestry has been practiced for thousands of years in many parts of the world, and despite all its benefits, the sector lacks favorable policies, legal frameworks, and coordination between different government mandates, such as agriculture, forestry, rural development, environment, and trade. Many governments (e.g., India, Indonesia, and Myanmar (Butler 2012)) do not allow free harvest and trade of timber, even for trees grown on private land, imposing several restrictive legal provisions for harvesting and transportation of lumbered trees. This has not been addressed sufficiently in policy formulation, nor has it been adequately integrated into land-use planning or rural development programs.

Thus, the potential of agroforestry to enrich farmers, communities, and by extension, national economies has not been fully exploited. Although motivated by a desire to increase tree cover, the Government of India has come out with a specific policy on Agroforestry (Down To Earth, 2014) which is likely to bolster research and development in the sector. India has been the first country in the world to adopt such a policy, which makes particular reference to the energy sector. It is expected that the policy will encourage other countries to follow suit in their developmental goals.

Unavailability of planting material

Research and development investments in trees have long lagged behind traditional agricultural crops. Several hybrid and suitable varieties of the latter have been developed, in addition to advanced techniques, ensuring that farmers obtain uniform stand and yield. In contrast with some horticultural crops (e.g., mango and citrus), most perennial species have so far relied on new domestication and selection efforts. The result has been that planting materials have a large genetic variation, are unregulated and their productivity is often very low and growers in remote areas lack access to them. Jatropha, for instance, was (Singh 2015) promoted in several countries without proper domestication or varietal development. In some cases, such as in India, large plantations of the crop were encouraged by government entities. As a result, most biofuel projects and policies focused on this crop failed worldwide, including in India. There is a need to have a mechanism in place, similar to that which exists for agricultural crops, which certifies planting material of tree species. Either an institutional mechanism needs to be developed for registration of nurseries or private players need to be encouraged to sell planting material under a "private label" or "self-certification". This could occur through accredited agencies which allow governments to recognize certification.

Research and extension services

One of the major constraints to the promotion and large-scale adoption of agroforestry innovations by farmers and communities is the lack of, or inadequate knowledge, of tree-growing practices. Agroforestry requires a careful study of all the factors that influence its adoption. Moreover, there are thousands of potentially useful trees and crop species that can be mixed in many different ways and local agroecological conditions can change the effects significantly. Considering the benefits trees provide for bioenergy, creative and challenging research is needed. Research results need to be used to confront and overcome the barriers to increased support from extension services. Several international organizations, including the World Agroforestry Centre, are engaged in research and development of trees and agroforestry systems. However, extension services tend to be undertaken by government agencies (in developing countries) which are often under-resourced or ineffective, and which seldom focus on trees. Furthermore, agroforestry is mostly about management techniques and not about products such as seed, fertilizer, and other inputs that can be sold. This way, few are lobbying for its wider application. The consequence is that poor farmers are left without proper advice. There are cases, however, in which these difficulties have been overcome with private sector support, for example, the pulp and paper industry in India has very strong R&D and extension services. There is a need to learn from these successful cases and understand how to replicate them.

Long gestation time

Another reason is that the benefits of integrating trees into landscapes are not fully known. Instead, farmers perceive the extra work and the long-growing periods.

The long lag time between planting and harvesting of tree products hinders the adoption of agroforestry by smallholders, especially those who struggle to meet their basic requirements on a day-to-day basis and prefer to invest in areas where returns are short term. There is a need to select and domesticate fast-growing native species that are coppiciable and also develop technologies for early flowering, such as grafting. Lessons from successful case studies, such as the cocoa project led by the World Agroforestry Centre, can be incorporated into the development of bioenergy feedstock strategies.

Access to market and value chains

Despite the fact that tree-based products, like fruits, medicines, fuelwood, and oilseeds etc., are part of day to day life in rural areas, scant information is available on their demand and potential supply in many developing countries. The market remains a cottage-style industry and middlemen often exploit growers. This leads to low profits to poor producers and production of low-quality products since there is no direct link between product and market. Except for a few examples, marketing systems for agroforestry products remains unorganized. Agroforestry forestry products frequently have not yet developed the same market linkages as for traditional agricultural crops as they are usually not a priority for farmers and communities. To ensure better development and utilization of forest and tree resources, there is a need for strong linkages between the various institutions in the sector. Marketing infrastructure similar to that for agricultural commodities - with assured markets by way of assured prices (e.g., minimum support price) - has to be introduced in order to mainstream agroforestry into agricultural systems.

For example, ITC (Indian tobacco Company) introduced a new concept named as "Agro-forestry Model". Under this model, farmers plant pulpwood trees on 25% of their land and on the remaining 75% of land, they grow food crops. The farmers can harvest the trees every 4 years and ITC provides them the market. In this way they generate income from both the food crops and the trees. This program started in April, 2010 and so far 2427 hectares of land has been covered under this model (Environmental performance, ITC Limited).

Access to credit

There are cumbersome procedures involved that hinder poor households accessing credit from the banking sector. One of the reasons why poor households cannot access credit is that they cannot present guarantees to the bank. Coupled with this, agroforestry investments do not attract the attention of the banking sector. Therefore, financial packages for rural energy programs, along with agroforestry, have to be included in comprehensive planning approaches if they are to have the desired impact. Credits are needed especially for agroforestry as the upfront investments can be higher than for annual crops. Capital is needed for buying seedlings and labor (acquiring and planting trees, caring for the young trees). Initially, trees can take space that was previously devoted to crops.

Land tenure

Secure land rights are necessary if farmers are to invest in agriculture and related activities. In Africa, customary tenure agreements exist in many countries, especially in Sub-Saharan countries. A family gains access to land as a social right; however, the land rights are governed by local and traditional authorities. Lack of permanent land tenure is a big deterrent to the adoption of long-term activities such as agroforestry. Without secure land rights, farmers might fear their land will be taken away by more powerful actors once its quality has been improved. Target groups will be more likely to be interested in agroforestry provided they have long-term access to land and rights to plant and use the products of trees. Land tenure issues also deter financial institutes from providing credit to farmers (see above).

Example of one such technique which takes care of land tenure issues is Farmer managed natural regeneration (FMNR). It is a land restoration technique, which involves the systematic regeneration and management of trees and shrubs from tree stumps, roots, and seeds. For successful implementation of this technique, FMNR ensures that farmers get control over their land and what they grow on it. For this, FMNR projects incorporate assistance to broker agreements between the relevant stakeholders – for example, farming communities, government agents, and other forms of local bodies (Francis et al. 2015).

Policy interventions are also required to change inadequate land tenure systems. USAID is investigating this important issue in Zambia, where a baseline survey of 4000 households from 315 villages is being carried out to see how strongly land tenure contributes to agroforestry interventions (USAID, 2014). It is expected that if land tenure issues are resolved, there would be increased uptake of agroforestry interventions for livelihood improvements.

Despite the above challenges, trees provide abundant opportunities for national economies, especially in the renewable energy sector. Similar to the agriculture sector, agroforestry also deserves more attention from governments, civil society, and the private sector to have its full potential realized.

Conclusions

Access to energy is crucial to realize development objectives. Bioenergy is an option to increase access to energy in rural areas of developing countries. It will however only be successful if the supply of feedstock is cost-effective and creates employment as well as additional incomes for smallholder farmers at the same time. The benefits of bioenergy feedstock production will be negated if agricultural land is compromised for production of nonfood crops or diversion of these crops for energy purposes. With dwindling agricultural resources in some developing countries and increased vulnerability of smallholders due to climate change, this remains a challenge.

Scientifically developed, locally adapted bioenergy systems may address a variety of challenges at the same time, food and energy security for a growing population, mitigating and adapting to climate change. Agroforestry and conservation agriculture should be key elements as they can improve the productivity of smallholder agricultural systems under climate change by strengthening their resilience.

To conclude, we would like to summarize the following findings of this article:

1. Among all types of modern renewable energy, bioenergy can be subject to adequate choices of technology and production models – an affordable option for smallholders in developing countries to locally produce and consume energy. It therefore has great potential for promoting sustainable agricultural growth, reducing energy poverty, driving rural development; creating jobs, and additional income for the poor.
2. Despite the significance of bioenergy in the livelihoods of rural populations, investments in this sector have largely been overlooked by international development players.
3. In many developing countries, bioenergy still has negative impacts such as forest degradation, indoor pollution, and food insecurity which are hindering its development.

4. Bioenergy can be sustainably produced and have positive environmental, social, and economic impacts. Solid scientific evidence and successful case studies in the developing world have proven this.
5. Smart agroforestry systems with integrated food and energy production are among the main solutions. Their potential benefits include:

   a. Production of different bioenergy sources within one comprehensive system, including sustainable charcoal, briquettes, straight vegetable oil, biodiesel, bioethanol, biogas, and electricity from the gasification of fuelwood. This leads to immediate energy poverty alleviation with cleaner alternatives to fossil fuels.
   b. Coproduction of organic fertilizers and biopesticides, as well as improved soil fertility with N_2 fixation from trees and reduced soil erosion, leading to enhanced food crop productivity.
   c. Creation of jobs, generation of additional income, and savings from fossil fuels, as well as integration of smallholders into new value chains.
   d. Strengthened food security due to the simultaneous production of food and through effects of fertilizer trees on productivity.
   e. Climate change mitigation with reduced GHG emissions, carbon sequestration, and storage in addition to adaptation services, with improved resilience of smallholder producers.
   f. No land-use change effects (either direct or indirect).
   g. Decreased risks for the farmer through the use of multiple species rather than monoculture plantations.
   h. Watershed protection and biodiversity conservation.

6. A number of bioenergy projects involving trees that have not taken into account a landscape approach in agroforestry models have not been successful. The main reasons for failure include poorly structured value chains, inability to ensure constant and economical supply of feedstock and lack of adequate market linkages.
7. Several successful agroforestry models have shown potential for sustainable charcoal, electricity and biofuel production using different feedstock such as *Gliricidia*, bamboo, and oil-bearing tree species.
8. Some bottlenecks remain for mainstreaming agroforestry for bioenergy production. These are mainly related to a lack of suitable policies, adequate planting material and extension services, land rights, long gestation periods, and difficulties with access to credit and markets.
9. With political will and improved investments in R&D and scaling-up, it is possible to achieve useful results. A few countries are already demonstrating this, for example, India has become the first nation to launch a national agroforestry policy and is accelerating its

investment in the sector. It is hoped other countries will be encouraged to promote agroforestry systems for bioenergy generation.

Acknowledgment

This study was financed by BMZ/GIZ.

Conflict of Interest

None declared.

References

Adebimpe, A. A. 2013. Population dynamics and infrastructure: meeting the millennium development goals in Ondo State, Nigeria. Afr. Popul. Stud. 27:229–237.

Ahlawat, S. P. 2014. Bamboo based agroforestry for livelihood security and environmental protection in semi-arid region of India, [Online]. Available at http://wca2014.org/abstract/bamboo-based-agroforestry-for-livelihood-security-and-environmental-protection-in-semi-arid-region-of-india/. (Accessed 19 August 2015).

Akinnifesi, F. K., W. Makumba, and F. R. Kwesiga. 2006. Sustainable maize production using gliricidia.maize intercropping in southern Malawi. Expl. Agric. 42:1–17.

Bayalaa, J., G. W. Sileshib, R. Coec, A. Kalinganirea, Z. Tchoundjeud, F. Sinclaire, et al. 2012. Cereal yield response to conservation agriculture practices in drylands of West Africa: a quantitative synthesis. J. Arid Environ. 78:13–25.

Behera, S. K., P. Srivastava, R. Tripathi, J. P. Singh, and N. Singh. 2010. Evaluation of plant performance of *Jatropha curcas* L. under different agro-practices for optimizing biomass – a case study. Biomass Bioenergy 34:30–41.

Bertrand, V., B. Dequiedt, and E. L. Cadre. 2014. Biomass for electricity in the EU-27: Potential demand, CO2 abatements and breakeven prices for co-firing. Energy Pol. 73:631–644.

Bi, L., and M. Haight. 2007. Anaerobic digestion and community development: a case study from Hainan province, China. Environ. Dev. Sustain. 9:501–521.

Bogdanski, A. 2012. Integrated food–energy systems for climate-smart agriculture. Agric. Food Secur. 1:9.

Budiadi, and H. T. Ishii. 2010. Comparison of carbon sequestration between multiple-crop, single-crop and monoculture agroforestry systems of Melaleuca in Java, Indonesia. J. Trop. Forest Sci. 22:378–388.

Butler, R. 2012. LOGGING: Timber certification, trade restrictions, [Online]. Available at: http://rainforests.mongabay.com/1010.htm.

Casillas, C. E., and D. M. Kammen. 2010. The energy-poverty-climate nexus. Science 330:1181–1182.

Chew, T. L., and S. Bhatia. 2008. Catalytic processes towards the production of biofuels in a palm oil and oil palm biomass-based biorefinery. Bioresour. Technol. 99:7911–7922.

Coimbra-Araújo, C. H., L. Mariane, C. B. Júnior, E. P. Frigo, M. S. Frigo, I. R. C. Araujo, et al. 2014. Brazilian case study for biogas energy: production of electric power heat and automotive energy in condominiums of agroenergy. Renew. Sustain. Energy Rev. 40:826–839.

Dale, V. H., K. L. Kline, L. L. Wright, and R. D. Perlack. 2011. Interactions among bioenergy feedstock choices, landscape dynamics, and land use. Ecol. Appl. 21:1039–1054.

Demirbas, A. 2011. Competitive liquid biofuels from biomass. Appl. Energy 88:17–28.

Denman, K. L. 2007. Couplings between changes the climate system and biogeochemistry climate change 2007: The physical science basis. Contribution of working group I to the fourth assessment report of the intergovernmental panel on climate change. University Press, Cambridge.

Dovie, D. B., E. T. Witkowski, and C. M. Shackleton. 2004. The fuelwood crisis in southern Africa – relating fuelwood use to livelihoods in a rural village. GeoJournal 60:123–133.

Down To Earth. 2014. India becomes first country to adopt an agroforestry policy. [Online]. Available at: http://www.downtoearth.org.in/content/india-becomes-first-country-adopt-agroforestry-policy.

Elevitch, C. 2008. Nitrogen fixing trees – the multipurpose pioneers. The Permaculture Research Institute [Online]. Available at: http://permaculturenews.org/2008/09/29/nitrogen-fixing-trees-the-multipurpose-pioneers/.

Endelevu Energy & Energy for Sustainable Development. 2008. A Road map for Biofuels in Kenya - Opportunities and Obstacles. Endelevu Energy & Energy for Sustainable Development, Africa, Kenya.

Energia. 2008. Turning Information into Empowerment: Strengthening Gender and Energy Network in Africa. TIE-ENERGIA Project, Netherland- Energia Secretariat, The Netherlands.

Environmental performance, ITC Limited. [Online]. Available at http://www.itcportal.com/sustainability/sustainability-report-2013/initiative-on-sustainable-agro-forestry.aspx.

Evans, A., V. Strezov, and T. J. Evans. 2010. Sustainability considerations for electricity generation from biomass. Renew. Sustain. Energy Rev. 14:1419–1427.

FAO. 2010. IFES developing countries. FAO, Rome.

FAO. Labour. Food and Agriculture Organization [Online]. Available at: http://www.fao.org/docrep/015/i2490e/i2490e01b.pdf.

FAO. BEFSCI (Bioenergy and Food Security Criteria Indicators). FAO [Online]. Available at: http://www.fao.org/bioenergy/31531-04f3908fc8f82ffc9a978592fc3bf1fa9.pdf.

Ferroukhi, R., A. Khalid, A. L. Pena, and M. Renner. 2015. Renewable energy and jobs annual review 2015. Int. Renew. Energy Agency.

Food and Agricultural Organisation of United Nations. FAO corporate documentary Repository: Bioenergy for development - Technical and environmental dimensions. Food and Agricultural Organisation of United Nations [Online]. Available at: http://www.fao.org/docrep/t1804e/t1804e06.htm. (Accessed 7 October 2014).

Francis, R., P. Weston, and J. Birch. 2015. The social, environmental and economic benefits of Farmer Managed Natural Regeneration (FMNR), World Vision.

Glover, J. D., J. P. Reganold, and C. M. Cox. 2012. Agriculture: plant perennials to save Africa's soils. Nature 489:359–361.

Gupta, V. K., R. Potumarthi, A. O'Donovan, C. P. Kubicek, G. D. Sharma, and M. G. Tuohy. 2014. Chapter 2-bioenergy research: an overview on technological developments and bioresources. Pp. 23–47 in V.G. Gupta, Maria Tuohy, Christian P Kubicek, Jack Saddler, Feng Xu, eds. Bioenergy research: advances and applications. Elsevier, Oxford, UK.

HLPE. 2013. Biofuels and food security: a report by the high level panel of experts on food security. Food and Agriculture Organization, Rome.

IEA. 2011. World energy outlook 2011. International Energy Agency, Paris Cedex, France.

IFAD. 2011. Rural poverty report. International Fund for Agricultural Development, Rome.

Iiyama, M., H. Neufeldt, P. Dobie, M. Njenga, G. Ngegwa, and R. Jamnadass. 2014. The potential of agroforestry in the provision of sustainable woodfuel in sub-Saharan Africa. Curr. Opin. Environ. Sustain. 6:138–147.

Ilyama, M., D. Newman, C. Munster, M. Nyabenge, G. W. Sileshi, V. Moraa, et al. 2013. Productivity of Jatropha curcas under smallholder farm conditions in Kenya. Agrofor. Syst. 87:729–746.

IREA. 2016. Remap:Roadmap for a renewable energy future. International Renewable Energy Agency, Abu Dhabi.

Jain, R. K. 1994. Fuelwood characteristics of medium tree and shrub species of India. Bioresour. Technol. 47:81–84.

Joshi, S., and S. Joshi. Simarouba glauca. University of Agricultural Sciences, Bagalore, India, [Online]. Available at: http://ageconsearch.umn.edu/bitstream/43624/2/Simarouba%20brochure,%20UAS%20Bangalore,%20India.pdf.

Karekezi, S., K. B. Banda, and W. Kithyoma. 2002. Improving energy services for the Urban Poor in Africa- A Gender Perspective: Energia. Energia News 5:4.

Koh, L. P., J. Miettinen, S. C. Liew, J. Ghazoul, and P. Ehrlich. 2011. Remotely sensed evidence of tropical peatland conversion to oil palm. Proc. Natl Acad. Sci. USA 108:5127–5132.

Kulatunga, G. 2012. Electricity from Gliricidia - an entirely Sri Lankan concept. Daily News. [Online]. Available at: http://archives.dailynews.lk/2012/01/13/fea03.asp.

Larsson, M., M. Gorling, S. Gronkvist, and P. Alvfors. 2013. Bio-methane upgrading of pyrolysis gas from charcoal production. Sustain. Energy Technol. Assess. 3:66–73.

Lin, B. B., I. Perfecto, and J. Vandermeer. 2008. Synergies between agricultural intensification and climate change could create surprising vulnerabilities for crops. Bioscience 58:847–854.

Lundgren, B. O., and J. B. Raintree. 1982. Sustained agroforestry. Pp. 37–49 in Nestel, B, eds. Agricultural research for development: potentials and challenges in Asia. International Service for National Agricultural Research, The Hague. Hague, Netherlands

Malla, S. 2013. Household energy consumption patterns and its environmental implications: assessment of energy access and poverty in Nepal. Energy Pol. 61:990–1002.

Millennium Institute. 2013. Global food and nutrition scenarios. Millennium Institute, Washington, DC.

Mwampamba, T. H., A. Ghilardi, K. Sander, and K. J. Chaix. 2013. Dispelling common misconceptions to improve attitudes and policy outlook on charcoal in developing countries. Energy Sustain. Dev. 17:75–85.

Ndegwa, G., V. Moraa, R. Jamnadass, J. Mowo, M. Nyabenge, and M. Liyama. 2011. Potential)for)biofuel) feedstock)in)Kenya. World Agroforetsry Centre, Geoffrey' Ndegwa,'Violet'Moraa,'Ramni'Jamnadass,'Jeremias'Mowo,'.

Nwanze, K. F. 2011. Smallholders can feed the world. International Fund for Agricultural Development (IFAD), Rome, Italy.

Ong, C. K., R. M. Kho. 2015. Tree-crop interactions: agroforestry in a changing climate. CABI, Wallingford, UK..

Openshaw, K.. 2000. A review of Jatropha curcas: an oil plant of unfulfilled promise. Biomass Bioenergy 19:1–15.

Orwa, C., A. Mutua, R. Kindt, R. Jamnadass, and S. Anthony. 2009. Agroforestree Database:a tree reference and selection guide version 4.0. [Online]. Available at http://www.worldagroforestry.org/treedb/AFTPDFS/Leucaena_leucocephala.PDF. (Accessed 19 August 2015).

Ouya, D. 2013. Unpacking the evidence on firewood and charcoal in Africa (Agroforestry World Blog). World Agroforestry Centre, [Online]. Available at: http://blog.worldagroforestry.org/index.php/2013/10/03/unpacking-the-evidence-on-firewood-and-charcoal-in-africa/.

Pandey, K. K., N. Pragya, and P. K. Sahoo. 2011. Life cycle assessment of small-scale high-input Jatropha biodiesel production. Appl. Energy 88:4831–4839.

Ryabov, G., D. S. Litun, E. P. Dik, and K. A. Zemskov. 2006. The prospects and problems of utilizing biomass and combustible wastes to produce heat and electricity. Therm. Eng. 53:559–565.

Sanchez, P. A., R. J. Buresh, R. R. Leakey, L. T. Evans, G. D. Anderson, P. Wood, et al. 1997. Trees, soils, and food security [and Discussion]. Philos. Transa. Bio. Sci. 352:949–961.

Scovronick, N., and P. Wilkinson. 2014. Health impacts of liquid biofuel production and use: a review. Glob. Environ. Change 24:155–164.

Selvarajah, R. 2013. The landscape approach: what are ICRAF scientists saying?. World Agroforestry Centre [Online]. Available at: http://worldagroforestry.org/newsroom/highlights/videos-landscape-approach-what-are-icraf-scientists-saying.

Sepp, S., C. Sepp, and M. Mundhenk. 2014. Towards sustainable modern wood energy development, Commissioned by the GIZ Sector Project for the Global Bioenergy Partnership (GBEP).

Shankar, A. V. 2012. Agroforestry - A blog on Agrihortisilviculture: Melia dubia for Biomass power production. [Online]. Available at: http://agrowmania.blogspot.in/2012/05/melia-dubia-for-biomass-power.html.

Singh, S. 2015. Revisting the Jatropha Debate, Forbes India. [Online]. Available at http://forbesindia.com/article/special/revisting-the-jatropha-debate/36187/1.

Siraj, M. A. 2014. Bamboo Power, The Hindu, 11 April 2014. [Online]. Available at: http://www.thehindu.com/features/homes-and-gardens/green-living/bamboo-power/article5900988.ece.

Sood, K. K. 2006. The Influence of household economics and farming aspects on adoption of traditional agroforestry in Western Himalaya. Mt. Res. Dev. 26:124–130.

Sosovele, H. 2010. Policy challenges related to biofuel development in Tanzania/Politische Herausforderungen in Bezug auf Biokraftstoffe in Tansania. Africa Spectr. 45:117–129.

The Secretary-General's High-Level Group. 2012. Sustainable energy for all: A global action agenda [Online]. Available at: http://www.un.org/wcm/webdav/site/sustainableenergyforall/shared/Documents/SEFA-Action%20Agenda-Final.pdf.

Thorlakson, T., and H. Neufeldt. 2012. Reducing subsistence farmers' vulnerability to climate change: evaluating the potential contributions of agroforestry in western Kenya. Agric. Food Secur. 1:15.

Tijani, A. J., M. O. Uguru, and O. A. Salawu. 2008. Anti-pyretic, anti-inflammatory and anti-diarrhoeal properties of Faidherbia albida in rats. Afr. J. Biotechnol. 7:696–700.

Ubrabio. Brazilian Program for Biodiesel Production, [Online]. Available at: http://www.ubrabio.com.br/1933/textos/BrazilianProgramForBiodieselProduction_29990/.

UNCTD. 2016. Second generation biofuel markets: state of play, trade and developing country perspectives. United Nations Conference on Trade and Development, Geneva.

UNDP. Sustainable development goals. United Nations Development Programme [Online]. Available at: http://www.undp.org/content/undp/en/home/sdgoverview/post-2015-development-agenda/goal-7.html. (Accessed May 2016).

USAID. 2014. Erc success story: when farmers feel secure-providing the evidence to invest in climate smart agriculture and agroforestry. USAID [Online]. Available at http://usaidlandtenure.net/content/erc-success-story-when-farmers-feel-secure.

Wani, S. P., and T. K. Sreedevi. Pongamia's journey from forest to micro-enterprise for improving livelihoods. International Crops Research Institute for the Semi-Arid Tropics (ICRISAT), Andhra Pradesh, India.

Weiland, P. 2010. Biogas production: current state and perspectives. Appl. Microbiol. Biotechnol. 85:849–860.

WHO. Indoor air pollution. World Health Organization [Online]. Available at: http://www.who.int/indoorair/en/. [Accessed May 2016].

Wildy, D., J. Pate, and J. Bartle. 2003. Silviculture and water use of short rotation Mallee eucalyptus. Rural Industries Research and Development Corporation, Macquarie street, Barton Act, Kingston Act.

World Agroforestry Centre. 2009a. Gliricidia sepium. World Agroforestry Centre [Online]. Available at: http://www.worldagroforestry.org/treedb/AFTPDFS/Gliricidia_sepium.PDF.

World Agroforestry Centre. 2009b. Pongamia pinnata. World Agroforestry Centre [Online]. Available at: http://www.worldagroforestry.org/treedb/AFTPDFS/Pongamia_pinnata.PDF.

World Agroforestry Centre. Faidherbia albida. World Agroforestry Centre [Online]. Available at: http://www.worldagroforestry.org/sites/default/files/F.a_keystone_of_Ev_Ag.pdf.

World Agroforestry Centre. Programme for the development of alternative biofuel crops. World Agroforestry Centre [Online]. Available at: http://worldagroforestry.org/research/alternative-biofuel-crops.

World Agroforestry Centre. Programme for development of alternative biofuel crop: Approach and activities. World Agroforestry Centre [Online]. Available at: http://worldagroforestry.org/research/alternative-biofuel-crops/approach_and_activities.

World Bank. 2011. Household cookstoves, environment, health, and climate change: anew look at an old problem. World Bank, Washington, DC.

8

Advances in shrub-willow crops for bioenergy, renewable products, and environmental benefits

Timothy A. Volk, Justin P. Heavey & Mark H. Eisenbies

State University of New York College of Environmental Science and Forestry (SUNY-ESF), Syracuse, New York

Keywords

Biomass quality, harvesting, multifunctional systems, *Salix*, short-rotation coppice, extension.

Correspondence

Timothy A. Volk, 1 Forestry Drive 346 Illick Hall Syracuse, NY 13210.

E-mail: tavolk@esf.edu

Funding Information

Empire State Development Division of Science Technology and Innovation (NYSTAR), (Grant/Award Number: 'C060041'); Honeywell International, (Grant/Award Number: '4500114419'); New York State Research and Development Authority (NYSERDA), (Grant/Award Number: '30713'); New York State Department of Transportation (NYSDOT), (Grant/Award Number: 'C-06-09'); United State Department of Agriculture National Institute of Food and Agriculture (USDA NIFA), (Grant/Award Number: '2012-68005-19703'); US Department of Energy Bioenergy Technologies Office, (Grant/Award Number: 'EE0001307').

Abstract

Short-rotation coppice systems like shrub willow are projected to be an important source of biomass in the United States for the production of bioenergy, biofuels, and renewable bio-based products, with the potential for auxiliary environmental benefits and multifunctional systems. Almost three decades of research has focused on the development of shrub willow crops for biomass and ecosystem services. The current expansion of willow in New York State (about 500 ha) for the production of renewable power and heat has been possible because of incentive programs offered by the federal government, commitments by end users, the development of reliable harvesting systems, and extension services offered to growers. Improvements in the economics of the system are expected as willow production expands further, which should help lower establishment costs, enhance crop management options and increase efficiencies in harvesting and logistics. Deploying willow in multifunctional value-added systems provides opportunities for both potential producers and end users to learn about the system and the quality of the biomass feedstock, which in turn will help overcome barriers to expansion.

Introduction

There is potential to sustainably produce over 1 billion Mg_{dry} of biomass annually in the United States from a combination of agricultural systems, forestry, and bioenergy crops. Short-rotation coppice (SRC) systems, like shrub willow (*Salix* spp.) and poplar (*Populus* spp.) are projected to supply 20–25% of this potential biomass (U.S. Department of Energy 2011). Shrub willow can be successfully grown on a wide array of agricultural land capabilities and drainage classes to produce bioenergy and bioproducts, with environmental and rural development benefits. Shrub willow has many characteristics that make it an ideal feedstock including high yields, the ability to resprout after coppice and be harvested every 3–4 years, ease of propagation from dormant stem cuttings, ease of breeding, a broad genetic base, and a feedstock composition similar to other sources of woody biomass (Volk et al. 2014). Research on shrub willow for biomass energy and alternative applications (bioremediation, vegetative covers, treatment of organic wastes, riparian buffers, living snow fences) has also been ongoing in the United States since 1986 and has included trials in 15 states across the Northeast and Midwestern United States and several provinces in Canada. Considerable collaborative efforts involving both private and public entities at the local, state and federal level and NGOs have been made to facilitate the commercialization of this system (Volk et al. 2014).

A breeding and selection program for shrub willows has been developed and is producing improved cultivars for both the biomass and agroforestry markets (Volk et al. 2014; Smart et al. 2008) with long-term studies of potential yields across a range of sites (Fabio et al. 2016) over multiple rotations (Sleight et al. 2015). Research has been conducted on various aspects of the production cycle including nutrient amendments and cycling, alternative tillage practices, the use of cover crops for weed and erosion control, plant spacing and density, growth characteristics important for biomass production, harvesting systems, and logistics. Environmental factors have also been studied such as the use of willow plantations by pollinators, birds and small mammals; changes in soil microarthropod communities under willow; changes in soil carbon, greenhouse gas balances; and water quality and quantity. Financial analysis and life cycle assessments have evaluated the overall system through multiple rotations and advanced sustainability studies are now being undertaken to evaluate the entire supply chain using multiple metrics and integrated assessments. Results from these and other initiatives in North America and Europe have provided a base from which to expand and deploy willow biomass crops, and willow projects are being developed as a sustainable cropping system for agricultural and open land (Volk et al. 2006).

Current Willow Biomass Production

Willow is typically planted using 20-cm-long dormant hardwood cuttings at a density of about 13,500 plants ha^{-1}. Competing vegetation is managed using a combination of chemical and mechanical controls over the first few growing seasons. The crop is coppiced (cut back) after the first year to promote the production of multiple stems, followed by the first harvest 3–4 years later using a single-pass cut-and-chip forage harvester. The willow crop resprouts the following spring and is harvested again in another 3–4 years. Seven or more harvests are anticipated to be possible from a single planting. Yields between 8 and 12 Mg$_{dry}$ ha^{-1} year^{-1} across a range of sites have been observed (Volk et al. 2011); or about 42–72 Mg$_{wet}$ ha^{-1} at harvest. Yield increases of 20–40% are anticipated from breeding and selection efforts for new willow varieties (Serapiglia et al. 2013, Volk et al. 2011).

Despite a variety of benefits possible from willow production, deployment has been restricted by high establishment costs, inconsistent markets, and perceptions about willow chip quality and feedstock characteristics. Several of these barriers have been addressed in recent years through the collaborative efforts of numerous organizations and support from federal and state agencies, as well as private companies (e.g., Honeywell International, Case New Holland, Double A Willow, Celtic Energy Farm). Harvesting costs were reduced by about 35% with the development of an effective single-pass cut-and-chip harvesting system based on a New Holland (NH) forage harvester (Eisenbies et al. 2014a). The system is commercially available at NH dealers across North America and Europe and is being used to harvest willow in central and northern New York State and throughout the Northeastern United States. This system also resolved issues with chip size and quality and produces material that is acceptable to the primary end user in New York State, ReEnergy Holdings, and other end users who have tested and utilized the material.

These collaborative efforts among universities and industry partners have contributed to an emerging willow industry in the Northeast, which was catalyzed in New York State by the successful application to the USDA Biomass Crop Assistance Program (BCAP) developed in 2012 by ReEnergy, Cato Analytics and SUNY-ESF. BCAP is designed to improve domestic energy security, reduce the greenhouse gas emissions that cause climate change, and create opportunities for rural development (Volk and Harlow 2014). The rollout of this program has addressed a number of the barriers associated with willow biomass crops. BCAP provides partial cost-share payments for some of the upfront expenses of site preparation and planting willow, as well as annual land rental payments based on soil conservation rates. The site preparation and establishment support in the 2012 program covered up to 75% of the establishment cost, or a maximum of $1853 ha^{-1}. Subsequent offerings for BCAP in 2015 reduced the cost-share establishment payment to 50% or a maximum of $1237 ha^{-1}. BCAP also paired producers with an end user for their material.

As part of the 2012 BCAP agreement, ReEnergy signed 11-year contracts with willow producers to purchase harvested biomass, providing producers with a known market for about half of the expected lifespan of these plantings. ReEnergy is mixing willow biomass with other regionally sourced biomass feedstocks such as forest residues to produce biopower at the Black River (60 MW) facility and biopower and industrial process steam for an adjacent paper mill at the Lyonsdale (22 MW) facility. In 2014, ReEnergy signed a 20-year supply agreement with the United States Defense Logistics Agency to provide secure,

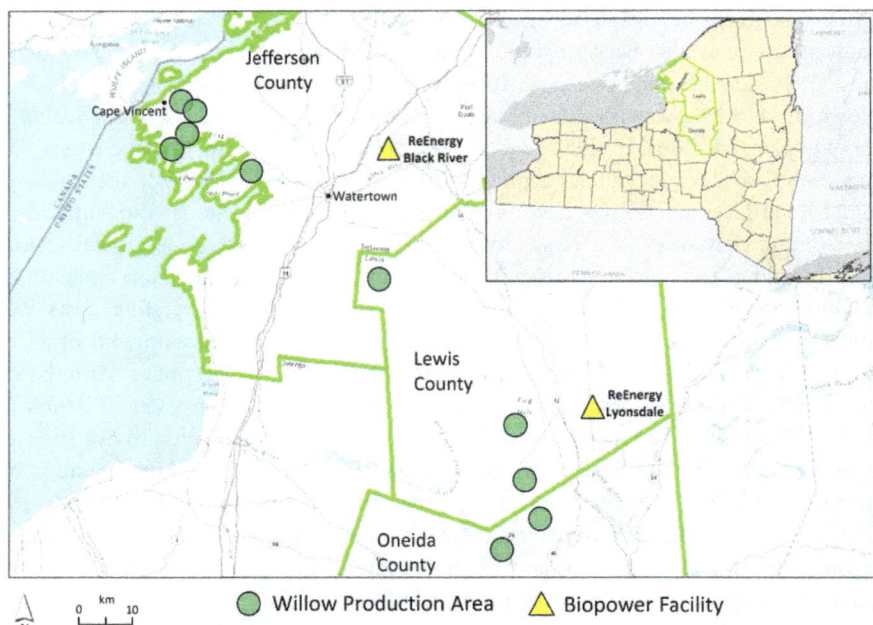

Figure 1. Willow production areas and two ReEnergy end-use facilities in central/northern region of New York State.

renewable electricity to the Fort Drum U.S. Army military base from the Black River facility, creating another level of assurance that this market for willow will remain in place. The window for the first round of BCAP signups in 2012 was limited to a two-month period, and 470 ha of willow biomass crops were enrolled in that time (Fig. 1). A second, one-month signup period was announced in late August 2015 with the potential to increase the area used to grow willow to about 1,000 ha under BCAP, but once again the window for signing up was very limited and this time no additional acreage was enrolled under the deadline, although several parties expressed interest and valuable connections with potential growers were made.

Extension Services

Since the first commercial scale-up of willow crops in 2012, SUNY-ESF (with support from NYSERDA – the New York State Energy Research and Development Authority) and the Northeast Woody/Warm-season Biomass Consortium (NEWBio) are providing a suite of extension services to producers and other stakeholders in New York and the Northeast. Nontechnical barriers to commercialization include a low level of awareness and understanding about the production and management among potential producers and support businesses; lack of understanding about the system among neighbors, policy makers, and broader public; and the lack of a functioning and organized biomass supply chain that meets the needs of the bioenergy system's stakeholders. If initial large-scale deployment of willow is

not successful, subsequent deployment in a region can be negatively impacted and delayed by years (Helby et al. 2006; McCormick and Kåberger 2007). To address these barriers and concerns, educational and outreach services are being provided by SUNY-ESF and NEWBio to the nascent willow industry in the Northeast including the development and delivery of educational materials such as brochures and fact sheets; training programs, field tours and webinars for producers and other stakeholders; newsletters, websites, social media, and other forms of information dissemination. Another element of current extension programming focuses on service provision including crop scouting; a willow equipment access program for specialized planting and harvesting machinery; and technical assistance in the field to assist with crop planting, management and harvesting. Analytical services such as soil sampling and interpretation of test results and the development of economic tools and analyses are also being provided. Extension staff are working with producers and end users to develop feedstock confidence and scale-up potential; providing insights from on-the-ground experience to supply chain and other analyses; and coordinating communication and joint efforts among university, public, government, NGO, and industry partners. These type of extension services have been shown to be critical to the adoption and success of novel bioenergy crops such as shrub willow, and were an integral component in each of seven Agriculture and Food Research Initiative (AFRI) Regional Bioenergy Coordinated Agricultural Projects supported by the United States Department of Agriculture National Institute of Food and Agriculture (USDA NIFA).

Economics

Many variables influence the profitability of willow biomass crops and a wide range of possible operating conditions and management strategies exist. Some of the most critical variables influencing profitability are biomass yield, the price received for delivered biomass, the cost of planting stock, efficiency of harvesting operations, the use and cost of fertilizers, and transport distances (Buchholz and Volk 2013). These factors are incorporated into a cash flow model developed by SUNY-ESF, EcoWillow 2.0. The model is a financial analysis tool for willow that encompasses all stages of the crop's life cycle over multiple harvest rotations. Data from research trials and commercial operations has been incorporated into the latest version of the model, along with several new features and a more user-friendly design. Users can download EcoWillow 2.0 and supporting documentation from the SUNY-ESF website (www.esf.edu/willow) for free and change input parameters to reflect the costs and operational realities or assumptions of their willow production systems.

A 2014 assessment of the economics of willow biomass crops in New York State is captured in a base case scenario representing conservative estimates of profitability. In order to assess how the economics of the system would change with improvements in yield and crop management practices (i.e., headlands and unplanted field area reduced from 20% to 10%, chip-collection vehicle capacity increased from 7 to 10 Mg_{wet}) as well as some reduction in input costs (i.e., 50% reduction in fertilizer use/costs, reduction in planting stock costs to $0.09 cutting^{-1}), an improved scenario was created. Each adjustment in this scenario is considered to be a realistic and achievable system improvement or best practice target based on current data, logistics, and management options of the crop.

The model can also assess the impact of incentive programs such as USDA BCAP, and two additional scenarios were created: an incentivized scenario that adds potential BCAP incentive payments to the base case, and an improved-incentivized scenario that adds both potential improvements and BCAP payments to the base case. For each scenario, the model provides outputs of net present value, internal rate of return (IRR), payback time and break-even price of biomass. All scenarios are based on a 22-year life cycle of the planting (including crop tear out). Prices are expressed in terms of Mg_{wet} for clarity from the producers' and end users' perspective. The expected moisture content of the crop is 45% for conversion into dry weight values, but as with other input parameters, this can be changed in the model by users.

The base case scenario indicates that the system is not currently profitable at the 2014 market price of woody feedstocks in the region of about $30.50 $Mg_{wet}$$^{-1}$, which is less than the base case break-even price of 33.00 $Mg_{wet}$$^{-1}$(Table 1, Heavey and Volk 2015). The improved scenario provides a positive IRR of 5% over 22 years and has a payback time of 13 years, or at the fourth harvest. The payback time is the same for the incentivized base case, 13 years or four harvests, but the IRR for that scenario is slightly higher at 7%. When the 2015 USDA BCAP incentive rates and the adjustments of the improved scenario are combined in the improved-incentivized scenario, the system has substantially higher 20% IRR and a payback time of 7 years, or just two harvests. The project cost distribution under all these scenarios is about 15% land costs, 20% establishment, 5% fertilizers, 35% harvest, 20% transport, and 5% stock removal. Future work will apply sensitivity analysis to these or similar scenarios and create combined techno-economic and life cycle analyses of willow biomass crops.

Harvesting Systems and Willow Chip Quality

Harvesting is the single largest cost component of willow biomass production and the single largest source of in-field fossil energy demand and related greenhouse gas emissions (Caputo et al. 2014). Efforts to reduce harvesting costs by improving the performance and reliability of the harvester and chip-collection system are essential to the profitability of willow biomass crops. In addition, having a reliable harvesting system that is commercially available and supported by a major agricultural equipment manufacturer increases the confidence level of potential project developers and producers that willow biomass crops can be grown and harvested effectively and efficiently.

The previous lack of a reliable harvesting system for willow biomass crops in North America had been a barrier to the deployment of the crop because landowners were unsure how their crop would be harvested. Many types of specialized machinery for harvesting SRC exist, including small and large single-pass cut-and-chip systems,

Table 1. Internal rate of return (IRR), payback times, and break-even prices for four different production scenarios of willow biomass crops grown in northern New York using EcoWillow 2.0 (Heavey and Volk 2015).

	IRR (%)	Payback	Break-even prices ($ $Mg_{wet}$$^{-1}$)
Base case	< 0	None	33.00
Improved	5	13 years	29.75
Incentivized base case	7	13 years	27.50
incentivized improved	20	7 years	22.75

whole-stem harvesters, and baling systems (Berhongaray et al. 2013; Ehlert and Pecenka 2013). However, due to the limited scale of willow and other SRC deployment, evolving technology, different operational scales, and management objectives, there has not been a dominant harvesting system in use in the United States. In New York State, several existing or modified harvesting platforms for SRC from Europe and North America were evaluated from 2001 to 2008 in SRC willow. Technical hurdles encountered on various harvesters tested during that time include the durability of equipment, low production rates, irregular feeding of stems into the harvester, limits on maximum stem sizes, and inconsistent size and quality of chips (Volk et al. 2010).

In 2008, Case New Holland and SUNY-ESF began developing and testing a prototype short-rotation coppice header (130FB) for their FR9000 and FR Forage Cruiser series of forage harvesters, specifically designed to cut and chip a range of SRC such as willow, poplar, and eucalyptus (Fig. 2). The header can be attached to a standard New Holland forage harvester in these series, although some modifications to the harvester itself are needed to harvest woody crops such as the use of forestry-grade tyres, an upgraded hydraulic system, and shielding below and across the front of the harvester. The performance objectives of the harvesting platform included the ability to harvest double rows of woody plants containing stems up to 120 mm in diameter at ground level, and to produce chips that are 10–45 mm long. Chipped material should be of a quality that allows it to be transported directly to a variety of end users for conversion to different forms of renewable energy and coproducts without requiring further processing.

Harvests of approximately 60 ha of willow biomass crops during late 2012 and early 2013 in New York State, and 20 ha of poplar biomass in Western Oregon, revealed important patterns in the operation of the New Holland

Figure 2. Harvesting willow biomass crops in New York State with a New Holland forage harvester and coppice header.

harvesting system (Eisenbies et al. 2014a). The throughput of the harvester is related to the quantity of standing biomass of the crop, but the pattern differs as the amount of standing biomass changes. At low levels of standing biomass, throughput increases in a linear trend until standing biomass reaches approximately 45–50 Mg_{wet} ha^{-1}. In this range of standing biomass, the throughput of the harvester is below its capacity because the speed of the harvester is limited by conditions in the field. If speeds are too high, the harvester becomes more difficult to operate and it begins to pull plants and roots out of the ground before the stems are cut. Beyond 45–50 Mg_{wet} ha^{-1} of standing biomass, the harvester throughput begins to plateau around 70–90 Mg_{wet} ha^{-1} (Eisenbies et al. 2014b). Operator experience, characteristics of the woody crop being harvested (such as stem morphology and size), and ground conditions also appear to be important factors that influence maximum throughput at various levels of standing biomass.

Over the past few years, the throughput from the single-pass cut-and-chip harvesting system has been improved from less than 20 wet Mg_{wet} h^{-1} with well over 25% downtime due to material jams or mechanical problems, to throughputs of 70–90 Mg_{wet} h^{-1} in willow biomass crops with standing biomass ranging from 20 to 65 Mg_{wet} ha^{-1} (Eisenbies et al. 2014a). The harvester can run consistently in these conditions with less than 10% downtime. Results from these harvesting trials and product development work have successfully led to New Holland making the 130FB short-rotation coppice header commercially available through its network of dealers.

One of the barriers associated with willow biomass in New York State has been the perception that the material is of substantially lower quality than forest residues that are available from the region. End users have expressed concern that willow biomass will have a higher ash and moisture content than forest residues, a lower energy content and a more inconsistent chip size and therefore result in a less desirable feedstock. To address this issue, samples were collected from over 200 truckloads of willow biomass that was harvested in New York State in 2012/2013. The results indicate that the mean ash content of 224 samples was 2.1% (CV 28%) and ranged from 0.8% to 3.5% (Eisenbies et al. 2015). Compared to samples that were hand-harvested from research plots (mean 1.7% CV 28%), the mean ash content of commercial scale samples was almost 0.5% higher but had a similar amount of variation. The ISO 17225-4 (ISO 2015) threshold for short-rotation coppice (B1) chips is < 3% ash content; this cutoff was met by 100% of the samples from one harvest location and 82% of those from a second site. Slight differences were found in the ash content between some willow cultivars that were grown at these sites. Most

Figure 3. Piles of forest residues (left) and a smaller pile of willow biomass crops (right) at the Lyonsdale ReEnergy facility in 2013. After using over 1,000 Mg$_{wet}$ of material in 2013, plant operators were comfortable enough with willow biomass to add it directly to their main wood chip pile in 2014.

notably, the average ash content of the cultivar Fish Creek (*Salix purpurea*) (1.3%) was significantly lower than other cultivars tested, which had mean ash contents up to 2.4%. The moisture and energy content of willow from these large scale harvesting trials is similar to debarked forest wood chips, but ash content of debarked chips was lower (0.6%) (Chandrasekaran et al. 2012).

The mean moisture content of the willow from the commercial harvests was 44% (CV 5%), ranging between 37% and 51% (Eisenbies et al. 2015). Moisture contents for hand-harvested samples from research plots were higher, ranging from 45% to 56%, with a mean moisture content of 46% (CV 6%). The only requirement in ISO standards is that the moisture content is reported for this kind of material.

The higher heating value of the commercial samples was 18.6 MJ kg^{-1} (CV 1%). In comparison, the hand-harvested samples had a higher heating value of 18.8 MJ kg^{-1} (CV 1%). The lower heating value, which accounts for the moisture content in the biomass, was 10.4 MJ kg^{-1} (CV 5%). Because the moisture content of the hand-harvested willow was slightly higher, the lower heating value of the hand-harvested material was slightly lower at 10.1 MJ kg^{-1} (CV 5%) (Eisenbies et al. 2015). Overall, the quality of willow feedstock in commercial trials is very similar to previous results from research trials, and has consistently low variability relative to other bioenergy feedstocks (Eisenbies et al. 2015).

Due to issues with inconsistent chips sizes from previous SRC harvesting systems, a focus of recent development work has been on producing a consistent size chip that meets the quality expectations of end users in the region. With the standard knife and machine configurations, the harvester is typically set to produce chips around 33 mm in size, which is the most fuel efficient mode. In willow crops from recent harvests this has resulted in particle size distributions where 40% of the mass is above 33 mm and 90% of the mass is above 19 mm, and the overall distribution of chips sizes meets the ISO P45S standard for particle size (Eisenbies et al., 2015).

While this data from commercial scale harvesting operations has been informative and helped build confidence in feedstock quality and variability, end users ultimately want to test large amounts of willow biomass in their facilities before they are really comfortable utilizing the feedstock. In 2013, about 1200 Mg$_{wet}$ of willow biomass was delivered to ReEnergy. All of this material was piled separately so plant operators could mix the willow in with other feedstocks in a controlled manner and understand how the material would work in their system (Fig. 3). After processing this material and having no problems in 2013, willow chips were added directly to the main chip piles at ReEnergy's wood yards in 2014, although harvests in 2014 were limited to about 16 ha due to weather and ground conditions which delayed operations at the primary harvest site for the season. In 2015, 36 ha were harvested at this same site, producing about 1,600 Mg$_{wet}$ of willow. Due to wet ground conditions that limited operations the previous season at this site, harvesting operations began in mid-August, prior to the normal harvest window, while leaves still persisted on the willow plants. Due to this fact, ReEnergy again piled willow feedstock separately at the wood yard, but did not encounter any issues mixing the willow with other feedstocks in 2015. Preliminary results from chip samples taken at the field edge and plant gate in 2015 showed that moisture and ash content of leaf-on willow were on the high end, but within the same range as previous commercial-scale trials with leaf-off willow conducted from 2012 to 2014. ReEnergy did not report any problem with the 2015 feedstock and is expected to handle willow in the same manner as other feedstocks in future years. Currently, there is about 80 ha of willow being harvested annually in the northeast using two New Holland FR9000 series harvesters equipped with the 130FB woody crops cutting head. From 2013 to 2015, over 3,500 Mg$_{wet}$ of willow was harvested and delivered to ReEnergy facilities and converted into renewable heat and power. Despite initial uncertainty, experience by operators at ReEnergy has increased overall confidence in the feedstock. Harvests over the next few seasons could reach 160 ha annually or more, as recently planted crops mature, new crops are planted, and the efficiency of harvesting logistics is further improved.

Developing and Deploying Shrub-Willow Systems

The components of a developing willow bioenergy system are now in place in New York State and the Northeast. Efforts are underway to expand willow biomass crop production to meet the demand for woody feedstocks by ReEnergy and other end users in the region and increase the adoption of willow for value-added multifunctional systems. There are several potential pathways to make willow biomass crops more economically feasible so that these systems can be expanded across the region. The first is to work within and improve traditional bioenergy systems. There is a stable long-term market for biomass for heat and power, but the current price being paid ($\sim$ \$30 Mg_{wet}^{-1}) does not provide a positive internal rate of return for growers without support from government programs and/or successfully achieving a suite of best practice targets to offset establishment and maintenance costs. The high establishment costs for willow ($\sim$ \$2,500 ha^{-1}) is also a barrier to many growers because positive returns are not generated for several years and multiple harvests. Reducing initial costs through programs such as USDA BCAP is one approach to improving economics over the short-term while more innovations are made. ReEnergy's commitment, following the program's initial success, to incorporate more willow into its feedstock supply, positions the region to increase the BCAP area up to 2,500–5,000 ha if future funding should become available. However, this expansion will be impacted by prices ReEnergy receives for electricity, which are currently at the low end of the range of the past few years. If the area planted with willow expands and demand for planting material is more consistent, improvements in the management of nurseries and cutting production can be made that will lower the cost of planting stock. In addition, expanding the area under willow will foster innovation and efficiency improvements in crop management and harvesting, further reducing costs.

Producing a wider array of products and/or higher-value products via a biorefinery pathway would increase the value for biomass feedstock and is another possible method for maximizing returns and expanding production. Trials have been conducted at SUNY-ESF with a biorefinery partner, Applied Biorefinery Sciences, using an incremental deconstruction approach based around a hot-water extraction process to recover hemicellulose and other chemicals from willow and other woody feedstocks (Amidon et al. 2011). Following this process, the remaining biomass can be used for the production of premium quality pellets that have lower ash content, higher energy content and more hydrophobic properties than unextracted willow. Alternatively, the processed material could be used as a source of cellulose sugars and lignin, although the most effective pathways to recover these products are still being developed. Other pathways are being explored that will generate multiple products from willow and other woody biomass to increase the value of willow feedstock.

A third potential opportunity for the expansion of shrub willow is multifunctional bioremediation/bioenergy systems. SUNY-ESF, Honeywell International and other organizations have worked together since 2004 to develop, deploy, and research an alternative shrub willow evapotranspiration (ET) cap on 50 ha of former industrial land near Syracuse, NY. The primary objective of this system is to address human health and environmental concerns related to chloride salts moving from the site into the watershed. The second objective is to produce biomass for renewable energy. Willows are able to tolerate the salty substrate of the site with minimal remediation efforts of incorporating 15 cm of organic wastes to the top 50 cm of substrate, combined with standard willow site preparation techniques (Mirk and Volk 2010). The willow on this site produce biomass with yields and quality similar to biomass plantings on mineral soils, while also effectively controlling the water budget of the site (Heavey et al. 2013). Life cycle assessments of the system have also shown the willow vegetative cap to be more cost effective than a traditional geomembrane cap, and require about one tenth the energy inputs and greenhouse gas emissions (Patel 2014). Honeywell and SUNY-ESF have engaged with state and local regulatory agencies to demonstrate the effectiveness of this system and the associated benefits, and there is potential to expand it to 250 ha.

A fourth potential avenue for willow expansion is development of multifunctional systems that balance willow establishment and management costs by providing other valuable environmental services. Recent studies of belowground biomass show that willow crops can store about 31 Mg_{dry} ha^{-1} in roots and stool (stump) material by the time they are 12–14 years old, which is equivalent to about 55 Mg CO_{2eq} ha^{-1} (Pacaldo et al. 2013). If a monetary value were attached to this carbon storage capacity, it would improve the economics of the system. Commercial-scale willow biomass planting can also be combined with wastewater and biosolid treatment systems, and other value-added bioremediation applications. Wastewater treatment is a particularly good option for willow plants, which can benefit from both the additional water and nutrient inputs, likely improving biomass yield, while providing a safe and effective means of processing of waste materials, a valuable environmental service (McCracken et al. 2014). These systems are typically done at smaller scales, but opportunities exist to implement them near larger municipalities and at many rural municipalities that lack waste water treatment infrastructure, and also have nearby sources of organic wastes such as livestock manure. Other potential multifunctional willow systems are being explored to increase the amount of willow being grown in the region, increase producers' experience with the crop, and provide end users opportunities to incorporate the biomass into their systems. Additional environmental benefits and ecosystem services from willow biomass crops include a high life cycle net-energy ratio, low or no pesticide and herbicide use once the crop is established, low potential for soil erosion, improved water quality, an abundant source of early pollen for bees and other pollinators, and the productive use of marginal and idle agricultural land for rural economic development and job creation (Rowe et al. 2009; Volk and Luzadis 2009; Caputo et al. 2014; Tumminello et al. 2015).

Aside from biomass plantings, willow can also be used in smaller scale plantings such as riparian buffers, streambank stabilization, and living fences. Willow living snow fences (LSF) are a promising alternative application that has been researched at SUNY-ESF since 2006. Like willow bioenergy/bioremediation projects or biorefinery pathways, willow LSF can provide a range of benefits including reduced cost of snow and ice control for transportation agencies, improved road safety for drivers, improved travel times, and a suite of environmental benefits (Heavey and Volk 2014). Willow LSF can also be more cost effective than structural snow fences and LSF of other species due to their rapid growth rates, multiple stems and other characteristics.

Conclusion

Research and development on willow biomass crops has been ongoing since 1986 in the United States and considerable progress has been made in understanding and improving the production system. In addition, as the level of understanding about shrub willow has increased, it has been tested and deployed in other applications including living snowfences, bioremediation projects, and other multifunctional systems. While the work over the past three decades has demonstrated a number of shrub willow's valuable attributes in various systems, deployment of the crop for biomass production and other applications is just beginning to develop. One of the largest barriers to deployment is the high establishment costs and the low rate of return in current energy markets. Efforts to improve crop management, harvesting, and logistics will reduce costs and help to improve returns. The development of biorefinery conversion pathways for multiple, higher-value products from each Mg of willow, or the valuation of some of the ecosystem services and environmental benefits provided

by shrub willow, may also help to improve revenues for producers and end users and make the economics more attractive. A suite of extension services is bridging the gap between ongoing research and adoption by the commercial industry for a sustainable bioeconomy. These and other methods will be researched and applied over the next few years in continued efforts to expand shrub willow in the United States. Integrative approaches that synergize these various factors and maximize economic, environmental, and social benefits at various scales will further advance the development, deployment, and utilization of shrub willow for multifunctional systems that produce bioenergy, renewable products, and environmental benefits.

Acknowledgments

We thank collaborators from partner organizations that are involved in developing willow biomass crops in the region including Cato Analyitcs, Celtic Energy Farm, Cornell University, Double A Willow, New Holland, O'Brien and Gere, Pennsylvania State University, ReEnergy Holdings, USDA Farm Services Agency, and USDA Natural Resources Conservation Service.

Conflict of Interest

This work was made possible by the funding under award #EE0001037 from the U.S. Department of Energy Bioenergy Technologies Office, the New York State Research and Development Authority (NYSERDA), the Empire State Development Division of Science, Technology and Innovation (NYSTAR), Honeywell International, The New York State Department of Transportation (NYSDOT), and through the Agriculture and Food Research Initiative Competitive Grant No. 2012-68005-19703 from the United State Department of Agriculture National Institute of Food and Agriculture (USDA NIFA). T.A. Volk is a co-inventor on the patents for the following willow cultivars that are included in these trials: Tully Champion (U.S. PP 17,946), Fish Creek (U.S. PP 17,710), Millbrook (U.S. PP 17,646), Oneida (U.S. PP 17,682), Otisco (U.S. PP 17,997), Canastota (U.S. PP 17,724), Owasco (U.S. PP 17,845).

References

Amidon, T., B. Bujanovic, S. Liu, and J. Howard. 2011. Commercializing biorefinery technology: a case for the multi-product pathway to a viable biorefinery. Forests 2:929–947.

Berhongaray, G., O. El Kasmioui, and R. Ceulemans. 2013. Comparative analysis of harvesting machines on an operational high-density short rotation woody crop (SRWC) culture: one-process versus two process harvest operation. Biomass Bioenergy 58:333–342.

Buchholz, T., and T. A. Volk. 2013. Profitability and deployment of willow biomass crops affected by different incentive programs. Bioenergy Res. 6:53–64.

Caputo, J., S. B. Balogh, T. A. Volk, L. Johnson, M. Puettmann, B. Lippke, et al. 2014. Incorporating uncertainty into a life cycle assessment (LCA) model of short-rotation willow biomass (Salix spp.) crops. Bioenergy Res. 7:48–59.

Chandrasekaran, S. R., P. K. Hopke, L. Rector, G. Allen, and L. Lin. 2012. Chemical composition of wood chips and wood pellets. Energy Fuels 26:4932–4937.

Ehlert, D., and R. Pecenka. 2013. Harvesters for short rotation coppice: current status and new solutions. Int. J. Forest Eng. 24:170–182.

Eisenbies, M., T. A. Volk, L. P. Abrahamson, R. Shuren, B. Stanton, J. Posselius, et al. 2014a. Development and deployment of a short rotation woody crops harvesting system based on a Case New Holland forage harvester and a SRC woody crops header. Final report to U.S. Department of Energy (Award DE-EE0001037) SUNY-ESF, Syracuse NY, 176pp.

Eisenbies, M. T. A., C. Volk, S. Foster, J. Karapetyan, and S. Shi. Posselius. 2014b. Evaluation of a single-pass, cut and chip harvest system on commercial-scale short-rotation shrub willow biomass crops. Bioenergy Res. 7:1506–1518.

Eisenbies, M., T. A. Volk, J. Posselius, S. Shi, and A. Patel. 2015. Quality and variability of commercial-scale short rotation willow biomass harvested using a single-pass cut-and-chip forage harvester. Bioenergy Res. 8:546–559.

Fabio, E. S., T. A. Volk, R. O. Miller, M. J. Serapiglia, H. G. Gauch, K. C. J. Van Rees, et al. (2016). Genotype by environment interactions analysis of North America shrub willow yield trials confirms superior performance of triploid hybrids. Glob. Change Biol. Bioenergy doi: 10.1111/gcbb.12344.

Heavey, J. P., and T. A. Volk. 2014. Living snow fences show potential for large storage capacity and reduced drift length shortly after planting. Agrofor. Syst. 88:803–814.

Heavey, J. P., and T. A. Volk. 2015. Willow crop production scenarios using EcoWillow 2.0. Shrub Willow Fact Sheet Series. The Research Foundation for SUNY.

Heavey, J. P., S. Shi, T. A. Volk, S. Lewis, E. Fabio, D. J. Daily, et al. 2013. Shrub Willow Sustainable Remediation for Solvay Settling Basin 14: Annual Report 2013. The Research Foundation for SUNY

Helby, P., H. Rosenqvist, and A. Roos. 2006. Retreat from Salix – Swedish experience with energy crops in the 1990s. Biomass Bioenergy 30:422–427.

McCormick, K., and T. Kåberger. 2007. Key barriers for bioenergy in Europe: economic conditions, know-how and institutional capacity, and supply chain coordination. Biomass Bioenergy 31:443–452.

McCracken, A., A. Black, P. Cairns, A. Duddy, P. Galbally, J. Finnan, et al. 2014. Use of Short Rotation Coppice (SRC) willow for the bioremediation of effluents and leachates: Project Report. European Regional Development Fund, Agri-Food and Biosciences Institute.

Mirk, J., and T. A. Volk. 2010. Seasonal sap flow of four Salix varieties growing on the Solvay wastebeds in Syracuse, NY, USA. Int. J. Phytorem. 12:1–23.

Pacaldo, R., T. A. Volk, and R. D. Briggs. 2013. Greenhouse gas potential of shrub willow biomass crops based on below- and aboveground biomass inventory along a 19 year chronosequence. Bioenergy Res. 6:252–262.

Patel, A. 2014. Life cycle analysis of shrub willow evapotranspiration covers compared to traditional geomembrane covers. Master's thesis, State University of New York College of Environmental Science and Forestry.

Rowe, R. L., N. R. Street, and G. Taylor. 2009. Identifying potential environmental impacts of large-scale deployment of dedicated bioenergy crops in the UK. Renew. Sustain. Energy Rev. 13:271–290.

Smart, L. B, M. J. Serapiglia, K. D. Cameron, A. J. Stipanovic, T. A. Volk and L. P. Abrahamson. 2008. Genetics of Yield and Biomass Composition of Shrub Willow Bioenergy Crops Bred and Selected in North America. In: Zalesny, Ronald S., Jr.; Mitchell, Rob; Richardson, Jim, eds. Biofuels, bioenergy, and bioproducts from sustainable agricultural and forest crops: proceedings of the short rotation crops international conference; 2008 August 19–20; Bloomington, MN. Gen. Tech. Rep. NRS-P-31. Newtown Square, PA: U.S. Department of Agriculture, Forest Service, Northern Research Station: 54.

Serapiglia, M. J., K. D. Cameron, A. J. Stipanovic, L. P. Abrahamson, T. A. Volk, and L. B. Smart. 2013. Yield and woody biomass traits of novel shrub willow hybrids at two contrasting sites. Bioenergy Res. 6:533–546.

Sleight, N., T. A. Volk, G. A. Johnson, M. H. Eisenbies, S. Shi, E. S. Fabio, et al. 2015. Change in yield between first and second rotations in willow (Salix spp.) biomass crops is strongly related to the level of first rotation yield. Bioenergy Res. 9:270–287.

Tumminello, G., T. A. Volk, S. H. McArt, and M. K. Fierke 2015. Pollinator diversity associated with willow biomass crop. Entomological Society of America National Conference, Minneapolis, MN Nov 15–18, 2015.

U.S. Department of Energy. 2011. U.S. Billion-Ton Update: Biomass Supply for a Bioenergy and Bioproducts Industry. R. D. Perlack, B. J. Stokes (Leads), ORNL/TM-2011/224. Oak Ridge National Laboratory, Oak Ridge, TN. 227p.

Volk, T. A., L. P. Abrahamson, C. A. Nowak, L. B. Smart, P. J. Tharakan, and E. H. White. 2006. The development of short-rotation willow in the northeastern United States for bioenergy and bioproducts, agroforestry and phytoremediation. Biomass and Bioenergy 30:715–727.

Volk, T. A., and S. Harlow. 2014. BCAP helps commercialize shrub willow for bioenergy in northern

New York. Extension research summary (https://www.extension.org/pages/71099/bcap-helps-commercialize-shrub-willow-for-bioenergy-in-northern-new-york#.Vfl8899VhBc)

Volk, T. A., and V. Luzadis. 2009. Willow biomass production for bioenergy, biofuels and bioproducts in New York. Pp 238–260 *in* B. Solomon and V. A. Luzadis ed. Renewable energy from forest resources in the United States. Routledge Press, New York, NY.

Volk, T. A., L. P. Abrahamson, and P. Castellano. 2010. Reducing the cost of willow biomass by improving willow harvest system efficiency and reducing harvesting costs. NYSERDA Report 10-23, Albany, NY.

Volk, T. A., L. P. Abrahamson, K. D. Cameron, P. Castellano, T. Corbin, E. Fabio, et al. 2011. Yields of biomass crops across a range of sites in North America. Asp. Appl. Biol. 112:67–74.

Volk, T. A., L. P. Abrahamson, T. Buchholz, J. Caputo, and M. Eisenbies. 2014. Development and deployment of willow biomass crops. Pp 201–217 *in* D. Karlen, ed. Cellulosic energy cropping systems. John Wiley and Sons, Chichester, UK.

9

Assessment of driving factors for yield and productivity developments in crop and cattle production as key to increasing sustainable biomass potentials

Sarah Gerssen-Gondelach[1], Birka Wicke[1] & Andre Faaij[2]

[1]Copernicus Institute of Sustainable Development, Utrecht University, Heidelberglaan 2, 3584 CS Utrecht, The Netherlands
[2]Energy and Sustainability Research Institute, University of Groningen, Nijenborg 4, 9747 AC Groningen, The Netherlands

Keywords
Biomass potential, cattle production, crop production, driving factors, sustainability, yield developments

Correspondence
Sarah Gerssen-Gondelach, Heidelberglaan 2, 3584 CS Utrecht, The Netherlands.

E-mail: s.j.gerssen-gondelach@uu.nl

Funding Information
This research was conducted within the research program "Knowledge Infrastructure for Sustainable Biomass", which is funded by the Dutch Ministries of 'Economic Affairs' and 'Infrastructure and the Environment'.

Abstract

The sustainable production potential of biomass for energy and material purposes largely depends on the future availability of surplus agricultural lands made available through yield improvements in crop and livestock production. However, the rates at which yields may develop, and the influence of technological, economic and institutional factors on these growth rates are key uncertainties in assessing the potentials and impacts of biomass production. This study analyzes the pace and direction of historical yield developments (1961–2010) of five major crops, beef and cow milk in Australia, Brazil, China, India, USA, Zambia, and Zimbabwe, and examines the driving factors behind these developments. In addition, it explores how future yields are modeled and how modeling efforts may be improved. Average yield growth rates over the investigated period ranged in most cases between 0.7–1.6% year^{-1} for crops, 1.0–1.5% year^{-1} for milk, and 0.4–0.8% year^{-1} for beef (relative to 2010). The role of different drivers is region specific. Yet, supporting agricultural policies have played an important role in increasing yields in all countries, especially for crops. In cattle production, a key factor was the importance of commercial beef and milk production for the national or export market. Based on regional differences in drivers and yield developments, models that assess biomass potentials and impacts should take into account regional drivers, yield gaps, and potential policy pathways.

Introduction

The use of biomass for energy, chemicals, and materials is considered an important alternative to fossil resources (Chum et al. 2011; Harvey and Pilgrim 2011). For biomass to deliver a sizeable contribution, the availability of sufficient sustainable and affordable biomass feedstock is crucial. Assessment studies that have evaluated the current and future global availability of biomass resources show that the largest future potential contribution can come from energy crops grown on various types of land (Hoogwijk et al. 2005; Smeets et al. 2007). The most important land type is surplus agricultural land, which can be released through increased production efficiency

of food, animal feed or pasture. There is, however, disagreement about the availability of surplus agricultural land. Key uncertainties in predicting this area of land are technological progress in agricultural production systems and the related increases in crop and livestock yields (Dornburg et al. 2010; Slade et al. 2011; Batidzirai et al. 2012).

Several studies have investigated the effects of yields on the availability of surplus agricultural land and biomass potentials and impacts. For example, van Vuuren et al. (2009) assessed the impact of food crop yield changes on the global woody biomass potential in 2050. They found that an additional yield improvement of 12.5% compared to the baseline scenario resulted in an increase of the biomass potential from 150 to 230 EJ. Erb et al.

(2009) found that the biomass potential in 2050 would be 79 EJ year^{-1} in the case of intermediate agricultural intensification and humane livestock rearing and 105 EJ year^{-1} in the case of greater intensification of crop and livestock production.[1] Dornburg et al. (2010) estimated that improvements in agricultural management could account for 140 EJ year^{-1} of the total biomass supply potential of 500 EJ year^{-1} in 2050. Slade et al. (2011) derived from a review study that more than 1 Gha of high yielding agricultural land, equal to about 20% of the global agricultural land area in 2010, could be made available for bioenergy crops in 2050 if food crop yields increase at a higher rate than food demand and if the consumption of livestock products is limited.

The degree of yield improvements also affects the environmental performance of biomass production. Without sufficient improvements in yields, there is a large risk of direct or indirect land use change (DLUC and ILUC, respectively), which can result in high greenhouse gas (GHG) emissions (Searchinger et al. 2008; Laborde 2011). In addition, advances in agricultural production systems may also improve the performance of the agricultural sector as a whole. For example, Tilman et al. (2011) show that there is a significant potential in agriculture to reduce global land clearing, GHG emissions and nitrogen use through improved technology and adaption and transfer of high-yielding technologies to underyielding regions. Also, Havlík et al. (2014) show that the transition of livestock production toward more efficient systems would significantly decrease livestock-induced GHG emissions. These emission savings are mainly a result of a reduction in land use change (Havlík et al. 2014).

Models that assess land availability, land use change induced by biomass demand and other impacts of biomass production, such as those used in the studies mentioned above, generally base their crop yield projections on historical developments. Many of these studies also account for (a limited number of) endogenous drivers of future yields. These factors are related to, for example, climate change (Jaggard et al. 2010), crop or land prices (Eickhout et al. 2008; Rosegrant et al. 2008; EPA 2010; Khanna et al. 2011) or management changes like the increased use of fertilizer and other production factors (Eickhout et al. 2008; Beach et al. 2010; Mosnier et al. 2012). This diversity of factors reflects that in reality yield developments depend on numerous factors of various origins (e.g., economic, technological, and ecological). The question arises as to what role these different driving factors play, how they relate to each other and if their impact varies between regions. Moreover, productivity developments in the livestock sector have received much less attention in literature and modeling efforts than agricultural crops – despite the fact that livestock production accounts

for 70% of the total agricultural land and one third of the arable land area is used for feed crop production (Steinfeld et al. 2006). For this sector, the lack of insight into the possibilities to increase yields, the rate at which this can be established and the role of different driving factors is even larger.

Recently, de Wit et al. (2011) discussed what growth rates and maximum (sustainable) yields could be achieved in European agriculture. They assessed agricultural yield developments in the past five decades and compared these to policy developments, structural changes and trends in the use of production factors (inputs). De Wit et al. found that yield developments were clearly correlated to agricultural policy, but yield growth did not always coincide with more efficient use of inputs (de Wit et al. 2011). De Wit et al. focused on Europe and did not investigate other regions that are of critical importance in future biomass supply such as Latin America and sub-Saharan Africa (Hoogwijk et al. 2005; Smeets et al. 2007). Given the importance of yield projections in determining biomass supply and impacts, the aim of this study is to assess for seven countries in different world regions (i.e., Australia, Brazil, China, India, United States, Zambia, and Zimbabwe):

1. what the historical agricultural developments and their drivers are,
2. to what extent and at what growth rate crop and livestock product yields can improve in the future, and
3. how different settings and drivers can influence future yield developments.

These insights contribute to several aspects identified as a key to improving the assessment of biomass potentials and impacts, such as (1) the use of bottom-up analyses to enhance the understanding of current (agricultural) systems, options for improvement, the degree to which yields can be increased, drivers, and regional differences, and (2) a more explicit discussion of assumptions (including yields) (Batidzirai et al. 2012; Wicke et al. 2014).

Methods

Selection of agricultural products and producing countries

The potential area of surplus agricultural land is expected to be largely influenced by efficiency developments in the production of major agricultural products. Therefore, the agricultural developments are assessed for five crops that are most dominant in terms of global production and cultivated area: wheat, corn, rice, sugarcane, and soybean (FAOSTAT). In addition, for livestock, we only take into account beef and cow milk production since cattle uses most of the agricultural area for grazing. Pig and chicken production are often landless, but land is required for producing feed crops (Seré and Steinfeld 1996). The area of cropland needed largely depends on the crop yields, which are already taken into account in this study. Therefore, pig and chicken production are not included in the assessment.

Agricultural developments are assessed in seven countries: Australia, Brazil, China, India, the United States, Zambia, and Zimbabwe (Fig. 1). There are two reasons for this selection. First, Brazil, China, India, and the United States are major producers of the selected crops and cattle products (FAOSTAT). Second, Australia, USA, Zambia, and Zimbabwe can potentially release a large area of agricultural land for biomass feedstock production (Hoogwijk et al. 2005; Smeets et al. 2007). de Wit et al. (2011) assessed agricultural developments in France, The Netherlands, Poland, and Ukraine. For comparison, we have included data about France in the results section. In addition, we compare the general findings from de Wit et al. with our own results.

Historical developments in driving factors

The analysis starts with a description of the current status of agriculture in the selected countries and developments in driving factors that have taken place since 1961 (this part of the study is presented in Appendices 1–8). Based

on literature, the drivers of yield developments are classified into three types (Anderson 2010; Hengsdijk and Langeveld 2010; Neumann et al. 2010; Piesse and Thirtle 2010; Smith et al. 2010; de Wit et al. 2011): technological/ management, economic, and institutional. Economic drivers are, for example, market developments and agricultural R&D investments. Institutional drivers include agricultural policies and governance systems. The discussion of economic and institutional drivers is based on literature review. For technological/management drivers, the following are assessed: labor intensity and level of mechanization, irrigation, nutrient, and pesticide use. These indicators are derived from time series data (1961–2010) collected from the UN Food and Agricultural Organization statistical division (FAOSTAT). These statistics are aggregated on a country level, for example, annual national consumption of fertilizers (tonne year^{-1}). To enable comparison of the management levels between countries, the factors are expressed in average intensity per hectare of agricultural land, for example, the national number of tractors used divided by the total area of agricultural land. For cattle, another indicator for the management level or production intensity is the proportion of ruminants to the area of meadows and pastures (hereafter the ruminant density). This is also derived from FAO statistics. It is not chosen to evaluate the cattle density, because this neglects the importance of other ruminants in the occupation of meadows and pastures for grazing

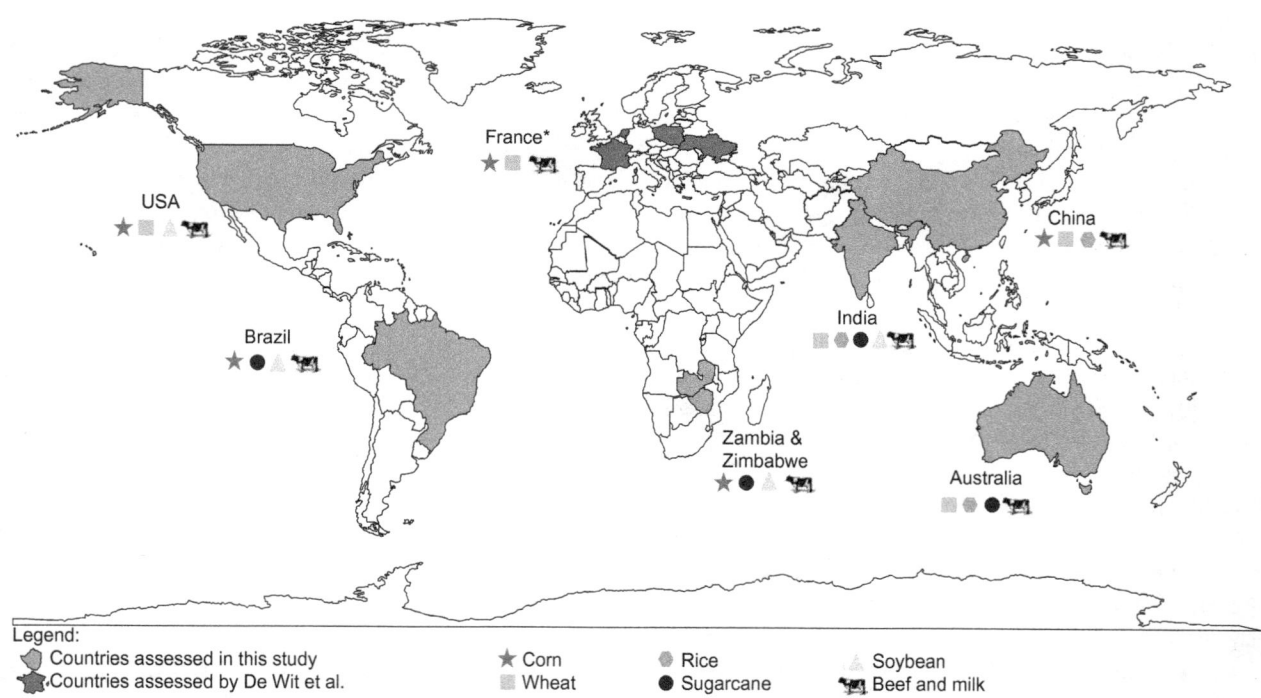

Figure 1. Selected countries and agricultural products. *France was earlier assessed by (de Wit et al. 2011), but for comparison, data for France are also presented in this study.

and hay. As a result of the feed requirements varying between ruminant species, the number of ruminants is expressed in livestock units (LU), where one unit represents the energy requirements for maintenance and production of a typical cow in North America. The livestock unit coefficients are obtained from the FAO (2011). The ruminant density is then calculated as the number of livestock units per hectare of meadows and pastures. The ruminants included are: buffaloes, camel, cattle, goats and sheep. The area of meadows and pastures consists of the total land area available for both for grazing and for the production of conserved forages. This approach thus also accounts for systems that combine grazing and confinement. In this study, we only consider a limited number of drivers that can be actively steered. Literature shows that more factors can influence yields, see the discussion.

Historical yield and productivity developments

To assess historical yield trends and yield growth rates for the selected products and countries, time series data (1961–2010) are collected from FAOSTAT. The crop yield is defined as the annual production quantity per hectare of area harvested (tonne ha^{-1} year^{-1}); the beef yield is given in terms of carcass weight (kg animal^{-1}); the milk yield is the annual milk production per cow (kg animal^{-1} year^{-1}). All numbers are national averages. The average beef and milk production per animal, however, are not the best indicators to study developments in livestock product yields. Beef and milk production can take place in different production systems ranging from pastoral to landless. But intensification does not always lead to higher beef or milk production per animal (e.g., because faster weight gain leads to shorter lifespans). Therefore, a better parameter for milk and beef yields would be the feed conversion efficiency (FCE, kg animal product per kg feed intake). The use of the FCE, however, has also limitations. These are considered in the discussion.

Average annual yield growth rates are obtained by applying linear regression to the historical yield data and are presented per product, per country, per decade and for the entire period. Growth rates are both expressed in absolute growth per year (e.g., t ha^{-1} year^{-2}) and in percentage per year (simple annual growth rate, relative to the initial year). Temporal shifts are identified for each product on country level and differences between products within a country are described. Explanations are sought by comparing the observed changes with the technological/management, economic, and institutional developments. In addition, developments in productivity of the total agricultural sector and of the total livestock sector are assessed and discussed. This productivity is defined as the proportion of aggregated outputs to aggregated inputs (output–input ratio). To derive the aggregated inputs and outputs, the trend of all inputs and outputs in physical units is calculated as an index (base year 1961). From these indices, an (unweighted) average of all inputs and all outputs is calculated for each year. For the agricultural sector. the included inputs are: agricultural land, fertilizer and tractors; the outputs are: crops, meat, milk, and eggs. For the livestock sector, the inputs are: feed crops, meadow, and pasture land; the outputs are: meat, milk, and eggs.

Future yield projections and the role of driving factors

Yield growth rates from projections in literature and models are compared to linear extrapolation of historical trends. It is discussed how yield projections are defined and how they can be improved based on the findings on historical driving factors. For each country, key factors are identified that may stimulate or limit future yield developments. To better understand what possible pathways could be defined for future yield developments, also the magnitude of yield gaps is taken into account. The yield gaps are derived from current yield levels and data on maximum attainable rain-fed or irrigated yields in 2020 as derived from the Global Agro-Ecological Zones database for the IPCC SRES B1 Scenario from the Australian Commonwealth Scientific and Research Organization (CSIRO) Mark 2 Mode (GAEZ Global Agriecological Zones).

Results

Yield and productivity developments

For each country, historical developments in agricultural inputs and yields are assessed. For each country, the status of agriculture and the developments in the different driving factors are discussed in detail in Appendices 1–8. Here a synthesis of historical yield developments and driving factors is presented for the case of Zimbabwe. Syntheses for the other countries can be found in Appendices 2–7. The key findings for each country are presented and compared after the example of Zimbabwe.

Example: developments and driving factors in Zimbabwe

Zimbabwe's first green revolution by commercial farmers (Eicher 1995; Langyintuo and Setimela 2009) is clearly represented by the increase of irrigated area and fertilizer use in the 1960s and early 1970s (Fig. 2). The yields of corn and soybeans only started to improve from the

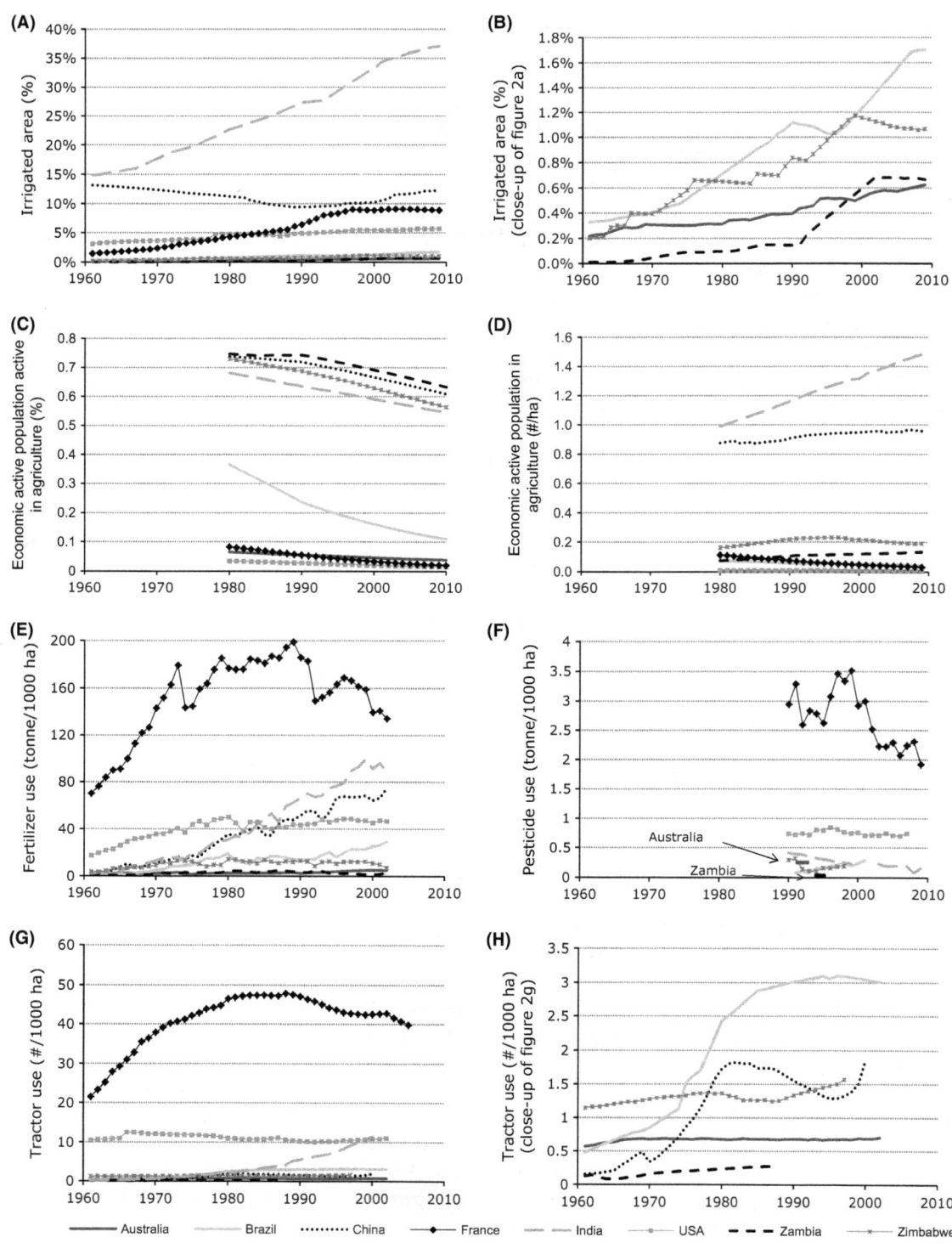

Figure 2. Development in agricultural inputs. All parameters are calculated from FAOSTAT data (FAOSTAT) according to the following definitions: (A) Agricultural area equipped for irrigation = Total area equipped for irrigation/Agricultural area. (B) Close-up of panel A, presenting a selection of the data to reveal differences between the countries at the lowest irrigation levels. (C) Labor share = Total economically active population in agriculture/ Total economically active population. (D) Labor intensity = Total economically active population in agriculture/Agricultural area. (E) Fertilizer = Total fertilizers/Agricultural area. (F) Pesticides = (Insecticides Total + Herbicides Total + Fungicides and Bactericides Total)/Agricultural area. (G) Tractors = Agricultural tractors/Agricultural area. (H) Close-up of panel G, presenting a selection of the data to reveal differences between the countries at the lowest levels of tractor use. Note: when no data are shown for a certain country and/or year, no data is available for this country or year. Note: for tractors, the capacity or size of machinery is not taken into account.

second half of the 1960s (Fig. 3), which coincides with the shift from tobacco production to other crops because of export sanctions imposed in 1965 (Whitlow 1988; Eicher 1995) (also see Appendix 8). During the 1970s, corn yields declined again, while soy yields only fell down in 1979. In addition, irrigation levels stagnated and fertilizer use dropped in the 1970s. Thus, management conditions seem to be affected by the civil war (Whitlow 1985) and corn production suffered more from

this than soybean cultivation. Sugarcane yields appear to be even less affected by economic and political changes in the 1960s and 1970s. Apart from significant fluctuations, sugarcane yields have increased from 1960 until 1986. The improvement rate of 0.9% year^{-1} in these years, however, is significantly lower compared to a growth rate of 3.1% year^{-1} for corn and 16.9% year^{-1} for soybeans between 1960 and 1980 (all relative to 1961), also see Table 1.

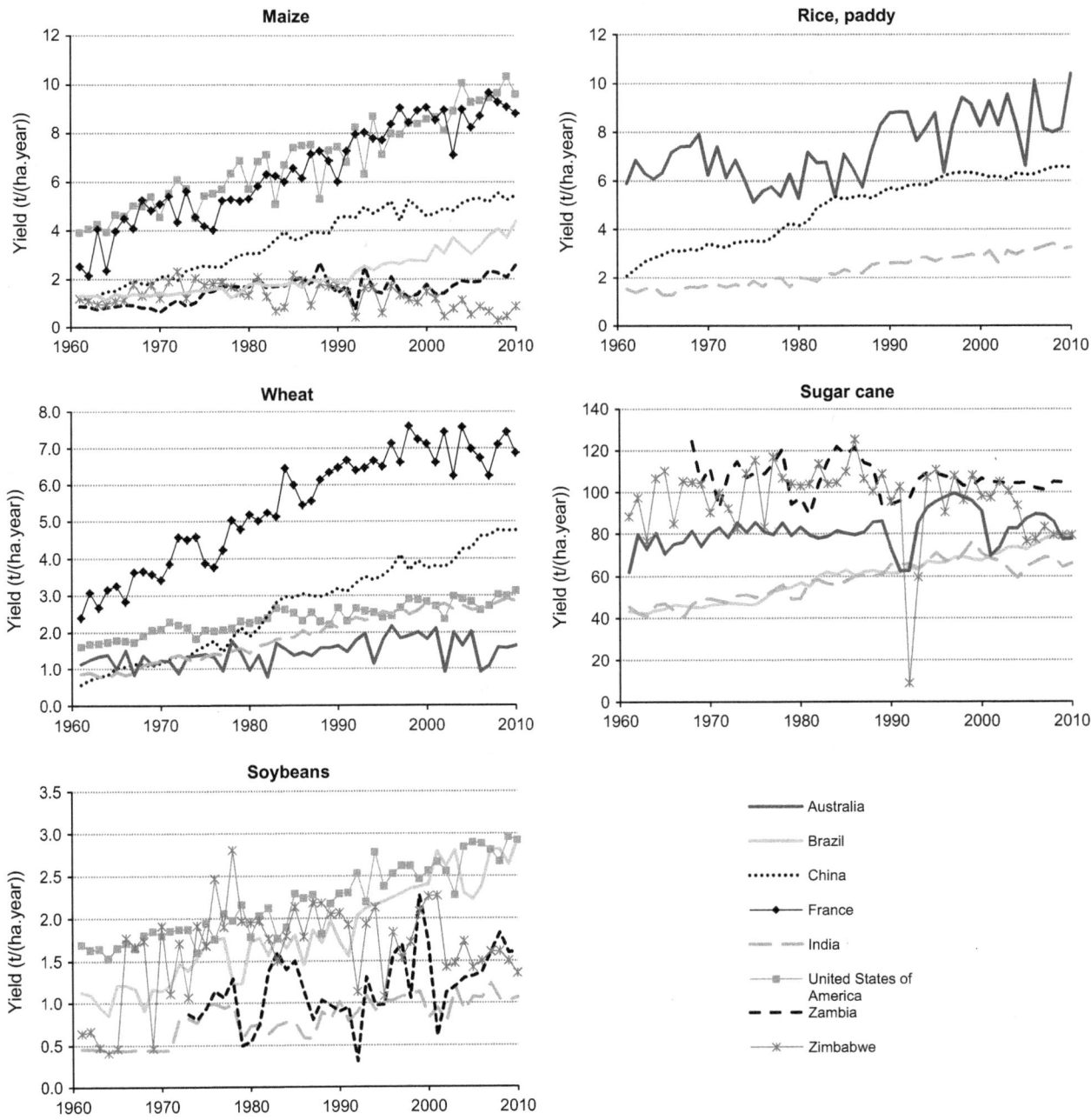

Figure 3. Historical yield developments (1961–2010) for the crops corn, paddy rice, wheat, sugarcane and soybeans (FAOSTAT).

Table 1. Absolute and relative growth in crop, beef and milk yields for the period 1961–2010 and per decade (based on FAO statistics).

Product	Country	Production Mt (kt) 2010	Yield t ha⁻¹ year⁻¹/kg animal⁻¹ year⁻¹ 1961	2010	Average annual yield change kg ha⁻¹ year⁻²/kg animal⁻¹ year⁻²/% year⁻¹ Per decade[3,4] 1961–70	1971–80	1981–90	1991–00	2001–10	Period 1961–2010[3]	1961–2010[5]
Corn	Brazil	56.1	1.3	4.4	12	21	23	85	101	55	
					0.9%	1.5%	1.3%	4.0%	3.2%	7.0%	1.6%
	China	177.5	1.2	5.5	95	111	116	24	79	93	
					8.0%	5.4%	3.6%	0.5%	1.6%	7.3%	1.6%
	France	14.0	2.5	8.8	325	33	83	179	101	131	
					13.4%	0.7%	1.4%	2.4%	1.2%	4.1%	1.4%
	USA	316.2	3.9	9.6	132	84	63	159	156	117	
					3.3%	1.6%	1.0%	2.2%	1.8%	2.9%	1.2%
	Zambia	2.8	0.9	2.6	−11	102	4	−4	117	25	
					−1.2%	11.2%	0.2%	−0.3%	8.3%	2.8%	1.2%
	Zimbabwe	1.2	1.2	0.9	54	−57	33	16	−39	−15	
					5.4%	−2.9%	2.4%	1.4%	−4.3%	−0.9%	−1.6%
Rice, paddy	Australia	0.2	5.9	10.4	127	−157	172	24	27	62	
					2.1%	−2.4%	2.8%	0.3%	0.3%	1.1%	0.7%
	China	197.2	2.1	6.5	127	105	108	83	56	93	
					5.5%	3.3%	2.3%	1.5%	0.9%	3.6%	1.3%
	India	120.6	1.5	3.3	18	30	80	30	48	43	
					1.2%	1.8%	4.2%	1.1%	1.6%	3.4%	1.3%
Soybeans	Brazil	68.8	1.1	2.9	8	18	15	70	20	37	
					0.8%	1.3%	0.9%	3.8%	0.8%	4.1%	1.4%
	India	12.7	0.5	1.1	−2	13	22	16	19	14	
					−0.4%	1.8%	3.4%	1.8%	2.0%	2.9%	1.2%
	USA	90.6	1.7	2.9	22	19	25	22	42	27	
					1.4%	1.1%	1.3%	0.9%	1.7%	1.8%	1.0%
	Zambia	0.0 (41.0)	0.9[1]	1.6	0	−33	−40	134	100	17	
					0.0%	−4.0%	−3.0%	19.8%	11.0%	2.0%	1.2%
	Zimbabwe	0.1 (57.3)	0.6	1.4	130	114	43	46	−45	12	
					30.0%	8.5%	2.5%	2.9%	−2.5%	0.9%	0.6%
Sugarcane	Australia	31.5	62.2	77.7	947	−3	211	3475	811	221	
					1.3%	0.0%	0.3%	4.8%	1.0%	0.3%	0.3%
	Brazil	716.2	43.4	79.2	399	1239	439	671	1081	775	
					0.9%	2.8%	0.7%	1.1%	1.5%	1.9%	1.0%
	India	277.8	45.6	66.1	530	329	738	930	94	593	
					1.2%	0.7%	1.3%	1.4%	0.1%	1.4%	0.8%
	Zambia	4.1	124.4[2]	106.1	−6198	−27	−344	747	4	−152	
					−5.2%	0.0%	−0.3%	0.7%	0.0%	−0.1%	−0.1%
	Zimbabwe	2.8	88.3	79.5	1071	1385	−649	4972	−2897	−270	
					1.2%	1.5%	−0.6%	7.5%	−2.9%	−0.3%	−0.3%
Wheat	Australia	22.1	1.1	1.6	−7	16	42	42	−27	12	
					−0.6%	1.4%	3.3%	2.6%	−1.6%	1.0%	0.7%
	China	115.2	0.6	4.7	65	82	90	79	126	88	
					10.0%	6.5%	3.7%	2.4%	3.0%	17.0%	1.8%
	France	38.2	2.4	6.4	114	90	139	97	−9	98	
					4.3%	2.2%	2.7%	1.5%	−0.1%	3.2%	1.3%
	India	80.7	0.9	2.8	43	26	58	45	19	47	
					5.8%	2.0%	3.5%	2.0%	0.7%	6.2%	1.5%
	USA	60.1	1.6	3.1	49	6	−3	51	46	25	
					3.1%	0.3%	−0.1%	2.1%	1.8%	1.5%	0.8%

Table 1. (Continued)

Product	Country	Production Mt (kt)	Yield t ha⁻¹ year⁻¹/kg animal⁻¹ year⁻¹		Average annual yield change kg ha⁻¹ year⁻²/kg animal⁻¹year⁻²/% year⁻¹						
					Per decade[3,4]					Period	
		2010	1961	2010	1961–70	1971–80	1981–90	1991–00	2001–10	1961–2010[3]	1961–2010[5]
Beef	Australia	2.1	150	254	1.7 / 1.1%	**−0.9** / **−0.5%**	4.0 / 2.3%	1.5 / 0.7%	2.5 / 1.1%	2.1 / 1.4%	0.8%
	Brazil	7.0	192	238	0.1 / 0.0%	**−2.4** / **−1.2%**	**−0.1** / 0.0%	3.2 / 1.6%	4.1 / 2.1%	0.8 / 0.4%	0.4%
	China	6.2	97	141	0.1 / 0.1%	0.2 / 0.2%	5.4 / 5.8%	**−2.4** / **−1.6%**	1.3 / 1.0%	1.2 / 1.3%	0.8%
	France	1.5	186	296	1.1 / 0.6%	2.7 / 1.3%	3.9 / 1.7%	0.1 / 0.0%	1.7 / 0.6%	2.7 / 1.5%	0.9%
	India	1.1	80	103	0.0 / 0.0%	0.9 / 1.1%	1.1 / 1.3%	0.1 / 0.1%	0.0 / 0.0%	0.6 / 0.8%	0.6%
	USA	12.0	215	341	4.2 / 2.0%	0.3 / 0.1%	3.4 / 1.3%	2.0 / 0.7%	2.1 / 0.6%	2.7 / 1.3%	0.8%
	Zambia	0.1 (60.8)	190	160	**−0.6** / **−0.3%**	**−2.7** / **−2.0%**	0.0 / 0.0%	0.0 / 0.0%	0.0 / 0.0%	**−0.7** / **−0.4%**	**−0.4%**
	Zimbabwe	0.1 (99.6)	167	225	0.0 / 0.0%	**−2.4** / **−1.4%**	6.2 / 4.1%	3.8 / 2.1%	0.0 / 0.0%	1.6 / 1.1%	0.7%
Cow milk	Australia	9.0	1985	5810	93 / 4.7%	7 / 0.3%	113 / 3.8%	80 / 1.9%	91 / 1.7%	75 / 4.0%	1.4%
	Brazil	30.7	707	1340	9 / 1.2%	**−7** / **−0.9%**	8 / 1.2%	51 / 6.9%	19 / 1.6%	12 / 2.0%	1.0%
	France	23.3	2671	6278	73 / 2.8%	47 / 1.5%	114 / 3.0%	93 / 1.8%	30 / 0.5%	84 / 3.5%	1.3%
	China	36.0	1208	2882	6 / 0.5%	55 / 4.6%	**−43** / **−2.3%**	6 / 0.4%	80 / 3.5%	29 / 3.0%	1.2%
	India	50.0	424	1284	2 / 0.4%	7 / 1.4%	19 / 3.4%	27 / 3.7%	36 / 3.8%	17 / 6.0%	1.5%
	USA	87.5	3307	9595	125 / 3.8%	95 / 2.1%	136 / 2.5%	147 / 2.1%	147 / 1.8%	128 / 4.1%	1.4%
	Zambia	0.1 (88.5)	300	300	0 / 0.0%	0 / 0.0%	0 / 0.0%	0 / 0.0%	0 / 0.0%	0 / 0.0%	0.0%
	Zimbabwe	0.4	406	430	**−1** / **−0.3%**	3 / 0.7%	**0** / **−0.1%**	1 / 0.1%	**0** / **0.0%**	1 / 0.2%	0.2%

Negative growth in bold.

[1]Yield in 1973.

[2]Yield in 1968.

[3]The average annual yield change in terms of percentage is given relative to the first year of the selected period, that is, the average growth rate is expressed as a percentage of the estimated yield in the first year of selected period (e.g., from 1971–1980, the average annual growth of corn yield in Brazil was 1.5% of the yield level in 1971 as derived from linear regression).

[4]Note that the yield growth rates per decade are calculated for a limited time period of 10 years. This means that the choice for a certain timeframe and outliers in the data can have significant influence on the result of the linear regression. The use of longer timeframes may show a very different trend in yield development.

[5]Relative to 2010.

The introduction of smallholder support in the early 1980s led to a second green revolution (Eicher 1995), but this is not clearly reflected in the statistics. A major factor is the severe drought in 1983, resulting in a significant drop of corn and soybean yields. Good climate conditions in 1985 led to high yields (Eicher 1995). But due to the reduction of smallholder support, and maybe the decline of agricultural R&D as well (Eicher 1995), fertilizer use and crop yields declined in the late 1980s and the 1990s. Remarkably, the area under irrigation increased in the same period. Due to a severe drought, crop yields plummeted in 1992 (Eicher 1995). With the fast-track land reform in 2000 (Matondi, 2012a), yields and input levels dropped further and continued declining in the following years. Overall, crop yields and also agricultural productivity (Fig. 4) have fluctuated considerably throughout the period 1961–2010.

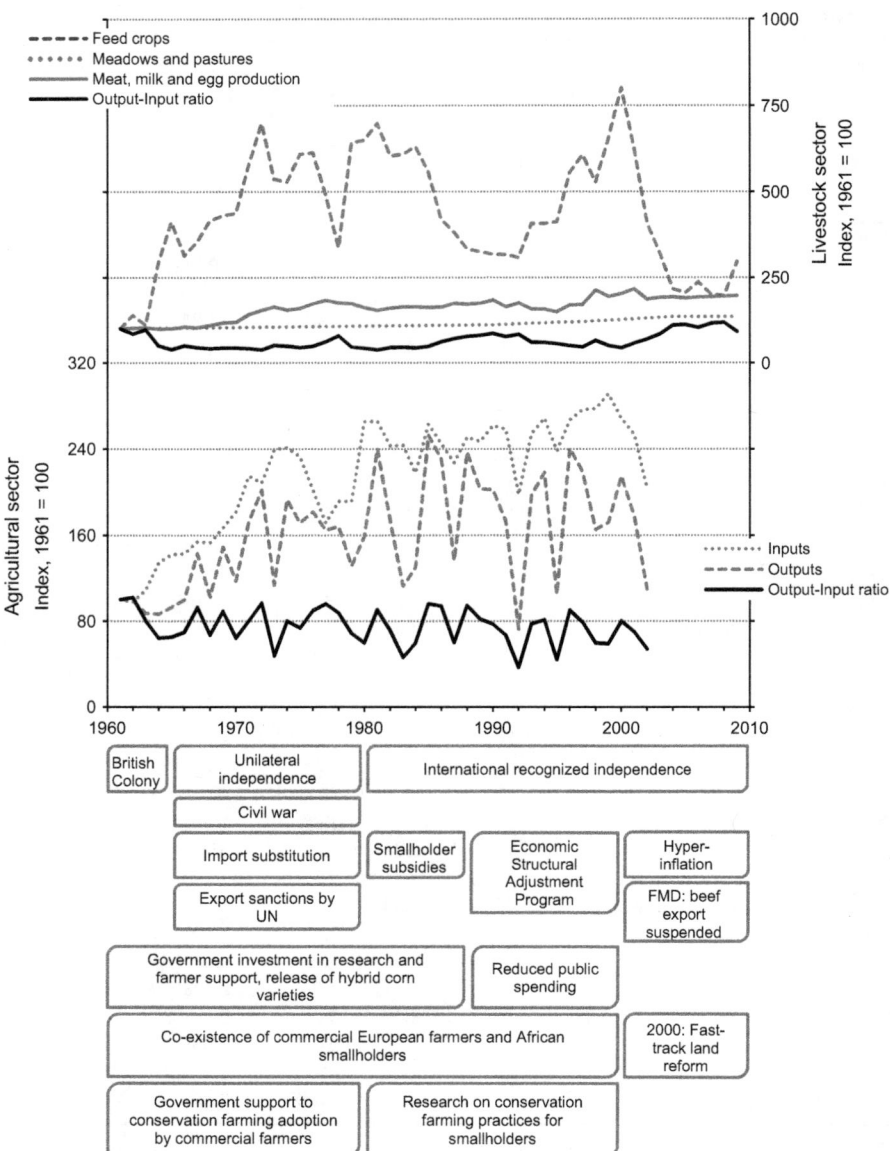

Figure 4. Productivity developments in the Zimbabwean agricultural and livestock sector and institutional, economic and technological/management developments. Because of limited data, agricultural tractors are not included in the inputs and in the output-input ratio for the agricultural sector. (FMD, foot-and-mouth disease).

Considering cattle production, beef yields were stable in the 1960s and early 1970s, but declined during the civil war in the late 1970s (see Fig. A9 in Appendix 9). In the 1980s and 1990s, the yields improved at an average rate of 1.8% year⁻¹ (relative to 1981), but dropped temporarily during the drought of 1992. According to the FAO statistics, beef yields have stagnated since the hyper-inflation and outbreak of foot-and-mouth disease (Marquette 1997; Matondi, 2012b). Between 1960 and 1990, milk yields increased at a very low rate of 0.2% year⁻¹ (relative to 1961). Thus, it seems that technological improvements were limited while economic and political changes did not significantly affect milk yields either. After a small drop in 1992, yields peaked in 1993 and have stabilized since the late 1990s.

Summary and comparison between countries

Over the past five decades, most crop yields showed an upward trend (Table 1). The yield growth rates, however, varied significantly between regions. Average yield growth rates over the period investigated (1961–2010) ranged in most cases between 0.7–1.6% year⁻¹ for crops, 1.0–1.5% year⁻¹ for milk and 0.4–0.8% year⁻¹ for beef (all relative to 2010). Highest rates were found for wheat in China (1.8% year⁻¹), milk in India (1.5% year⁻¹), and beef in France (0.9% year⁻¹). The lowest rates for a crop are found for sugarcane (−0.3 to 1.0% year⁻¹), for any one country the rates are lowest for Zimbabwe (−1.6 to 0.7% year⁻¹). For comparison, in the European countries studied by de Wit et al. (2011), the average growth rates of wheat are 1.0% year^{-1} for Poland to 1.3% year^{-1} for France (relative to 2010). This is lower compared to the wheat growth rates in China and India, but higher compared to the figures for wheat in Australia and the USA. Absolute wheat yield growth in France and the Netherlands (approximately 100 kg ha^{-1} year^{-2} (de Wit et al. 2011)), however, was higher compared to the four countries producing wheat in the present study. For beef, the absolute growth in France and Poland is comparable to the USA, but average growth rates in these European countries are higher than in all non-European countries assessed in this study (0.9% year^{-1} for France and 1.2% year^{-1} for Poland). Absolute and relative yield growth figures for beef in the Netherlands are comparable to Brazil.

In this study, the most observed trend over five decades for crop yield growth is linear. This is in accordance with other studies (Hafner 2003; Ray et al. 2013; Fischer et al. 2014). Yet, in each case, the analysis revealed periods during which yields improved at a higher rate compared to the long-term average as well as periods during which yield growth rates were lower than this average. In each case, technological as well as economic and institutional factors have played a role and these drivers often influenced each other. Yet, the importance and the effect of a driving factor varied from case to case. Table 2 gives an overview of the most important factors behind yield and productivity developments in each country (for more details, see Appendices 2–8; for France see de Wit et al. (2011)).

Improvements in agricultural technology and management have often led to considerable yield growth (Figs 2 and 3). Especially in China and India, large-scale adoption of new technologies (including high-yielding crops) resulted in high average yield growth rates (Appendices 4 and 5). In France and the United States, improved technologies resulted in considerable absolute yield improvements (Appendix 6). The technological improvements, however, included a significant increase in the use of inputs like fertilizer and often caused a decline in agricultural productivity (Fig. 5). In the cattle sector, yield improvements were often achieved through the increased use of feed crops. In Australia and Zimbabwe, however, the consumption of feed crops grew faster than the production of meat and milk, which led to a reduction in the productivity of the livestock sector. Other countries, like India and the USA were able to compensate for the higher input levels by increasing the production output levels of the livestock sector at a similar or even higher rate.

Economic factors often play a vital role in the improvement of agricultural technology. Investments in R&D enabled the development of new technologies. In most countries, these were mainly public investments. In the USA, also private investments were very important (Appendix 6). These investments had already started in the period of industrial and agricultural protectionism. Only in Australia, industrial protectionism indirectly biased the agricultural sector and hindered improvements in production practices and crop yields (Appendix 2). The introduction of economic liberalization provided an incentive for many farmers to (further) improve yields and increase or stabilize agricultural productivity. In Australia, yields and yield improvement rates improved quickly after liberalization started. In Brazil and the USA, yield improvement rates reduced in the first instance but increased again after about ten years (Appendices 3 and 6). Agricultural production in China diversified after the reforms and yield improvements of predominant crops slowed down (Appendix 4). Commercial farmers in Zambia profited from liberalization as they were able to improve soybean yields at high rates, while corn yields of smallholders decreased (Appendix 7). For the cattle sector, the importance of commercial beef and milk production for the domestic or export market was found to be a key factor for yield improvements. In Australia and the USA, such markets already existed during the period of protectionism, while the beef market in Brazil has especially grown after economic liberalization (Appendices 2, 3 and 6). In France (and the EU as a whole), dairy markets got saturated and a quota on milk production was introduced. The number of dairy cows reduced significantly, but milk yields continued to increase through improved management (Huyghe 2012). This, however, led to a reduction in livestock productivity in terms of the output–input ratio.

In accordance with the findings of de Wit et al. (2011) for European countries, policies are found to be an important instrument in steering changes in the agricultural sector in the seven countries investigated in the present study. New technologies could be adopted by farmers because of farmer support programs, for example, in India (Appendix 5). Market liberalization policies created new markets for agricultural products, for example, in Australia (Appendix 2). In some cases, yield improvements have been attained by policies that were focused on a specific commodity: for example, in Zambia, the focus of policies on corn production during a long period resulted in significantly higher yield growth rates for corn compared to other crops (Appendix 7). In Brazil, the ProÁlcool program positively affected sugarcane yields (Appendix 3). The import substitution policy for edible oils in India, especially stimulated the increase of soybean yields. In contrast to the successful implementation of policies in the above examples, Zimbabwe also shows the impact of a lack of good governance and stimulating policies. A civil war in the 1970s and economic reforms around the 1990s disrupted agricultural production and the economy. Due to

Table 2. Key driving factors behind historical yield and productivity development[1].

	Driving factors	Effect on other driving factors	Effect on crops	Effect on cattle
Australia	• Market reforms, trade liberalization, opening of export markets		+ Rice and wheat yields + Agricultural productivity	+/– Beef and milk yields
	• Cattle: growth of export markets	+ Cattle management and technology		+ Milk and beef yields – Livestock productivity
	• Introduction of agri-environmental policies	+ Agricultural production, fertilizer use, irrigation	+/– Agricultural productivity	+ Decline in livestock productivity slowed down
Brazil	• Industrial protection and fertilizer subsidies	+ Fertilizer use	+ Corn and soybean yields – Agricultural productivity	+/– Beef and milk yields
	• Economic reforms	– & +/– Fertilizer use	– Corn and soybean yield growth rates +/– Agricultural productivity	+/– Beef and milk yields
	• Opening of agricultural export markets	+ Fertilizer use	+ Yield growth rates corn and soybean	+ Beef and milk yield growth rate & Livestock productivity increase
	• ProÁlcool program	+ Tractor use, irrigation	+ Sugarcane yields	
China	• Agricultural reforms, public investments in infrastructure and R&D	+ Irrigation, fertilizer use, mechanization	+ Yields	
	• Economic reforms	+ Agricultural diversification	– Yield growth rates	+ Beef and milk yields + Milk yields
	• Increased consumption of milk and dairy products			
France	• Protection of agricultural markets, stimulation of mechanization and fertilizer use	+ Inputs	+ Yields – Agricultural productivity	+ Milk and beef yields
	• Stimulation of modernization and scaling-up, land reforms	+/– Fertilizer use	+ Yields & Agricultural productivity	+ Milk and beef yields
	• Shift to high-yielding crops		+ Yields & Agricultural productivity	
	• Agri-environmental policy, reform of farmer support programs and stimulation of organic farming	– Inputs	– Yield growth rates	– Yield growth rates
	• Quotation milk production			– Livestock productivity
India	• Public investments and subsidies	+ Inputs	+ Yields – Agricultural productivity	
	• Increased milk consumption			+ Milk yields
USA	• Investment in R&D, biotechnology	+ Livestock technology and management	+ Yields	+ Beef and milk yields
	• Trade liberalization and reform of farmer support policies		– Yield growth rates (during reforms) + Yield growth rates (after reforms)	
	• Growing milk market			+ Milk yields
	• Agri-environmental programs	– Fertilizer use	+ Agricultural productivity	
Zambia	• Fertilizer subsidies	+ Fertilizer use	+ Corn yields	
	• Economic liberalization, elimination of fertilizer subsidies	– Fertilizer use + Irrigation	– Corn yields + Soybean yields	
	• Conservation farming technologies		+ Agricultural productivity	
	• Fertilizer Support Program	+ Fertilizer use +/– Irrigation	– Agricultural productivity + Corn yields	
Zimbabwe	• R&D (commercial farmers)	+ Irrigation & Fertilizer use		
	• Civil war	– Irrigation & Fertilizer use	– Yields	– Beef yields
	• Economic reforms, reduction of smallholder support and of agricultural R&D	– Fertilizer use	– Yields	
	• Economic crisis and fast track land reform	– Inputs	– Yields	
	• FMD and ban on beef export			– Beef yields

FMD, foot-and-mouth disease.

[1]Effect: +, increase; – decrease; +/– stabilization.

the long-term unstable situation, crop yields and agricultural productivity have fluctuated heavily. In addition to agricultural policies aimed at economics and production, most countries have also introduced agri-environmental policies which aimed at, for example, enhanced quality of degraded agricultural lands (Australia, China, Zambia, Zimbabwe), balanced use of inputs (China, France) and controlled use and management of natural resources (Australia, India, the United States). In the USA, China, and Zambia, this led to improved agricultural productivity (Appendices 4, 6, and 7). In Australia, the productivity did not improve considerably compared to previous years (Appendix 2). In India, the productivity continued to decline due to weak enforcement of agri-environmental policies (Appendix 5).

Yield projections

Crops

Models that assess biomass potentials and/or impacts of biomass production apply either only exogenous yield projections (determined by factors outside the model) or a combination of exogenously and endogenously (determined by the model based on internal factors) defined yield projections. The exogenous yield projections are based on historical trends. As mentioned in the previous section, the analysis of historical yield developments shows that the most observed trend for crop yield growth is linear. Yet, longer historical time series show that in, for example, the USA crop yield growth has not always followed the current linear trend (see

Fig. A11 in Appendix 9). Also, over shorter time frames, variability in the trend is found with periods of decline, stagnation and/or strong growth. Therefore, yield projections based on historical trends depend on the historical time frame taken into account. For example, Fischer et al. (2014) find global yield growth rates of 1.0% year^{-1} for wheat, rice, and soybean and 1.5% year^{-1} for corn based on the linear trend for 1991–2010. For the period 1961–2010, the present study finds global growth rates that are slightly higher for wheat, rice, and soybean (1.1–1.3% year^{-1}) and lower for corn (1.3% year^{-1}, all relative to 2010), see Table A1 in Appendix 9. Although these differences seem small, they may have considerable impact on future biomass potentials and impacts. This is illustrated by Fischer et al. (2014) who state that, in order to meet the projected food demand in 2050 with limited increase of real prices of crops, the minimum global yield growth rate for staple crops between 2010 and 2050 is 1.1% year^{-1} relative to 2010 (Fischer et al. 2014). Higher growth rates of, for example, 1.3% year^{-1} are preferred to account for factors that may influence supply and demand of crops, including increasing biofuel demand (Fischer et al. 2014). In model assessments, it is therefore important to make explicit what historical time frame is considered to define exogenous yield projections.

Although exogenous yield projections are based on historical trends, some models assume extrapolation of this trend (e.g., in (Jaggard et al. 2010)), while most models assume the overall future yield growth to slow down compared to the historical trend (e.g., in (Laborde 2011; OECD 2012)). Van Dijk and Meijerink (2014)

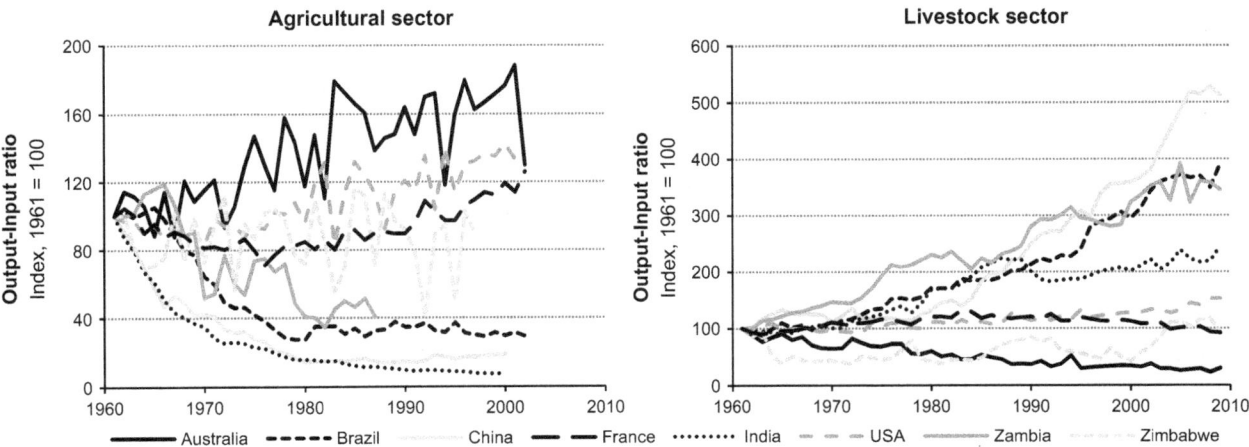

Figure 5. Comparison of developments in productivity (output-input ratio) of the agricultural and livestock sector in the seven selected countries. The input-output ratio is indexed to 100 for the year 1961. This means that when the ratio is higher than 100 in 1 year, the productivity has improved compared to 1961; when the ratio is lower than 100, the productivity has declined. For the agricultural sector the included inputs are: agricultural land, fertilizer and tractors; the outputs are: crops, meat, milk and eggs. For the livestock sector, the inputs are: feed crops, meadow and pasture land; the outputs are: meat, milk and eggs.

give several reasons for assuming decreasing yield growth. First, the opportunities for increasing yields and exploiting existing yield gaps are more and more exhausted (Searchinger et al. 2013). Also, investments in agricultural R&D have declined and considerable socioeconomic constraints in many developing countries are considered to remain a limiting factor for yield growth (Alston et al. 2009; McIntyre et al. 2009; Alexandratos and Bruinsma 2012). Although these motivations are reasonable, these are mainly expectations about how different factors are likely to develop. There may, however, also be other possible pathways. Indeed, the present study shows that in the past 50 years, significant developments in technology, for example, mechanization, fertilizer use, and crop breeding, have driven yield improvements. But although these technologies are wide-spread now, there still exist considerable differences in technology level and yield gaps between regions (see first part of the results and Table A2 in Appendix 9). The present study and the study by de Wit et al. (2011) show that in regions where historical yield growth was high, the development and adoption of new technologies was primarily driven by policies (e.g., subsidies for farmers, trade liberalization and public investments in R&D). In regions where there is still room for considerable yield improvements, stimulating policies could thus play a vital role in materializing this potential. Similarly, other factors could also have a significant effect on future yield developments. The presence of different potential pathways shows that, in the assessment of future biomass potentials and the impacts of biomass production, it is important to investigate different scenarios and to include various endogenous driving factors.

To determine the endogenous yield projections, models relate yield change to driving factors like land or crop prices, climate change, or management. According to Dietrich et al. (2014), technological change is considered to be the key driver for yield change. There is, however, little consensus about the drivers of technological change and the influence of these drivers on yield change (Dietrich et al. 2014; Robinson et al. 2014). Therefore, the number of endogenous factors is generally limited in models, often to only one or two factors. As a result, the different technical, economic, and institutional driving factors are not covered well. Furthermore, the findings from the historical analysis make clear that future yield developments largely depend on how the different driving factors develop in each region. As the number of endogenous factors is generally limited, models do not properly distinguish different driving factors between regions. Important examples of how different drivers affect yield growth potentials are the following:

- In the case of corn in the USA, current yields have attained almost 80% of the *maximum attainable yield*[2] (see Table A2 on yield gaps, Appendix 9). Thus, it is likely that the current technological limits will constrain and slow down yield growth in the near future. Continuation of the historical yield improvement trend would require significant *technological progress*, for example, increase in the potential agroclimatic yield through biotechnology.

- On the other side of the spectrum, yields of rice and soy in India and corn and soy in Zambia and Zimbabwe are less than 40% of the maximum attainable yield. In various other cases, the yield gap is smaller but still leaves room for considerable yield improvements as well. In such situations, the historical analysis shows that accelerated yield growth compared to the longer term trend is possible under favorable circumstances with regard to, for example, governance.

- Zambia and Zimbabwe are examples of cases where *stable agricultural and trade policies* are needed to improve and support the agricultural sector and market and the economy in general. Under such conditions, farmers may be able to adopt improved technologies and management practices. As more advanced technologies and practices already exist, farmers could realize a significant acceleration in yield growth compared to the average trend over the past five decades.

- In Southern India, rice yields could be increased significantly if *management conditions* and *market access* would be improved (see Appendix 5 on agricultural characteristics). In Northern India, rice (and also wheat) is mainly produced on irrigated lands and yields are higher compared to Southern India. Ground water depletion, however, poses a risk to future yield improvements.

- An important measure to attain yield improvements in an environmentally sound way is *sustainable agricultural intensification*. In Australia, for example, management levels have been low relative to most other countries. The same is true for yield growth rates. Although intensification could have significant environmental impact, it is found that this can also be realized while improving the productivity, that is, increasing the crop production (output) per unit resource use (input), and reducing negative effects like emissions and water pollution (see also (Fischer et al. 2014; Hochman et al. 2013; de Wit et al. 2014)).

Thus, for future modeling work, detailed regional assessment of the most important driving factors for yield development and the implementation of more drivers are needed. For this purpose, Table 3 identifies some key drivers that may either limit or stimulate future yield improvements in each country assessed in this study.

Table 3. Key threats and opportunities for future yield improvements.

	Threats to future yield improvements	Opportunities for future yield improvements
Australia	Climate and climate change	Sustainable intensification of crop and livestock sector (see e.g., (Hochman et al. 2013))
Brazil	Weak enforcement or mitigation of land conservation policies	Intensification of cattle production
China	Decreasing water availability, loss of fertile land and land degradation	Expansion of region-specific policies, continuation or expansion of policies to improve productivity, increase of mechanization
India	Land degradation, decreasing water availability	Improvement of market access, management and production of smallholders in Southern India, enforcement of agri-environmental policies, improvement of productivity
USA	Lack of new advances in biotechnology: no significant improvements in the maximum attainable yields	Significant funding for and advances in biotechnology: shift of maximum attainable yields
Zambia	Soil erosion, climate variability	Stimulating policy; e.g., improvement of market access of smallholders (investment in infrastructure), increase in the adoption of conservation farming
Zimbabwe	Continuation of unstable economic situation, climate variability	Re-establishment of (beef) export markets: improvement of knowledge and production of smallholders

The factors are identified based on the historical assessment presented in the first part of the results section.

We highlighted the importance of stimulating policies, especially for Zambia and Zimbabwe earlier. The historical analysis showed that in all countries agricultural and trade policies play an important role in steering yield developments. Van Dijk and Meijerink (2014) show that in economic models, policies and institutions are included, but it seems that their effect on yields is not considered yet. Given our findings, it is important to include policies as a driving factor for yield changes in model assessments. For example, Dietrich et al. (2014) have attempted to implement endogenous yield change related to investments in R&D. Linking yield developments to R&D and other policy-related drivers could be used to define scenarios for different policy pathways and to evaluate their impact on yield changes. In addition, yield gap figures are a good indicator both for the degree of technological progress that can still be attained, and for the potential yield growth rates. It is useful to apply yield gaps in the models to define the yield development projections. For that, more research is needed on, for example, the (crop and region specific) correlation between yield gaps and yield growth rates. In the historical analysis, it was also found that yields have often fluctuated in the past and this affected the average yield growth rate. In current assessment models, projections are based on historical trends and the influence of fluctuations on the long-term trend is neglected. The regional identification of key drivers for yield developments could include an assessment of risk factors for yield fluctuations (e.g., the occurrence of extreme climate conditions), which can be taken into account in the yield projections.

A comparison of yield projections from global outlook studies in van Dijk and Meijerink (2014) shows that the projected yield growth varies significantly and depends on the underlying assumptions made. The influence of underlying assumptions on yield projections is also illustrated by a comparison of yield projections from the integrated assessment model IMAGE (which is used for the assessment of biomass potentials and impacts; see van Vuuren et al. (2009)) and the economic model MIRAGE models (which is used for analyses on, land use change induced by biofuel targets; see Laborde (2011)). In Table A1 (Appendix 9), growth rates derived from the yield projections in IMAGE and MIRAGE are compared to the extrapolation of historical trends. Both models combine exogenous yield projections with endogenously determined yield changes. The exogenous yield projections used in IMAGE and MIRAGE assume that, on the global level, yield growth rates will slow down compared to the historical linear trend (Laborde 2011; OECD 2012). Nevertheless, the two models do not always agree on whether yields in a certain region or even globally will improve at a pace higher or lower compared to the linear growth trend. Large differences in projections between the models are found for corn and soybean in Brazil and rice in China. Also, the projected global growth rates from MIRAGE for corn, rice, and soybean are higher compared to the projections based on linear extrapolation of the historical trends from 1961–2010. In IMAGE, the global projections for wheat, corn, rice, and soybean result in lower yield growth rates compared to linear extrapolation. It is most likely that these contrasting results can be explained by differences in how endogenous yield changes are modeled. The insights from this and other studies would help to make the underlying assumptions for endogenous yield projections more explicit and detailed, and help to assess how yield projections are influenced by different assumptions.

In the introduction, several studies were mentioned that assess the influence of increased yields on the biomass potential. Van Vuuren et al. (2009), Erb et al. (2009) and Dornburg et al. (2010) used yield projections from the FAO (Bruinsma 2003), the presented yield projections from the IMAGE model are in line with these projections (OECD 2012)). Van Vuuren et al. (2009) and Dornburg et al. (2010) take this FAO scenario as baseline and assess the extra biomass potential from additional yield increases. Erb et al. (2009) however, assume that the FAO scenario represents a high intensification scenario and baseline yield improvements are lower. This shows that the perception of the baseline varies. This again underlines the importance to make assumptions more explicit. In addition, it is important to discuss each scenario and address under what conditions the projected yields and the resulting biomass potentials and impacts can be attained; the degree to which investments have to be increased, or required changes in policy. This is considered to be highly valuable for decision making.

Beef and milk

As opposed to historical yield growth trends for crops often being linear, for beef and milk production, we only found linear trends for Australia, India, and the United States; these are countries where we found that yields had significantly improved over a longer timeframe because of the existence of a commercial market. The absence of a yield trend in the past makes it more difficult to define yield growth scenarios for the future. Also, compared to crops, less information can be found about the projections used in the models. Several studies, e.g. IMAGE (Bouwman et al. 2005; Eickhout et al. 2008), apply yield projections from the FAO, which are presented for aggregated world regions level in Wirsenius et al. (2010). A comparison of these projections with historical growth figures indicates that in developed regions, the average annual yield growth rate is projected to be significantly lower than in the last five decades. For example, in North America and Oceania, the increase in beef yield is projected to decline from 1.0% year^{-1} in the period 1961–2005 to 0.2% year^{-1} from 1997/99–2030. Also on the global level, yield growth of beef and milk will be slower compared to the historical trend. Acceleration of yield improvements is projected to especially take place in sub-Saharan Africa. In this region, the yield growth rates of milk are projected to increase from −0.4% year^{-1} (1961–2005) to 0.8% year^{-1} (1997/99–2030). In addition to the FAO projections, Wirsenius et al. (2010) defined an improved livestock production (ILP) scenario, assuming faster intensification of livestock production in low- and

medium income regions as a result of increased competition for land and stricter policies related to land use and livestock production. In this scenario, more regions (e.g., Asia) will realize accelerated yield increases compared to the past. Also on the global level, yield growth will increase from 0.9% year^{-1} to 1.5% year^{-1} for beef and from 0.5% year^{-1} to 2.2% year^{-1} for milk. The scenarios from FAO and Wirsenius et al. again illustrate regional differences, which should be taken into account in the models.

To define yield projections, it is again helpful to consider the potential role of different driving factors per region. In the historical assessment, it was found that the role of commercial livestock production for the domestic or export market is an important factor for explaining yield improvements. For modelling purposes, several scenarios could be defined for market development based on assumptions regarding the size and location of beef and milk consumption and production. In addition, the speed of yield improvements may be based on other driving factors like the possibilities for technological developments and the introduction of agri-environmental policies. Similar to crops, the technological improvement potential could be assessed through yield gap analysis. For livestock, however, no standardized methods exist to assess the yield gap (ILRI). One approach is similar to the conceptual framework for crops and is based on three groups of production factors; production defining (climate and animal genetic characteristics), production limiting (water and feed intake) and growth-reducing (diseases, pollutants) (van de Ven et al. 2003). This method is still new; the first calculations of potential beef production were recently conducted by van der Linden et al. (2013).

Discussion

FAO data

This study analyses a large amount of statistical data which is obtained from the FAOSTAT database. The quality of FAO data, however, can vary significantly. When available, the FAO presents *official data*, which means that the data are collected directly from the states. Yet, the data collection capacities and practices vary between countries and affect the reliability of the data. In addition, the concepts, definitions, coverage, and classifications used by the countries are not uniform and require harmonization to enable international comparison. When no official data is available, the FAO gives figures from secondary semiofficial or unofficial datasets or own estimations (FAOSTAT). With regard to crop yields used in this study, the amount of underlying data that is nonofficial data is limited and mainly restricted to Zambia and Zimbabwe. In the case of milk and beef yields, secondary and estimated data are more common and also presented on a regular basis for Brazil, China and India (FAOSTAT). Sometimes these data seem to be artificial as yields remain constant over one to five decades (see for example yields from beef and milk production in Zambia and Zimbabwe, Fig. A9).

To get an impression of the reliability of FAO yield statistics, their consistency with USDA data was analyzed (Fig. A12, Appendix 9). In some cases, FAO and USDA diverge significantly. For example, in Zambia, corn yield development between 1961 and 1982 is highly uncertain; figures from the FAO show an increasing yield trend, while the USDA data presents a downward trend. This has significant impact on the interpretation of historical developments. The FAO data suggests that corn yield improvements started in the early 1970s with the introduction of fertilizer subsidies for smallholder farmers, while the USDA data implies that the yields were negatively affected by the new pricing and subsidy policies and only started to increase after the introduction of the first new corn varieties in the late 1970s. Differences between the two statistical sources were also found for soybean in Zimbabwe, milk in Brazil and China, and beef in China and India. Not all data sets could be compared as the USDA database had no statistics on milk and beef production in Zambia and Zimbabwe and on sugarcane yields. The varying quality and reliability must be taken into account when interpreting the results. Still, the FAO database is the most complete source for yield figures currently available.

Yield indicators for cattle

With regard to cattle production, it is preferred to assess yield developments in terms of changes in feed conversion efficiency (FCE, kg animal product per kg feed intake) instead of beef or milk production per animal. The main problem of using the production level per animal is that this figure does not always reflect technological advancements. For example, an improved beef cattle production system may achieve faster weight gain and be able to reduce the cattle lifetime. As a result, the beef production per animal may remain constant or even reduce, while the total production can be increased. Also, de Wit et al. (2011) showed that beef yields in the Netherland decreased in the 1990s and 2000s because of the large share of dairy cows that are optimized for milk and not for meat production. In the present study, the historical data show that the average yield per animal has continued to increase in the main cattle producing countries. The rates at which these yields have increased, however, are likely to differ from improvement rates in feed conversion efficiency.

Another reason why the use of the feed conversion efficiency is preferred is the underlying idea of this study that yield improvements have an important role in making land available for biomass production without increasing overall land use. While crop yields are directly related to land use, figures of beef or milk production per animal give no indication of the related land use. As feed consumption can be linked more directly to land use, the feed conversion efficiency would give a better insight in how developments in the cattle sector would influence land requirements.

Ideally, the FCE is measured over the lifetime of an animal because its value is not constant over time. To analyze historical developments in average FCE, a more simple but less accurate way is to calculate the feed conversion efficiency by dividing the produced amount of beef or milk by the gross feed intake. This feed intake is based on estimated energy requirements and the amount of energy supplied by feed inputs, factors that are highly depending on the production system. Therefore, the examination of developments in feed conversion efficiencies over time would require the allocation of animal populations and production quantities to the different production systems for at least several points in time, for example, building on previous work by Seré and Steinfeld (1996) and Bouwman et al. (2005). This was not feasible for the present study. As the carcass weight and annual milk production per animal are the best available data over a longer historical time period, these figures were used here to assess beef and milk yield developments. The same data were used by de Wit et al. (2011) and Wirsenius et al. (2010) to study livestock yield growth rates.

Yield projections and assessment of biomass potentials

This study investigated historical developments in yields and their drivers and provides suggestions for how potential studies can better account for yield developments and driving factors. The assessment focused on three types of driving factors: technological/management, economic, and institutional. Other factors, however, may also influence yield developments. Climate change, for example, may either have a positive or negative effect on yield growth depending on the location (see e.g., Jaggard et al. (2010)). Also, several studies indicate that yield improvement rates of crops are related to the GDP level of a country (Hafner 2003; Powell and Rutten 2013). This correlation, however, does not necessarily mean that GDP itself is a driver of yield development. It is more likely that GDP is an indicator of other driving factors, such as market conditions and technology levels, which are included in the present study.

As shown in the section about yield projections, historical yield growth rates depend on the time frame considered. This is also seen when comparing the yield growth rates for France as calculated in this study and in de Wit et al. (2011). For wheat, for example, the present study found a growth rate of 4.3% year^{-1} (114 t ha^{-1} year^{-1}) for the period 1961 to 1970 while de Wit et al. found a growth rate of 5.2% year^{-1} (136 t ha^{-1} year^{-1}) for the period 1961 to 1969. Thus, the yield growth rates are highly sensitive to the timeframe applied, especially in the case of short time frames. The growth rates should thus be considered with great care and only be used as an indicator of the extent of yield growth or decline. Nevertheless, both studies show that the yield growth rates are very useful to assess the impact of driving factors on yield developments.

Although the historical assessment gives important insights into how different factors may influence future yield developments, it is not possible to predict future yield growth rates. The insights can thus only be used to assess how yields may develop under certain conditions. Particularly the application of endogenous factors and scenarios is useful to assess how yield developments change under different assumptions and how this affects biomass potentials. As mentioned in the section about yield projections, it is important to translate each scenario to conditions for meeting the projected yield developments. This can help identify (regional) strategies for increasing yields.

Finally, in addition to yields, there are also other factors that may affect biomass production potentials. For example, market developments and incentives could influence the balance between crop and livestock production on the one hand and biomass production on the other hand. Also, sustainability criteria could affect the area of land that is excluded from biomass production. An overview of more key factors is provided by Dornburg et al. (2010). Similar to the drivers for yield developments, it is important to make the assumptions regarding these factors explicit. Also, the application of scenarios could be useful.

Conclusions

Global, sustainable biomass production potentials of energy crops largely depend on the future availability of surplus agricultural lands made available through yield improvements in crop and livestock production. This study analyzed the pace and direction of historical yield developments between 1961 and 2010 in Australia, Brazil, China, India, the United States, Zambia, and Zimbabwe. Furthermore, it assessed the technological, economic, and institutional driving forces behind these developments and explored how the insights gained can help to improve the modeling of future yields.

This study showed that historical yield growth (especially of crops) has often followed a linear trend. Mainly, the average yield improvement rates for crops and milk were between 0.7% and 1.6% year^{-1}. For beef, the rates were lower (maximum of 0.8% year^{-1} in Australia; all relative to 2010). In all cases, yields and yield growth rates have fluctuated to various degrees. Large fluctuations were especially found for crops when driving factors changed strongly (e.g., extreme climate conditions in Australia). Also, in each case, the analysis revealed periods during which yields improved at a higher rate compared to the long-term average as well as periods during which yield growth rates were lower than this average. The periods of high yield growth, for example, 8.5% year^{-1} for soybean in Zimbabwe in the 1970s, show that relatively fast improvements can be attained in cases where the yield gap is still large. Such significant improvements can especially be realized under favorable conditions with regard to economics and governance that stimulate improvements in agricultural technology and management. The future development of yields depends on how driving factors will change in each region.

The historical assessment shows that all three types of driving forces have influenced yield changes. The importance and the effect of each factor, however, is country- and even regional- specific. Overall, supporting agricultural policies have played an important role in increasing yields. Examples of successful policies are subsidies to stimulate adoption of new technologies, trade liberalization (resulting in increased demand for agricultural products which stimulated investments and innovations in the agricultural sector) and public investments in R&D. In some periods and countries, such policies were absent or eliminated (e.g., Australia in the 1960s and Zambia in the 1990s). As a result, yields stagnated or declined. Although agricultural policies led to yield increases in many cases, they failed to improve output-input ratios (i.e., unsuccessful to realize more efficient use of resources like fertilizers). Some countries like the USA and (to a lesser extent) China were able to increase this productivity by implementing specific agri-environmental policies. Other countries adopted such policies as well, but the result largely depended on the success to enforce these policies (e.g., productivity stabilized in Australia, but no effect was seen in India). The importance of policies in steering yields was especially high for crops. With regard to yield improvements in cattle production, a key factor was the importance of commercial beef and milk production for the national or export market. But policy and market can be closely related: in many cases, trade liberalization created new markets, which stimulated investments and resulted in improved yields as demonstrated in, for example, Brazil.

Current models that assess biomass potentials and impacts only take into account one or a limited number of endogenous factors influencing yields. Also, an explicit discussion of the assumptions behind yield projections is lacking, which hampers a comparison of yield projections between the models. Several suggestions are made to improve the models and thereby our understanding of potential future pathways for agricultural yield developments and for sustainable biomass production. First, scenarios based on regional assessment of key factors for yield development, as conducted in this study, could help to gain more insight in potential pathways and regionally differentiated effects. Second, to define such scenarios, yield gap figures are an important indicator of possible technological progress and the potential rate of yield improvement. Also, different policy strategies should be included and tested in the scenarios. Finally, the assessment of important factors for yield development could help to make the underlying assumptions of yield projections more explicit. The implementation of these suggestions will help to identify policy options and preconditions for specific development pathways.

Acknowledgments

The authors thank Vassilis Daioglou (Utrecht University and PBL Netherlands Environmental Assessment Agency) for sharing yield data from the IMAGE model. David Laborde is thanked for sharing data from the MIRAGE model. This research was conducted within the research program "Knowledge Infrastructure for Sustainable Biomass", which is funded by the Dutch Ministries of 'Economic Affairs' and 'Infrastructure and the Environment'.

Conflict of Interest

None declared.

Notes

[1] Assuming continuation of current trends in diet and crop land area expansion.

[2] Maximum attainable yield: the yield resulting from combining (1) the constraint-free potential agroclimatic yield with regard to temperature, radiation, and soil moisture conditions prevailing in the specific region and (2) reduction factors related to climate (e.g., pests and diseases), soil and terrain conditions, and assumptions regarding the management level (GAEZ Global Agri-Ecological Zones). Generally, it is assumed that there is a minimum yield gap where the actual yield level is equal to the economically attainable yield. Fischer et al. (2014) consider this economically attainable yield to be about 23%

below the maximum attainable yield. Larger yield gaps are assumed to be 'economically exploitable yield gaps'. These caps could (largely) be closed with existing technologies (Fischer et al. 2014).

[3] The high growth rate for soybean yields is also related to the drought induced yield drop in 1992. Without this outlier in the dataset, however, the improvement rate is still 12.2 % year^{-1}. Sugarcane yields were not considerably affected by the drought.

References

Alexandratos, N., and J. Bruinsma. 2012. World agriculture towards 2030/2050: the 2012 revision. Food and Agricultural Organisation of the United Nations, Rome, Italy. ESA working paper no. 12-03.

Alston, J. M., J. M. Beddow, and P. G. Pardey. 2009. Agricultural research, productivity, and food prices in the long run. Science 325:1209–1210.

Alston, J. M., J. S. James, M. A. Andersen, and P. G. Pardey. 2010a. A Brief History of U.S. Agriculture. Pp. 9–21. in J. M. Alston, J. S. James, M. A. Andersen and P. G. Pardey, eds. Persistence Pays. Springer, New York, NY.

Alston, J. M., J. S. James, M. A. Andersen, and P. G. Pardey. 2010b. Research funding and performance. Pp. 137–185 in J. M. Alston, J. S. James, M. A. Andersen and P. G. Pardey, eds. Persistence Pays. Springer, New York, NY.

Alston, J. M., J. S. James, M. A. Andersen, and P. G. Pardey. 2010c. The federal role. Pp. 187–236 in J. M. Alston, J. S. James, M. A. Andersen and P. G. Pardey, eds. Persistence Pays. Springer, New York, NY.

Anderson, K. 2010. Globalization's effects on world agricultural trade, 1960–2050. Philos. Trans. Royal Soc. B: Biol. Sci. 365:3007–3021.

Anderson, K., and W. Martin. 2009. 9 China and Southeast Asia. Pp. 359–387 in K. Anderson, ed. Distortions to agricultural incentives: A global perspective, 1955–2007. The International Bank for Reconstruction and Development/The World Bank, Washington, DC.

Anderson, K., and A. Valdés. 2009. 7 Latin America and the Caribbean. Pp. 289–322 in K. Anderson, ed. Distortions to Agricultural Incentives: A Global Perspective, 1955–2007. The International Bank for Reconstruction and Development / The World Bank, Washington, DC.

Anderson, K., R. Lattimore, P. J. Lloyd, and D. MacLaren. 2009. 5 Australia and New Zealand. Pp. 289–322 in K. Anderson, ed. Distortions to Agricultural Incentives: A Global Perspective, 1955–2007. The International Bank for Reconstruction and Development/The World Bank, Washington, DC.

ASTI Data Tool version 1.1 [Internet]. ASTI Agricultural Science and Technology Indicators [accessed 2013 23 August]. Available at: http://www.asti.cgiar.org/data/.

Baer, W. 1972. Import Substitution and Industrialization in Latin Amercia: experiences and Interpretations. Latin Am. Res. Rev. 7:95–122.

Banerjee, O., A. J. Macpherson, and J. Alavalapati. 2009. Toward a policy of sustainable forest management in Brazil: a historical analysis. J. Environ. Develop. 18:130–153.

Bao, J. 2011. Dairy production in diverse regions | China. Pp. 83–87. *in* J. W. Fuquay, ed. Encyclopedia of Dairy Sciences (Second Edition), Academic Press, San Diego.

Barros, S.. 2013. Brazil Biofuels Annual: Annual Report 2013. USDA Foreign Agricultural Service, Sao Paulo; BR13005. Available at: http://gain.fas.usda.gov/Recent%20 GAIN%20Publications/Biofuels%20Annual_Sao%20 Paulo%20ATO_Brazil_9-12-2013.pdf.

Batidzirai, B., E. M. W. Smeets, and A. P. C. Faaij. 2012. Harmonising bioenergy resource potentials— Methodological lessons from review of state of the art bioenergy potential assessments. Renew. Sustain. Energy Rev. 16:6598–6630.

Beach, R. H., D. Adams, R. Alig, J. Baker, G. S. Latta, and B. A. McCarl, et al. 2010. Model Documentation for the Forest and Agricultural Sector Optimization Model with Greenhouse Gases (FASOMGHG). RTI International. Available at: http://www.cof.orst.edu/cof/fr/research/tamm/ FASOMGHG_Model_Documentation_Aug2010.pdf

Blandford, D., and R. N. Boisvert. 2006. 16 Policy for agricultural adjustment in the United States. Pp. 237–253 *in* D. Blandford and B. Hill, eds. Policy reform and adjustment in the agricultural sectors of developed countries. CAB International, Wallingford, Oxfordshire, UK.

Bonaglia, F. 2009. Zambia: Sustaining agricultural diversification. OECD J. Gen. Pap. 2009:103–131.

Bouwman, A. F., K. W. van der Hoek, B. Eickhout, and I. Soenario. 2005. Exploring changes in world ruminant production systems. Agric. Syst. 84:121–153.

Bowman, M. S., B. S. Soares-Filho, F. D. Merry, D. C. Nepstad, H. Rodrigues, and O. T. Almeida. 2012. Persistence of cattle ranching in the Brazilian Amazon: a spatial analysis of the rationale for beef production. Land Use Policy 29:558–568.

Bruinsma, J.. 2003. World agriculture: towards 2015/2030. An FAO perspective. Earthscan Publications Ltd., London, UK.

Carmo Oliveira, J. D.. 1986. Trade policy, market 'distortions', and agriculture in the process of economic development Brazil, 1950–1974. J. Dev. Econ. 24:91–109.

Carpentier, C. L., and D. E. Ervin. 2002. USA. Pp. 95–139 *in* F. Brouwer and D. E. Ervin, eds. Public concerns, environmental standards and agricultural trade. CAB International, Wallingford, Oxfordshire, UK.

Carvalho, J. 1991. Agriculture, industrialization and the macroeconomic environment in Brazil. Food Policy 16:48–57.

Cederberg, C., D. Meyer, and A. Flysjö. 2009. Life cycle inventory of greenhouse gas emissions and use of land and energy in Brazilian beef production. Swedish Institute for Food and Biotechnology (SIK), Göteborg, Sweden. 792.

Chambers, W. B.. 2002. World trade and concerns for the human environment. Pp. 39–55. *in* F. Brouwer and D. E. Ervin, eds. Public concerns, environmental standards and agricultural trade. CAB International, Wallingford, Oxfordshire, UK.

Chum, H., A. Faaij, J. Moreira, G. Berndes, P. Dhamija, H. Dong, et al. 2011. Bioenergy. Pp. 209–332 *in* O. Edenhofer, R. Pichs-Madruga, Y. Sokona, K. Seyboth, P. Matschoss, S. Kadner, T. Zwickel, P. Eickemeier, G. Hansen, S. Schlömer, and von Stechow C., eds. IPCC Special Report on Renewable Energy Sources and Climate Change Mitigation. Cambridge University Press, Cambridge, UK and New York, NY.

Dietrich, J. P., C. Schmitz, H. Lotze-Campen, A. Popp, and C. Müller. 2014. Forecasting technological change in agriculture—an endogenous implementation in a global land use model. Technol. Forecast. Soc. Chang. 81:236–249.

Van Dijk, M., and G. W. Meijerink. 2014. A review of global food security scenario and assessment studies: results, gaps and research priorities. Global Food Sec. 3:227–238.

Dixon, J., A. Gulliver, and D. Gibbob. 2001. Farming Systems and Poverty: Improving farmers' livelihoods in a changing world. FAO and World Bank, Rome and Washington, DC.

Djurfeldt, G., E. Aryeetey, and A. C. Isinika. 2011. African Smallholders: Food Crops, Markets and Policy. CAB International, Wallingford, Oxfordshire, UK.

Dornburg, V., D. van Vuuren, G. van de Ven, H. Langeveld, M. Meeusen, M. Banse, et al. 2010. Bioenergy revisited: key factors in global potentials of bioenergy. Energy Environ. Sci. 3:258–267.

Doyle, P. T., and C. R. Stockdale. 2011. DAIRY FARM MANAGEMENT SYSTEMS | Seasonal, Pasture-Based, Dairy Cow Breeds. Pp. 29–37. *in* J. W. Fuquay, ed. Encyclopedia of Dairy Sciences (Second Edition), Academic Press, San Diego.

Eicher, C. K. 1995. Zimbabwe's maize-based Green Revolution: preconditions for replication. World Dev. 23:805–818.

Eickhout, B., J. C. M. van Meijl, A. A. Tabeau, and E. Stehfest. 2008. The Impact of Environmental and Climate Constraints on global food supply. Netherlands Environmental Assessment Agency (MNP)/Agricultural Economics Research Institute (LEI), Bilthoven/The Hague, The Netherlands.

EPA. 2010. Renewable Fuel Standard Program (RFS2) Regulatory Impact Analysis. Environmental Protection Agency, Washington, DC. EPA-420-R-10-006.

Erb, K., H. Haberl, F. Krausmann, C. Lauk, C. Plutzar, and J. K. Steinberger, et al. 2009. Eating the Planet: Feeding and fuelling the world sustainably, fairly and humanely–a scoping study. Commissioned by Compassion in World Farming and Friends of the Earth UK. Institute of Social Ecology, Vienna, Austria. Social Ecology Working Paper No. 116.

FAO. 2011. Guidelines for the preparation of livestock sector reviews. Animal Production and Health Guidelines. FAO, Rome. Report No 5.

FAOSTAT [Internet]. FAO [updated 2014; accessed 2012-2014]. Available at: http://faostat.fao.org.

Ferreira, J., R. Pardini, J. P. Metzger, C. R. Fonseca, P. S. Pompeu, G. Sparovek, et al. 2012. Towards environmentally sustainable agriculture in Brazil: challenges and opportunities for applied ecological research. J. Appl. Ecol. 49:535–541.

Fischer, R. A., D. Byerlee, and G. O. Edmeades. 2014. Crop yields and global food security: will yield increase continue to feed the world? Australian Centre for International Agricultural Research, Canberra, Australia.

GAEZ Global Agri-Ecological Zones [Internet]. Food and Agriculture Organisation of the United Nations (FAO) and International Institute for Applied Systems Analysis (IIASA) [updated 2014; accessed 2014 19 February]. Available at: http://gaez.fao.org/Main.html#.

Gulati, A., and G. Pursell. 2009. 10 India and other South Asian Countries. Pp. 389–415 in K. Anderson, ed. Distortions to Agricultural Incentives: A Global Perspective, 1955–2007. The International Bank for Reconstruction and Development / The World Bank, Washington, DC.

Hafner, S. 2003. Trends in maize, rice, and wheat yields for 188 nations over the past 40 years: a prevalence of linear growth. Agric. Ecosyst. Environ. 97:275–283.

Haggblade, S., and G. Tembo. 2003. Conservation farming in Zambia. International Food Policy Research Institute, Washington, DC, USA. EPTD Dicussion Paper No. 108. Available at: http://www.ifpri.org/sites/default/files/publications/eptdp108.pdf

Harvey, M., and S. Pilgrim. 2011. The new competition for land: food, energy, and climate change. Food Policy 36:S40–S51.

Havlík, P., H. Valin, M. Herrero, M. Obersteiner, E. Schmid, M. C. Rufino, et al. 2014. Climate change mitigation through livestock system transitions. Proc. Natl Acad. Sci. USA 111:3709–3714.

Hawkes, C., and S. Murphy. 2010. 2. An overview of global food trade. Pp. 16–32. in C. Hawkes, C. Blouin, S. Henson, N. Drager and L. Dubé, eds. Trade, Food, Diet and Health: Perspectives and Policy Options. Wiley Blackwell, Oxford.

Hawkes, C., S. Friel, T. Lobstein, and T. Lang. 2012. Linking agricultural policies with obesity and noncommunicable diseases: A new perspective for a globalising world. Food Policy 37:343–353.

Hengsdijk, H., and J. W. A. Langeveld. 2010. Yield trends and yield gap analysis of major crops in the world. Statutory Research Tasks Unit for Nature & the Environment(WOT Natuur & Milieu), Wageningen, The Netherlands. WOt-werkdocument 170.

Hochman, Z., P. S. Carberry, M. J. Robertson, D. S. Gaydon, L. W. Bell, and P. C. McIntosh. 2013. Prospects for ecological intensification of Australian agriculture. Eur. J. Agron. 44:109–123.

Hoogwijk, M., A. Faaij, B. Eickhout, B. de Vries, and W. Turkenburg. 2005. Potential of biomass energy out to 2100, for four IPCC SRES land-use scenarios. Biomass Bioenergy 29:225–257.

Howard, J. A., and C. Mungoma. 1996. Zambia's stop-and-go revolution: the impact of policies and organizations on the development and spread of maize technologies. International Development Working Paper No. 61. MSU Agricultural Economics, Michigan.

Huyghe, C. 2012. Country Pasture/Forage Resources Profiles. France [Internet]. FAO [updated 2012; accessed 2015 5 February]. http://www.fao.org/ag/agp/AGPC/doc/Counprof/France/france.htm#4.RUMINANT.

ILRI 2014. Closing livestock yield gaps in the developing world: imperatives for people and the planet [Internet]. ILRI [updated 2014; accessed 2014 26 June]. Available at: http://www.slideshare.net/ILRI/smith-gfsc-apr2014.

Jaggard, K. W., A. Qi, and E. S. Ober. 2010. Possible changes to arable crop yields by 2050. Philos. Trans. Royal Soc. B: Biol. Sci. 365:2835–2851.

Jayne, T. S., J. Govereh, M. Wanzala, and M. Demeke. 2003. Fertilizer market development: a comparative analysis of Ethiopia, Kenya, and Zambia. Food Policy 28:293–316.

Karplus, V. J., and X. W. Deng. 2008a. Transformation in China's Agriculture in the Twentieth Century. Pp. 27–44 in V. J. Karplus and X. W. Deng, eds. Agricultural biotechnology in China: origins and prospects. Springer, New York.

Karplus, V. J., and X. W. Deng. 2008b. Agricultural biotechnology takes root in China. Pp. 55–77 in V. J. Karplus and X. W. Deng, eds. Agricultural biotechnology in China: origins and prospects. Springer, New York.

Khanna, M., C. L. Crago, and M. Black. 2011. Can biofuels be a solution to climate change? The implications of land use change-related emissions for policy. Interface Focus 1:233–247.

Laborde, D.. 2011. Assessing the land use change consequences of European biofuels policies. International Food Policy Research Institute, Washington, DC. Available at: http://trade.ec.europa.eu/doclib/docs/2011/october/tradoc_148289.pdf.

Lamers, P., C. N. Hamelinck, M. Junginger, and A. P. C. Faaij. 2011. International bioenergy trade—A review of past developments in the liquid biofuel market. Renew. Sustain. Energy Rev. 15:2655–2676.

Langyintuo, A. S., and P. Setimela. 2009. Assessing the effectiveness of a technical assistance program: the case of maize seed relief to vulnerable households in Zimbabwe. Food Policy 34:377–387.

van der Linden, A., G. W. J. van de Ven S. J. Oosting, van Ittersum M. K., and de Boer I. J. M.. 2013. Can we extend yield gap analysis to livestock production? First International Conference on Global Food Security (poster presentation); 29 Sept-2 Oct 2013; Noordwijkerhout, The Netherlands.

Marongwe, L. S., I. Nyagumbo, K. Kwazira, A. Kassam, and T. Friedrich. 2012. Conservation agriculture and sustainable crop intensification: A Zimbabwe Case Study. Int. Crop Manag. 17:1–29.

Marquette, C. M. 1997. Current poverty, structural adjustment, and drought in Zimbabwe. World Dev. 25:1141–1149.

Martinelli, L. A., R. Naylor, P. M. Vitousek, and P. Moutinho. 2010. Agriculture in Brazil: impacts, costs, and opportunities for a sustainable future. Curr. Opin. Environ. Sustain. 2:431–438.

Matondi, P. B. 2012a. 1. Understanding Fast Track Land Reforms in Zimbabwe. Pp. 1–17 in P. B. Matondi, ed. Zimbabwe's fast-track land reform, 1st edn. Zed Books, London.

Matondi, P. B. 2012b. 5. Complexities in understanding agricultural production outcomes. Pp. 130–160 in P. B. Matondi, ed. Zimbabwe's fast-track land reform, 1st edn. Zed Books, London.

Matondi, P. B. 2012c. Zimbabwe's fast-track land reform. Zed Books, London, UK; New York.

McCormick, M. E.. 2011. Dairy farm management systems | non-seasonal, pasture optimized, dairy cow breeds in the United States. Pp. 38–43 in J. W. Fuquay, ed. Encyclopedia of dairy sciences (Second Edition). Academic Press, San Diego.

McIntyre, B. D., H. R. Herren, J. Wakhungu, and R. T. Watson. 2009. Agriculture at Crossroads. International Assessment of Agricultural Knowledge, Science and Technology for Development - Global Report. IAASTD, Washington, DC.

Mosnier, A., P. Havlík, H. Valin, J. S. Baker, B. C. Murray, S. Feng, et al. 2012. The Net Global Effects of Alternative U.S. Biofuel Mandates: Fossil Fuel Displacement, Indirect Land Use Change, and the Role of Agricultural Productivity Growth — Nicholas Institute. Nicholas Institute for Environmental Policy Solutions, Durham, NC. NI R 12-01. Available at: http://nicholasinstitute.duke.edu/climate/policydesign/net-global-effects-of-alternative-u.s.-biofuel-mandates.

Mukherjee, S., and D. Chakraborty. 2012. Editors' introduction: the Indian growth story: towards a sustainable development? Pp. 1–18 in S. Mukherjee and D. Chakraborty, eds. Environmental scenario in India: successes and predicaments: Routledge Studies in Ecological Economics. UK; Routledge, Abingdon, Oxon.

Neumann, K., P. H. Verburg, E. Stehfest, and C. Müller. 2010. The yield gap of global grain production: a spatial analysis. Agric. Syst. 103:316–326.

OECD. 2012. OECD Environmental Outlook to 2050. OECD publishing. Available at: http://dx.doi.org/10.1787/9789264122246-en.

Pelletier, N., R. Pirog, and R. Rasmussen. 2010. Comparative life cycle environmental impacts of three beef production strategies in the Upper Midwestern United States. Agric. Syst. 103:380–389.

Piesse, J., and C. Thirtle. 2010. Agricultural R&D, technology and productivity. Philos. Trans. Royal Soc. B: Biol. Sci. 365:3035–3047.

Pink, B. 2012. 2012 Year book Australia. Australian Bureau of Statistics (ABS), Canberra.

Pletcher, J. 2000. The politics of liberalizing Zambia's Maize Markets. World Dev. 28:129–142.

Powell, J. P., and M. Rutten. 2013. Convergence of European wheat yields. Renew. Sustain. Energy Rev. 28:53–70.

PSD Online [Internet]. USDA Foreign Agricultural Service [accessed 2013 September]. Available at: http://www.fas.usda.gov/psdonline/psdQuery.aspx.

Quick Stats [Internet]. USDA National Agricultural Statistics Service [accessed 2013 September]. Available at: http://quickstats.nass.usda.gov/?source_desc=CENSUS.

Ray, D. K., N. D. Mueller, P. C. West, and J. A. Foley. 2013. Yield trends are insufficient to double global crop production by 2050. PLoS ONE 8(6):e66428.

Robinson, S., van Meijl H., D. Willenbockel, H. Valin, S. Fujimori, T. Masui, et al. 2014. Comparing supply-side specifications in models of global agriculture and the food system. Agricult. Econ. 45:21–35.

Rosegrant, M. W., C. Ringler, S. Msangi, T. B. Sulser, T. Zhu, and S. A. Cline. 2008. International Model for Policy Analysis of Agricultural Commodities and Trade (IMPACT): Model Description. International Food Policy Research Institute, Washington, D.C. Available at: http://www.ifpri.org/sites/default/files/publications/impactwater.pdf

Searchinger, T., R. Heimlich, R. A. Houghton, F. Dong, A. Elobeid, J. Fabiosa, et al. 2008. Use of U.S. Croplands for biofuels increases greenhouse gases through emissions from land-use change. Science 319:1238–1240.

Searchinger, T., C. Hanson, J. Ranganathan, B. Lipinski, R. Waite, and R. Winterbottom, et al. 2013. Creating a Sustainable Food Future. A menu of solutions to sustainably feed more than 9 billion people by 2050, World Resources Report 2013–14: Interim Findings. World Resources Institute (WRI), Washington, DC.

Seré, C., and H. Steinfeld. 1996. World Livestock Production Systems: Current status, issues and trends. Food and Agriculture Organization of the United Nations, Rome. FAO Animal Production And Health Paper 127. Available at: http://www.fao.org/docrep/004/W0027E/W0027E00.HTM.

Shah, A. 2012. Agriculture and environment in India Policy implications in the context of North-South trade. Pp. 219–242 in S. Mukherjee and D. Chakraborty, eds. Environmental scenario in India: successes and predicaments. Routledge, Abingdon, Oxon, UK.

Silvis, H., and C. van Rijswick. 2002. Agricultural policies and trade liberalisation. Pp. 11–37. in F. Brouwer and D. E. Ervin, eds. Public concerns, environmental standards and agricultural trade. CAB International, Wallingford, Oxfordshire, UK.

Slade, R., R. Saunders, R. Gross, and A. Bauen. 2011. Energy from biomass: the size of the global resource. Imperial College Centre for Energy Policy and Technology and UK Energy Research Centre, London, UK.

Smeets, E. M. W., A. P. C. Faaij, I. M. Lewandowski, and W. C. Turkenburg. 2007. A bottom-up assessment and review of global bio-energy potentials to 2050. Prog. Energy Combust. Sci. 33:56–106.

Smith, P., P. J. Gregory, D. van Vuuren, M. Obersteiner, P. Havlík, M. Rounsevell, et al. 2010. Competition for land. Philos. Trans. Royal Soc. B: Biol. Sci. 365:2941–2957.

Stattman, S. L., O. Hospes, and A. P. J. Mol. 2013. Governing biofuels in Brazil: a comparison of ethanol and biodiesel policies. Energy Pol. 61:22–30.

Steinfeld, H., P. Gerber, T. Wassenaar, V. Castel, M. Rosales, and C. de Haan. 2006. Livestock's long shadow: environmental issues and options. FAO, Rome, Italy.

Stringer, R., and K. Anderson. 2002. Australia. Pp. 181–214 in F. Brouwer and D. E. Ervin, eds. Public concerns, environmental standards and agricultural trade. CAB International, Wallingford, Oxfordshire, UK.

Tauger, M. B. 2011. Agriculture in World History, 1st edn. Routledge, New York, NY.

Tilman, D., C. Balzer, J. Hill, and B. L. Befort. 2011. Global food demand and the sustainable intensification of agriculture. Proc. Natl Acad. Sci. USA 108:20260–20264.

Trewing, D.. 2004. Chapter 14 – Agriculture. Pp. 417–456. in D. Trewing, ed. 2004 Year book Australia, Volume 86. Australian Bureau of Statistics (ABS), Canberra.

Veeck, G., C. W. Pannell, C. J. Smith, and Y. Huang. 2011. China's Geography: Globalization and the Dynamics of Political, Economic, and Social Change, 2nd ed.. Rowman & Littlefield Publishers, Plymouth, UK.

van de Ven, G. W. J., N. de Ridder, H. van Keulen, and M. K. van Ittersum. 2003. Concepts in production ecology for analysis and design of animal and plant–animal production systems. Agric. Syst. 76:507–525.

van Vuuren, D. P., J. van Vliet, and E. Stehfest. 2009. Future bio-energy potential under various natural constraints. Energy Pol. 37:4220–4230.

Whitlow, R. 1985. Conflicts in land use in Zimbabwe: political, economic and environmental perspectives. Land Use Policy 2:309–322.

Whitlow, R. 1988. Soil erosion and conservation policy in Zimbabwe: past, present and future. Land Use Policy 5:419–433.

Wicke, B., F. van der Hilst, V. Daioglou, M. Banse, T. Beringer, S. Gerssen-Gondelach, et al. 2014. Model collaboration for the improved assessment of biomass supply, demand, and impacts. GCB Bioenergy.

Wint, G. R. W., and T. P. Robinson. 2007. Gridded livestock of the world 2007. FAO, Rome, Italy. Available at: http://www.fao.org/docrep/010/a1259e/a1259e00.htm.

Wirsenius, S., C. Azar, and G. Berndes. 2010. How much land is needed for global food production under scenarios of dietary changes and livestock productivity increases in 2030? Agric. Syst. 103:621–638.

de Wit, M., M. Londo, and A. Faaij. 2011. Productivity developments in European agriculture: relations to and opportunities for biomass production. Renew. Sustain. Energy Rev. 15:2397–2412.

de Wit, M. P., J. P. Lesschen, M. H. M. Londo, and A. P. C. Faaij. 2014. Greenhouse gas mitigation effects of integrating biomass production into European agriculture. Biofuels, Bioprod. Biorefin. 8:374–390.

Appendix 1

Historical Developments in Global Agricultural Sector

In the first half of the 20th century, global agriculture had been affected by weather crises (droughts, floods, famines, plant, and animal diseases), the great depression in 1930, and two world wars (Tauger 2011; Hawkes et al. 2012). Later, after a decolonization process during the 1950s–1960s, the agricultural sector in former colonies was also underperforming (Hawkes et al. 2012). At the same time, agriculture had to fulfill an important role in supporting the industrialization process by providing cheap food to urban work force (Hawkes and Murphy 2010; Hawkes et al. 2012). As a result, during the 1940s–1970s, governments aimed to increase their production and continue the modernization and mechanization of agricultural systems which had started in the 19th century (Tauger 2011). To support and protect agricultural production and prices, states adopted agricultural protection policies. Since the 1950s, many developed countries imposed strong state controls on agriculture and trade through instruments like import tariffs, export subsidies and producer support (Hawkes and Murphy 2010; Smith et al. 2010; de Wit et al. 2011). These policy instruments guaranteed a minimum return to farmers and compensated for price differences between the internal and global market (Hawkes and Murphy 2010; de Wit et al. 2011). Developing and industrializing countries heavily taxed agricultural exports and protected producers from import competition (Hawkes et al. 2012). Rapid economic growth facilitated investments in agricultural R&D, which led to breeding of high-yielding crop varieties, increasing application of fertilizers and irrigation, and mechanical innovations (Piesse and Thirtle 2010; Tauger 2011; de Wit et al. 2011). This is referred to as the Green Revolution. By the 1970s, the Green revolution and state intervention resulted in overproduction in developed countries. Some less developed countries like China and Brazil were also able to increase agricultural production at a higher rate than population growth. But other developing countries (especially in Africa) lagged behind and suffered from underproduction (Hawkes et al. 2012).

In the last 25 years, both developed and developing countries have begun to reform their agricultural policies, which has resulted in growing international trade of agricultural products.

Trade liberalization started with the first General Agreement on Tariffs and Trade (GATT) in 1947. Although the trade of agricultural products was subject of international negotiations, it was typically excluded from multilateral trade agreements until 1990 (Hawkes and Murphy 2010; Hawkes et al. 2012). In 1995, the Uruguay Round Agreement on Agriculture (AoA) came into effect and imposed measures on signatory countries to open their agricultural markets (Hawkes and Murphy 2010; Hawkes et al. 2012). Another issue that remained unresolved in negotiations was the question how policies for sustainable development and environmental protection could be aligned with and integrated in trade regulations. Although environmental stewardship is a global issue, the interests in multilateral negotiations are diverse. While developed countries emphasize the need for an environmental reform of trade regulations, developing countries are mainly concerned with questions related to market access, dumping, and agricultural subsidies (Chambers 2002). As will be shown in the next sections, this diversity in interests is also reflected in the varying degree of adoption and enforcement of agri-environmental policies in the selected countries.

Appendix 2

Australia

Agricultural characteristics

Australia is a dry continent. Its climatic zones range from a tropical Northern region, through an arid interior, to a temperate Southern region (Pink 2012). The wet northern summer conditions allow beef cattle grazing and sugarcane production (east coast). The drier southern summer conditions favor wheat production, and grazing of sheep and dairy and beef cattle (Pink 2012). Rice is mainly grown in the Southeast of Australia (Pink 2012). About 10% of the agricultural area is cultivated (cropping and sown pastures and grasses) (Pink 2012; FAOSTAT). The remaining area consists of permanent pastures and meadows for livestock grazing (FAOSTAT). The majority of the farms are engaged in either livestock farming or grain growing (Pink 2012).

The management levels are low by OECD standards. Until the early 1980s, about 3.5% of the cultivated land (arable land) was equipped for irrigation. From the late 1980s, this share increased to about 5.5% in recent years (FAOSTAT). Irrigation is mainly applied for vegetables and fruits, rice, and also sugarcane (Stringer and Anderson 2002; Pink 2012). Because low rainfall limits the returns on fertilizer expenditure, fertilizer use is relatively low (Stringer and Anderson 2002), see Figure 2. The highest rates of fertilizer and pesticide are applied in horticulture (fruits and vegetables) (Stringer and Anderson 2002).

The average ruminant density on pastures is very low (see Fig. A10 in Appendix 9). Beef cattle are mainly held in Northern Australia, where production is extensive and the technology level low. In the South, production is more intensive. This is illustrated by higher stocking rates per hectare, improved pastures, and the use of fodder crops and animal health products (Pink 2012).

Dairy production mainly takes place in the south eastern high-rainfall coastal areas and is based on year round pasture grazing. Feedlot-based dairying is expanding, but is still uncommon (Pink 2012). Between 1980 and 2000, farmers switched to another dairy cattle breed (Doyle and Stockdale 2011).

Economic and institutional developments

Compared to other OECD countries, Australia has been more protective to industry for most of the 20th century. The country did not participate in the General Agreement on Tariffs and Trade (GATT) between 1947–1979. In the 1950s and 1960s, Australia's policies were focused on industrial protectionism, characterized by price support, and trade protection (e.g., import restrictions on manufacturing products). These policies isolated farmers from national and international market signals (Stringer and Anderson 2002) and resulted in indirect disincentives for agriculture (Anderson et al. 2009). The subsidies and protection provided to the agricultural sector were limited and could not offset these disincentives (Anderson et al. 2009).

From the 1970s, Australia has been reforming its trade policies (Anderson et al. 2009). The past two decades have been a period of especially rapid total factor productivity (TFP) growth (Anderson et al. 2009). One important factor explaining the increase is the openness of the Australian economy to trade and investment (Fischer et al. 2014).

In 2002, Stringer and Anderson (2002) mentions that "four-fifths of the Australian agricultural production is exported". The main export markets are the United States and Asia (Stringer and Anderson 2002).

Until the 1990s, agricultural policies were led by socio-economic objectives. Since the 1990s, more emphasis has been put on sustainable agricultural development (Stringer and Anderson 2002). Current agricultural policies aim to improve market responsiveness, and encourage sustainable agricultural practice (i.e., approaches that combine economic, environmental, and social aspects). Projects also focus on food quality. Agri-environmental policies concentrate on water, soil erosion, salinity, and biodiversity loss (Stringer and Anderson 2002). Measures to achieve this include research, education, voluntary adoption of best practice, and development of guidelines in collaborations between governments, NGOs, industries, and communities. Over time, regulatory approaches have been complemented, or even substituted, with market oriented mechanisms like for example the polluter pays principle. The basis of these measures is that prices reflect social costs and benefits, that is, positive and negative externalities are taken into account (Stringer and Anderson 2002).

Yield developments

In Australia, a small average change in wheat and rice yields between 1961 and 1980 was followed by a significant increase in the 1980s (Fig. 3 and Table 1). As this coincides with the participation in the General Agreement on Tariffs and Trade (GATT) after 1979, it is likely that the improvements are a result of the trade policy reforms which opened the international market (Anderson et al. 2009). Notably, the input use levels remained stable. Thus, it seems that the reforms motivated farmers to use their resources more efficiently, which resulted in higher production levels and improved productivity in the early 1980s (Fig. A1). Agri-environmental policies were first implemented in the 1990s and aimed to improve market responsiveness and encourage sustainable agricultural practice (Stringer and Anderson 2002). Since their introduction, agricultural production has further increased. But, fertilizer use and irrigation levels increased as well and the productivity of the agricultural sector did not improve compared to the 1980s (Fig. A1). Also, wheat and rice yields stagnated in the 1990s. Sugarcane yields have been relatively steady and did not significantly improve in the 1980s. A considerable drop in yield in the period 1990–92 was followed by a peak in the second half of the 1990s and another plunge in 2001–02 due to drought. Then, sugarcane yields seem to have stabilized again around previous levels.

Beef and milk yields have almost continuously improved, except for a period of stagnation in the 1970s and early 1980s (Fig. A9 and Table 1). In this period of market reforms, also the production and export of beef and milk temporarily declined (FAOSTAT). Between 1980 and 2000, beef and milk yields improved while their export markets grew, feed crop consumption increased enormously (Fig. A1), and a shift in dairy cattle breed from British breeds to Holstein–Friesian animals was made (Trewing 2004; Doyle and Stockdale 2011). Although the majority of cattle production is still extensive, the rise in feed crop consumption is likely to be related to the intensification of beef production in South Australia and the expansion of feedlot-based dairying (Pink 2012). The above findings suggest that the development of export markets has been an important driver for changes in the production systems

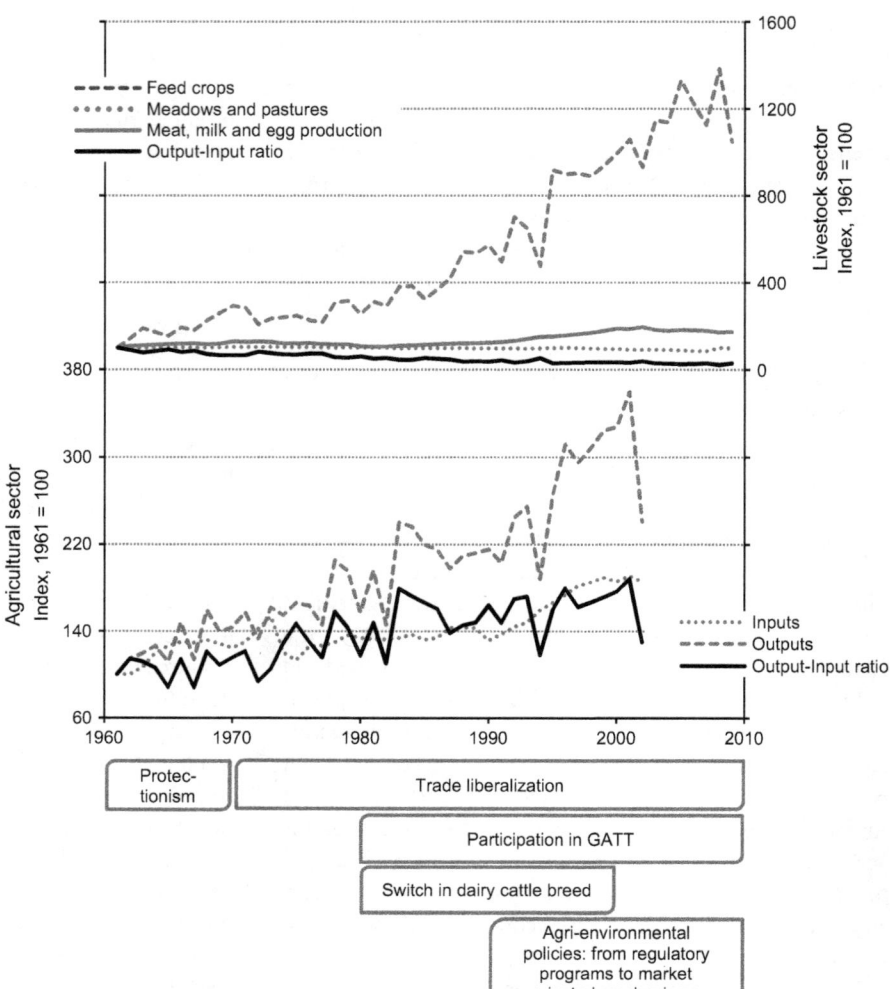

Figure A1. Australian agricultural and livestock productivity developments and institutional, economic and technological/management developments. (GATT, general agreement on tariffs and trade).

of beef and milk, and all these factors together have contributed to yield improvements. Due to the massive increase in feed crop use, however, the output–input ratio of the livestock sector has decreased significantly. The decline has slowed down since the mid-1990s, which may be related to the introduction of agri-environmental policies.

Appendix 3

Brazil

Agricultural characteristics

Agriculture in Brazil is characterized by concentrated land ownership; medium- and large-scale commercial farms contribute the bulk of agricultural output (Anderson and Valdés 2009). Crop and livestock are combined in mixed

farming systems (Dixon et al. 2001). The majority of cattle is held in *extensive mixed* (rain-fed) production systems (Wint and Robinson 2007). These are found in the wooded and open savannah areas (the Cerrados) in the Central West of Brazil, and also in the Southeast (Dixon et al. 2001; Cederberg et al. 2009). Extensive ranching is the primary activity, but cultivation of soya and corn is increasing. In addition to this farming type, *intensive mixed farming* takes place in Eastern and Central Brazil. This system produces most of the sugarcane, which is mainly cultivated in the Central South of Brazil. *Dryland mixed farming* (which is mainly semisubsistence farming) is the major system in northeastern Brazil (Dixon et al. 2001).

A significant expansion of the agricultural sector in the last decades was accompanied by a considerable increase in deforestation, replacement of native vegetation, and biodiversity loss (Martinelli et al. 2010; Ferreira et al. 2012). Also, the use of fertilizers and other inputs, and the ruminant density on meadows and pastures have risen significantly since the 1970s (Figs 2 and A10). The ruminant density in Brazil is high compared to the other selected countries, except to India. The largest share of the cattle population is being held for beef production; about 10% are dairy cows (Cederberg et al. 2009; FAOSTAT). Beef and milk production are mainly based on extensive systems, in which cattle grazes on pastures all year round (Cederberg et al. 2009). In the emerging semiextensive systems, herds also receive supplemental feed from crops and various concentrates. Feedlot-based, intensive systems are still rather uncommon (Cederberg et al. 2009).

Economic and institutional developments

Prior to 1950, the Brazilian market was concentrated around the export of food and raw materials, and the import of industrial products (Baer 1972). Between 1950 and the mid-1970s, the focus shifted to national industrialization, and policies aimed at replacing foreign imports with domestic production (Carvalho 1991). This is called Import Substitution Industrialization (ISI) (Carmo Oliveira 1986). To protect the industry, wage rates were kept low by restrained food prices (Carvalho 1991). In order to realize low food prices, Brazilian agriculture was heavily and increasingly taxed (Carmo Oliveira 1986). Levies consisted partly of direct export taxes, but were dominated by indirect taxation resulting from industrial protection policies (Anderson and Valdés 2009). Due to the industrial protection, also input prices increased. Therefore, the government provided credit and fertilizer subsidy to promote the use of fertilizer and other inputs (Carvalho 1991).

Trade liberalization started in the 1980s, and continued to the mid-1990's (Anderson and Valdés 2009). Reforms included the removal of import and export restrictions, and the redistribution of resources from import-competing to export-oriented sectors (Anderson and Valdés 2009). This transformation took also place in the agricultural sector (Anderson and Valdés 2009). Today, major exports of agricultural products like soybean, sugar, beef, and ethanol contribute to Brazil's positive balance of trade. Thus, the agricultural sector plays an important role in the economic development of Brazil (Martinelli et al. 2010; Ferreira et al. 2012).

During the last years of industrial protectionism, the oil crisis in 1973 prompted the Brazilian government to phase out fossil fuels. The ProÁlcool program was launched to promote the sugarcane industry and bio-ethanol production (Stattman et al. 2013). Blending ethanol to fossil fuel was already introduced in 1931, but the ProÁlcool program brought about a major increase in ethanol consumption and production (Stattman et al. 2013). Today, most of the ethanol produced is still intended for the domestic market. In 2009, almost 14% of the production was exported (Lamers et al. 2011). To improve Brazil's energy diversity and independence, a second biofuel program was implemented in 2004: the National Program of Production and Use of Biodiesel (PNPB). Due to the abundance of soy and the search for new soy markets, biodiesel production is largely based on the conversion of soybean oil (Stattman et al. 2013).

The need to control deforestation was already recognized in the 1920s. The first Forestry Code, which dates from 1934, regulated the conservation of forests on private land (Banerjee et al. 2009). In 1965, a second code expanded the land dedicated to preservation from forests to other sensitive areas. Also, it created conservation areas outside the private rural properties (Banerjee et al. 2009). Due to economic priorities, however, enforcement of the codes was weak. Between 1974 and 1987, when the focus shifted from protectionism to trade liberalization, the government promoted livestock production, forestry, and mining in the Brazilian Amazon (Banerjee et al. 2009). The markets for Brazilian beef have been growing since the 1970s and led to considerable expansion of extensive cattle ranching on cleared forest land in this region (Bowman et al. 2012). In addition, the more recent expansion of soy production on previous pastures causes further expansion of cattle ranching into the Amazon (Bowman et al. 2012). Policies and other initiatives that aim to intensify cattle production are in early stages yet (Bowman et al. 2012).

Yield developments

In Brazil, the developments in the yield of corn and soybeans are comparable. From the 1960s to 1980s, yield growth rates were moderate and highest in the 1970s (Fig. 3 and Table 1). In the 1990s, high improvement rates of more than 3.5% year^{-1} were attained (relative to 1991). Fertilizer use increased substantially in the late 1960s and the 1970s, declined and stagnated in the 1980s and increased again in the 1990s (Fig. 2). Thus, it appears that fertilizer subsidies provided during industrial protection (1950s–70s) (Carvalho 1991) led to an increase of fertilizer use and of yields, especially in the 1970s. The economic reforms in the 1980s (Anderson and Valdés 2009) temporarily hindered agricultural development, but the opening of agricultural export markets stimulated further improvements in the 1990s. Notably, after agricultural productivity declined in the 1960s and 1970s, the output-input ratio has remained fairly constant from the 1980s until 2002 (Fig. A2). Despite the introduction of the biodiesel program in 2004 (Stattman et al. 2013), soybean yields stagnated in this decennium. Yet, the share of soybeans used for biodiesel production has been rather small in the first years (0.5% in 2006 compared to 12% in 2010) (Barros

2013; FAOSTAT). The development of sugarcane yields is similar to the trend for corn, but the first period of major growth is found between 1975 and 1985. This is clearly related to the introduction of the PróAlcool program in the early 1970s (Stattman et al. 2013). Tractor use and irrigation also grew significantly after 1973, thus are probably related to the rise of sugarcane production.

Considering beef and cow milk production, no significant yield improvements were attained from 1961 until the 1980s (Fig. A9 and Table 1). This period of relatively stable yields was interrupted by a few years of decline in the mid-1970s. Major yield increases were only attained in the early 1990s. This suggests that the promotion and expansion of cattle ranching during the period of liberalization (Banerjee et al. 2009; Bowman et al. 2012) did not directly stimulate yield improvements. Only when the export markets were fully opened in the early 1990s, yields significantly improved. For milk, this growth continued

in the late 1990s and 2000s, but at a lower rate. The initial increase in beef yields was first followed by a decline in the late 1990s, before yields increased again in the 2000s. As the production of beef and milk has grown significantly faster than the use of feed crops and the area of pasture land, the output–input ratio of the livestock sector has continuously increased over the past five decades, especially since the mid-1990s (Fig. A2).

Appendix 4

China

Agricultural characteristics

Agriculture in China is characterized by large environmental diversity, and large diversity in agricultural products.

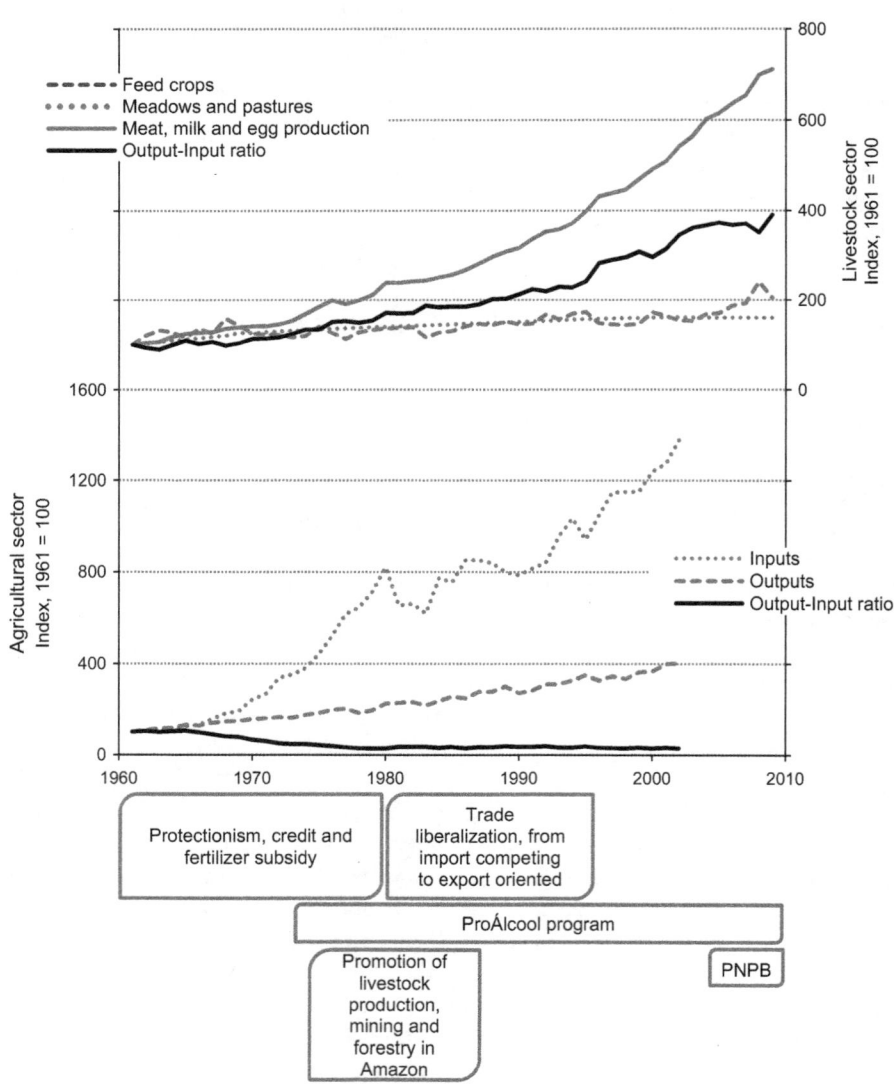

Figure A2. Brazilian agricultural and livestock productivity developments and institutional, economic and technological/management developments (PNPB, National Program of Production and Use of Biodiesel).

The Qinling mountains divide China into water-deficit (North, West) and water-surplus regions (South, Northeast) (Veeck et al. 2011). There are four major farming systems (Dixon et al. 2001). *Lowland rice production* is found in humid and moist sub humid areas in South and Central East China. Rice production is rain-fed, with supplementary irrigation where available. Important livelihoods are, besides rice, subsidiary crops like corn and soybeans, livestock, and off-farm work. *Upland Intensive Mixed Farming* is found in upland and hill areas with humid and subhumid climate (South East and North China). A significant area, mainly rice, is irrigated. Livestock contributes to draught power, meat, income and savings. Also off-farm work is an important income source. *Temperate mixed farming* is found in moist and dry subhumid areas in Central Northern China. The major crops are wheat and corn. Livestock is also an important livelihood.

Pastoral farming is located in semiarid and arid temperate climates in Western China. Pastoralism is based on extensive grazing of mixed herds (camels, cattle, sheep and goats) on native pasture. In local suitable areas, farmers apply irrigated crop production (e.g., cotton, barley, wheat). Characteristic for the agricultural sector of China are the majority of small-scale farms and the application of multiple cropping systems, that is, the production of more than one crop per year on the same land (Veeck et al. 2011).

China aims to be largely self-sufficient in grain production. Because of its large population, land availability for agricultural production is an important issue (Veeck et al. 2011). Economic growth has caused a major increase in population and demand for housing, transport, and industry in the Eastern coastal area. Much land in this region, however, is fertile and highly productive agricultural land which is lost due to the urbanization process (Karplus and Deng, 2008a; Veeck et al. 2011). Due to this pressure, even marginal lands (e.g., with very limited precipitation or extreme slope) are cultivated (Veeck et al. 2011).

Without irrigation, the dry areas are of marginal use for intensive agriculture. Water shortages hamper the improvement of agricultural production in these regions. Therefore, irrigation has expanded at a high rate (Fig. 2). This, however, is causing severe water shortages as water consumption outpaces replacement through precipitation; there are major concerns that groundwater reserves are being depleted, especially in arid areas in Eastern and Western China (Karplus and Deng, 2008a; Veeck et al. 2011). Fertilizer use has increased dramatically in the past five decades (Fig. 2). Fertilizer use is especially high in lowland rice and in temperate mixed farming systems (Dixon et al. 2001). The overuse of fertilizers is associated with land degradation, air pollution, and eutrophication of water sources (Veeck et al. 2011).

Economic and institutional developments

The period from 1949–1976 in China is called the Maoist era. During this era, the Chinese Communist Party (CCP) was in power. Governance was characterized by a strong inward orientation and self-imposed isolation (Veeck et al. 2011). Also, policies were relatively homogenous for the nation as a whole, and the use of capital and resources was regulated centrally (Veeck et al. 2011). In this era, farming took place in large farming communes (Tauger 2011). These communes were difficult to manage. In 1959, this caused a collapse of farm production and a huge famine. In order to solve the problems, agricultural reforms were introduced which included the reparation and construction of irrigation systems and the distribution of high-yield seeds (Karplus and Deng, 2008a; Tauger 2011). Programs to develop improved crop varieties had already started in the early 1950s. The communes were broken up when the market reforms started in 1978. Since then, individual farmers have been leasing land from the local authorities. The resulting diversification of crop production and farmers' activities, income, and education level became visible in the second half of the 1990s (Veeck et al. 2011).

Since the end of the Maoist era, the CCP has still been in charge. The central government continues to play an important role in planning and guiding the direction of development (e.g., economic decision making), but the role for local governments is increasing (Veeck et al. 2011). Also, from 1978, the focus of the market shifted from import substitution industrialization toward export-oriented development strategies (Anderson and Martin 2009). This resulted in an export-led industrial growth, and also a restructuring of the economy away from agriculture and heavy industry toward light manufacturing and service activities (Anderson and Martin 2009; Veeck et al. 2011). The taxation of agricultural exports has been reduced, but the protection of import-competing agriculture, especially of rice, has been increased (Anderson and Martin 2009). Because of the importance of the agricultural sector, the government increased investments in agricultural R&D and started to fund research in biotechnology. This support was continued in the following decades (Karplus and Deng, 2008b).

Environmental protection laws were first introduced in the late 1980s. These laws aimed to prevent the loss of high-productivity cropland caused by the expansion of urban and industrial areas in Eastern China (Veeck et al. 2011). Also, the increased awareness about environmental problems and the need for more efficient agriculture led to the implementation of the Comprehensive Agricultural Development (CAD) program in 1988 (Veeck et al. 2011). The CAD program was introduced because of the low productivity of a large share of arable land and increasing grain imports. The program aimed to enhance the quality

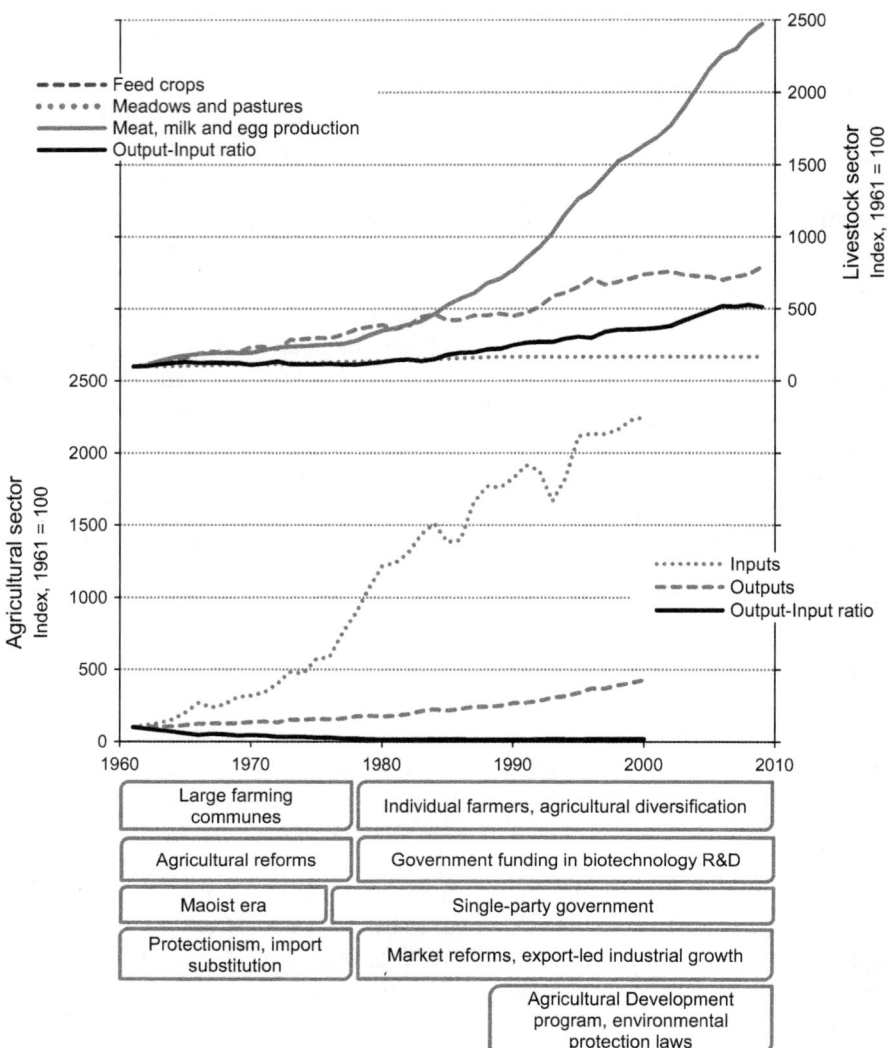

Figure A3. Productivity developments in the Chinese agricultural and livestock sector and institutional, economic and technological/management developments.

of agricultural land through better land management, including improved fertility and drainage, balanced use of inorganic fertilizer, irrigation and water storage and conservation. The CAD still exists (Veeck et al. 2011).

Yield developements

Since the agricultural reforms in the early 1960s (Karplus and Deng, 2008a; Tauger 2011), irrigation, fertilizer use, and mechanization have been increasing almost continuously (Fig. 2). In addition, fast yield growth is found for all crops (corn, rice and wheat), Table 1. On average, the average yield growth rate was the highest for wheat (1.8% year^{-1} relative to 2010), but the highest absolute improvement was achieved by corn and rice (93 kg ha^{-1} year^{-2}). These considerable gains can mainly be attributed to the introduction of new technology, which was realized by significant public investments in

infrastructure and research (Karplus and Deng, 2008a). In the 1990s, however, yield improvements dropped. This may reflect the diversification of farmers' crop production after the economic reforms (Veeck et al. 2011). In the following decade, the growth rates of corn and wheat yields increased again, while rice yield improvements continued to slow down.

The yield growth rates for beef and milk have been lower than for crops. This may be explained by the major importance of cattle for delivering draught power, as agricultural production is still labor-intensive in China (Fig. 2). The market reforms may have resulted in some improvements in cattle production; the beef and milk yield shortly increased around the 1980s. Afterwards, beef yields stabilized again. Milk yields also stagnated for some years, but have been increasing at a rate of 3.5% year^{-1} since the late 1990s (relative to 2001, Fig. A9 and Table 1). At the same time, the consumption of milk and dairy products by urban residents soared, which was caused by China's growing prosperity (Bao 2011).

After the output-input ratio of the agricultural sector had declined rapidly in the 1960s and 1970s, it stabilized in the 1980s and improved gradually in the 1990s (Fig. A3). It is likely that both market liberalization and agri-environmental policies, which aim to improve agricultural land quality (Veeck et al. 2011), have contributed to this reversal of the downward trend in productivity.

Appendix 5

India

Agricultural characteristics

India has two major farming systems: rice-wheat and rain-fed mixed. The rice-wheat system in Northern India is characterized by wetland rice production in summer (monsoon season), and irrigated wheat production in winter (cool, dry season). A significant amount of livestock is held in this system, where bovines produce draft power, milk and manure for composting (Dixon et al. 2001).

The rain-fed mixed system occupies the largest area in India (Central and Southern India). It is mainly rain dependent, but according to Dixon et al. (2001), about 16 percent of the area cultivated under this system was equipped with simple, small-scale irrigation techniques around 2000. Infrastructure and market access are poor, and agricultural activities are oriented toward subsistence. The main livelihoods are cereals, legumes, fodder crops, livestock and off-farm activities (Dixon et al. 2001).

The input-intensity of the agricultural sector has increased substantially since 1961 (Fig. 2). Yet, as the increase in input use outpaced the growth in total production, the output-input ratio has declined seriously (Fig. A4). This has caused major environmental issues. Large-scale irrigation in Northern India has inflicted soil salinization and groundwater depletion (Shah 2012). Also, groundwater is polluted due to intensive use of fertilizers and rudimentary processing of livestock wastes. In addition, large livestock populations cause soil degradation through the conversion of natural vegetation (Shah 2012). In Southern India, soil erosion is the main problem. The vegetative cover and organic matter content of soils are low. Yet, farmers continue to cultivate crops on marginal lands to meet their basic needs (Shah 2012).

Economic and institutional developments

India is a former colony of the United Kingdom and gained independence in 1947. In order to prevent famines,

and to ensure affordable prices for basic foods, the Indian government has been intervening in the food market since its independence in 1947. In the public distribution system, which was established in 1958 and is still present, basic foods are sold at subsidized prices (Gulati and Pursell 2009).

In response to droughts and famines in 1965–66, policies aimed at food grain self-sufficiency and agricultural imports began to be replaced by domestic production (Gulati and Pursell 2009; Tauger 2011). Green revolution technologies played an important role, as the government implemented many programs to modernize agriculture at a high speed. This included the development and planting of high-yielding wheat and rice varieties and large subsidies for electricity and fertilizers (Gulati and Pursell 2009; Tauger 2011). According to Dixon et al. (2001), however, agricultural development during India's Green Revolution did mainly take place in 10 percent of India's districts which had adequate local infrastructure for water management, transport and electricity (for tubewells). In the 1970s and early 1980s, the import of edible oils expanded significantly. This led to policies which aimed to decline these imports and substitute them with domestically produced oils (Gulati and Pursell 2009). Import substitution was abandoned in the 1990s and the focus shifted to an export oriented economy. Trade policies were reformed through the structural adjustment program (SAP), which was introduced 1991 (Mukherjee and Chakraborty 2012).

A significant number of environmental policies exist that aim to control the use and management of natural resources. Enforcement of these regulations, however, is weak (Shah 2012).

Yield developments

In India, crop yields have almost continuously grown in the last five decades, but at different rates. The highest rates are found for wheat and rice in the period 1961–1990 (Table 1). Explanations for these achievements can be found in the major investments by the government in the modernization of agriculture (in Northern India), the development and adoption of high-yielding rice and wheat varieties and agricultural subsidies (Dixon et al. 2001; Gulati and Pursell 2009; Tauger 2011). After 1990, the absolute and relative yield growth of these crops has decreased. Soybean yields started to increase in 1972. This coincides with the introduction of the import substitution policy for edible oils (Gulati and Pursell 2009). Since 1961, sugarcane yields have increased at a moderate rate until 1999. After a decline in 2000–2004, yields returned to the level of the mid-1990s. For all crops, there is no clear relation between market reforms and yield developments in the 1990s. Although the linear regression data suggest that yield growth of rice, wheat and soybeans slowed down in this decade, the graphs in Figure 3 do not show a significant deviation from an earlier trend which can be linked to the liberalization process.

Regarding cattle product yields, there is a significant difference in developments between milk and beef. Cow milk yields have increased all five decades, and growth rates increased as well. Similar to China, this is likely to be related to increased milk consumption (FAOSTAT). Beef yields are low compared to the other six countries. This is likely to be related to the protected status of cows in Hinduism, the major religion in India. Beef yields have been rather constant and only increased in the 1970s and 1980s. This temporary improvement may be explained by reduced need for draft power due to the mechanization process (Fig. 2). Mechanization, however, has continued in 1990s, but this is not reflected in further improvement of beef yields.

Due to the enormous increase in inputs, the agricultural productivity has continuously decreased between 1961 and 2000 (Fig. A4). Although the decline has slowed down, the lack of productivity improvements confirms the weak

enforcement of agri-environmental policies in India (Shah 2012).

Appendix 6

USA

Agricultural characteristics

In the United States, agricultural production is mainly concentrated in the Pacific and Central (Midwestern) regions and the Southern plains (Alston et al., 2010a). Over the past decades, the total number of farms in the United States has declined but the number of large scale farms has increased. Still, large-scale farms are a minority of all US farms, but they produce more than two-third of agricultural output (Anderson and Valdés 2009; Alston et al., 2010a). The production practices depend on the farm size and the natural resource base (e.g., soil moisture and fertility). For example, the major practice in the corn belt (Midwestern USA) is dryland farming. In the Central

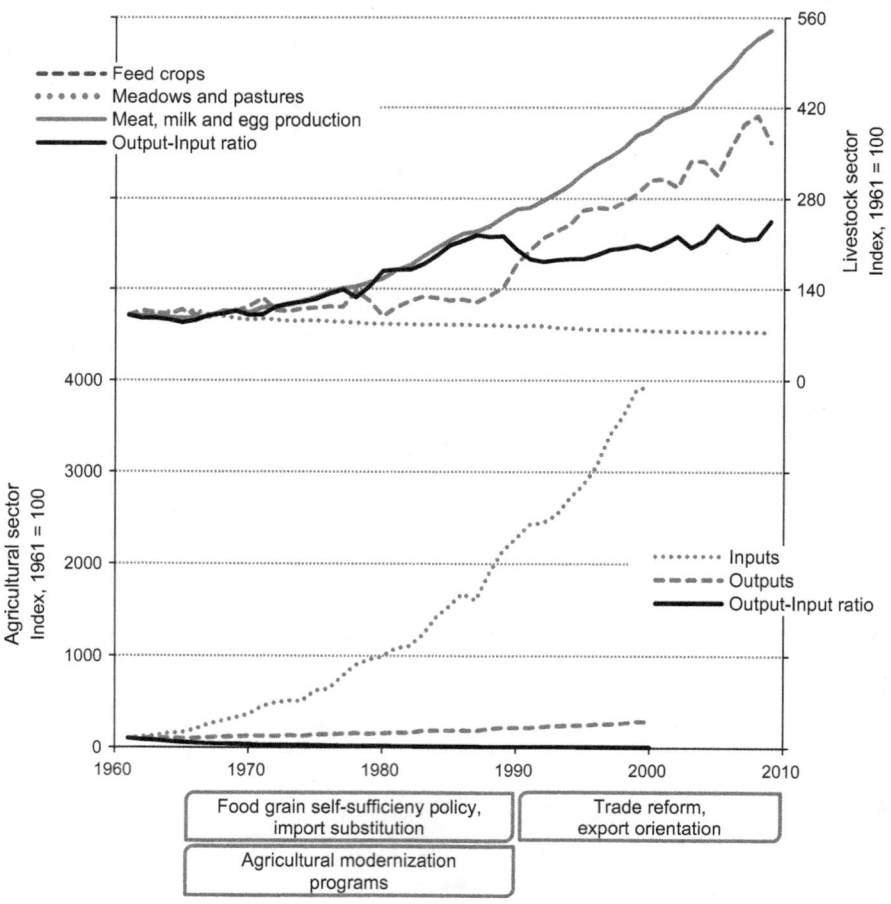

Figure A4. Productivity developments in the Indian agricultural and livestock sector and institutional, economic and technological/management developments.

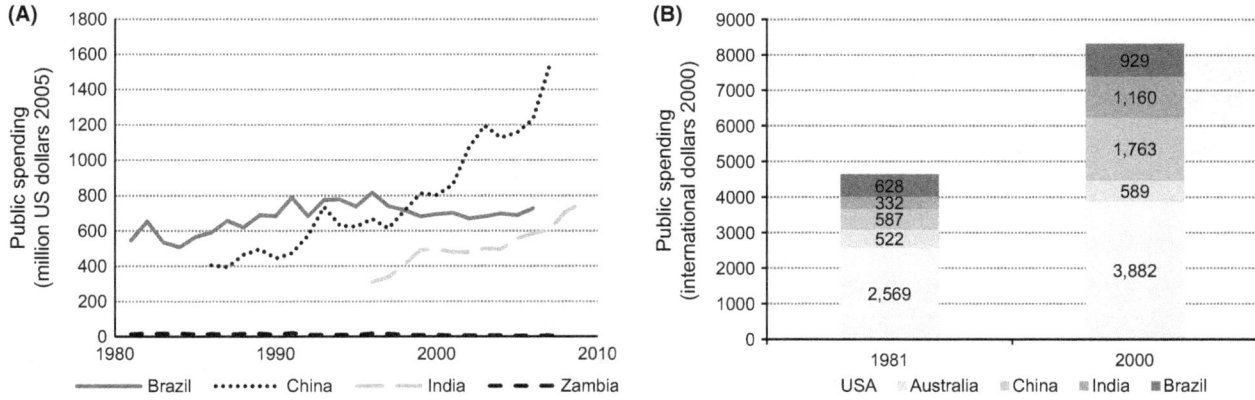

Figure A5. Public agricultural R&D spending: (A) Historical progress in spending in developing countries (ASTI Data Tool version 1.1); (B) Spending in OECD and developing countries (Alston et al., 2010b).

valley of California (Eastern USA), irrigation is applied (Carpentier and Ervin 2002).

The agricultural production of the last five decades can be characterized by intensive management, especially with regard to mechanization (Fig. 2). It seems, however, that the intensity of input uses has stabilized since the 1970s and 1980s. Also, the ruminant density on pastures has declined (Fig. A10). In the livestock sector, dairy production is mainly confinement based (McCormick 2011). Pasture use, however, has grown since the early 1990s (McCormick 2011). In pasture based production systems, dairy cows may be at pasture during parts of the year. At the same time, and in winter months, they receive stored forage along with varied levels of supplemental concentrates throughout the year (McCormick 2011). Beef production is mainly characterized by cow-calf herds on pasture and (winter) hay (Pelletier et al. 2010). Beef cattle are finished in feedlots where they receive a mixed, high concentrate feed ration. Less than 1% of beef cattle are finished in pastures (Pelletier et al. 2010). On pastures, no housing is provided for cow-calf herds. Hormone implants are employed in the feedlot stage. Calves can also be sent to feedlots directly. Pelletier et al. (2010) mentions that this is common practice in the US Upper Midwest.

The high intensity of agricultural management in the United States has led to a wide range of environmental issues. The most important problems are soil erosion (i.e., loss of the fertility and water-holding capacity of the soil) and contamination of water sources by agricultural chemicals and livestock manure (Carpentier and Ervin 2002).

Economic and institutional developments

From their introduction in the 1930s, agricultural policies in the United States have been differing significantly in composition from the EU and other OECD countries. The focus of agricultural policies has been on providing food aid and nutrition assistance. Assistance to farmers in the form of commodity support programs is placed second (Silvis and van Rijswick 2002). These commodity programs consisted of price support and direct income payments (Silvis and van Rijswick 2002; Blandford and Boisvert 2006). To limit payments by the government, crop programs, such as corn and wheat, placed limits on production. For other commodities, like milk, import restrictions were applied (Blandford and Boisvert 2006). Market liberalization and multilateral trade agreements have changed the programs for farm support. Since 1985, income support has shifted to payments that are decoupled from prices and production. Also, production limitations have been replaced by more planting flexibility, enabling farmers to make market-based decisions (Blandford and Boisvert 2006).

The United States have been dominant in agricultural R&D expenditures (Alston et al., 2010b). For example, Figure A5 shows that American public spending in 2000 was twice as high compared to the investments made by China. In addition, agricultural R&D in the United States has been funded extensively by the private sector. The public and private sector contribute both about half of the total investments (Alston et al., 2010b). In other countries, especially the developing countries, the share of private spending has been much smaller (Fig. A6). In addition, innovations in agricultural technology have been stimulated by intellectual property rights. Until the 1970s, however, this protection excluded inventions related to living organisms like plants and animals (Alston et al., 2010c). This changed in the 1980s and the legalization of patents on life forms cleared the way for biotechnology to rapidly expand (Alston et al., 2010c).

Agri-environmental policies in the US have been implemented from about 1970 (Carpentier and Ervin 2002). Traditionally, broad programs were implemented in each state. More recently, individual state and local programs

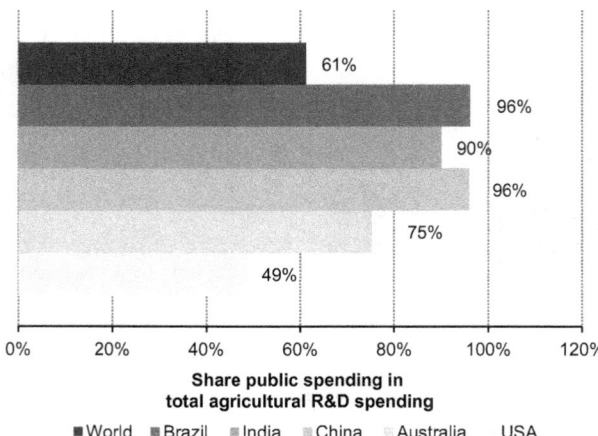

Figure A6. Public share in total agricultural R&D spending in 2000 (Alston et al., 2010b).

have emerged (Carpentier and Ervin 2002). The majority of agri-environmental policies have been voluntary-payment programs (Carpentier and Ervin 2002). The application of regulatory approaches which define input or performance standards for agriculture was limited by two factors. First, agriculture in the United States is characterized by a high variety in production practices and local circumstances (ecosystems, quality and sensitivity of resources). Regulation of this diversity in operations is technically difficult and expensive. Second, a strong agricultural lobby exists in the United States, and political influence on environmental protection has been modest (Carpentier and Ervin 2002). As a result, regulatory programs and environmental standards for agriculture have only been introduced since the second half of the 1990's. These regulatory programs focus on the quantity and quality of production inputs (especially water quality in livestock production, and pesticide use) (Carpentier and Ervin 2002). The most important environmental policies in the United States are the Conservation Reserve Program (CRP, farm bill 1985) and the Environmental Quality Incentives Program (EQIP, farm bill 1996). The CRP is a voluntary program that provides payments to farmers who apply conservation practices on environmentally sensitive lands (Silvis and van Rijswick 2002). The EQIP provides financial and technical assistance to farmers to improve and protect the environmental quality of their properties (e.g., soil and water) (Silvis and van Rijswick 2002).

Yield developments

In the United States, the crop yield growth trends have been positive for most of the period 1961–2010 (Fig. 3). It is very likely that the substantial investments in agricultural R&D (Alston et al., 2010b) have played an important

role in achieving these improvements. From the mid-1980s, growth of corn and wheat yields slowed down for about a decade. Probably, this deceleration is related to the reforms of trade and farmer support policies in the same period (Blandford and Boisvert 2006). Afterwards, however, absolute growth reached a record high in the 1990s (Table 1). For soybeans, growth accelerated in the 2000s. In addition to the effect of trade liberalization, these significant increases in yield growth may be attributed to the rise of biotechnology since the 1980s (Alston et al., 2010c). Improvements in technology and management have also driven yield growth in beef and cow milk production. This technological progress is likely to be stimulated by investments in R&D and growing domestic milk consumption (Alston et al., 2010b; FAOSTAT). Although absolute and relative yield growth of beef was highest in the 1960s, yields have also been increasing considerably since the mid-1980s after a period of stagnation in the 1970s. Cow milk yields have almost continuously increased and while the relative growth rate has been fairly constant, the absolute growth accelerated in the 1980s and 1990s.

Regarding agricultural management, tractor use peaked in 1966 and fertilizer use reached the highest level in 1980 (Fig. 2). Also, the output–input ratios of the agricultural and livestock have been improving since the 1970s (Fig. A7). As agri-environmental programs were introduced in the same decade, the developments in input use and agricultural productivity may well be related to these policies.

Appendix 7

Zambia

Agricultural characteristics

The agricultural sector in Zambia exists of a small number of large-scale commercial farmers who have good access to input and output markets, a few medium-scale commercial farmers for whom market access is difficult, and a majority of smallholders who are often engaged in subsistence farming (Howard and Mungoma 1996; Bonaglia 2009). The large-scale farmers produce and sell wheat, soybean, coffee, milk and other livestock products. Corn, however, dominates the agricultural sector and is mainly produced by the smallholders and medium-scale commercial farmers (Howard and Mungoma 1996).

The major farming systems in Zambia are maize mixed (Central and East Zambia) and cereal-root crop mixed (West Zambia) (Dixon et al. 2001). Maize mixed systems are found in plateau and highland areas with a dry subhumid to moist subhumid climate. Besides corn, principal livelihoods are tobacco, cotton, cattle, goats, poultry, and off-farm work.

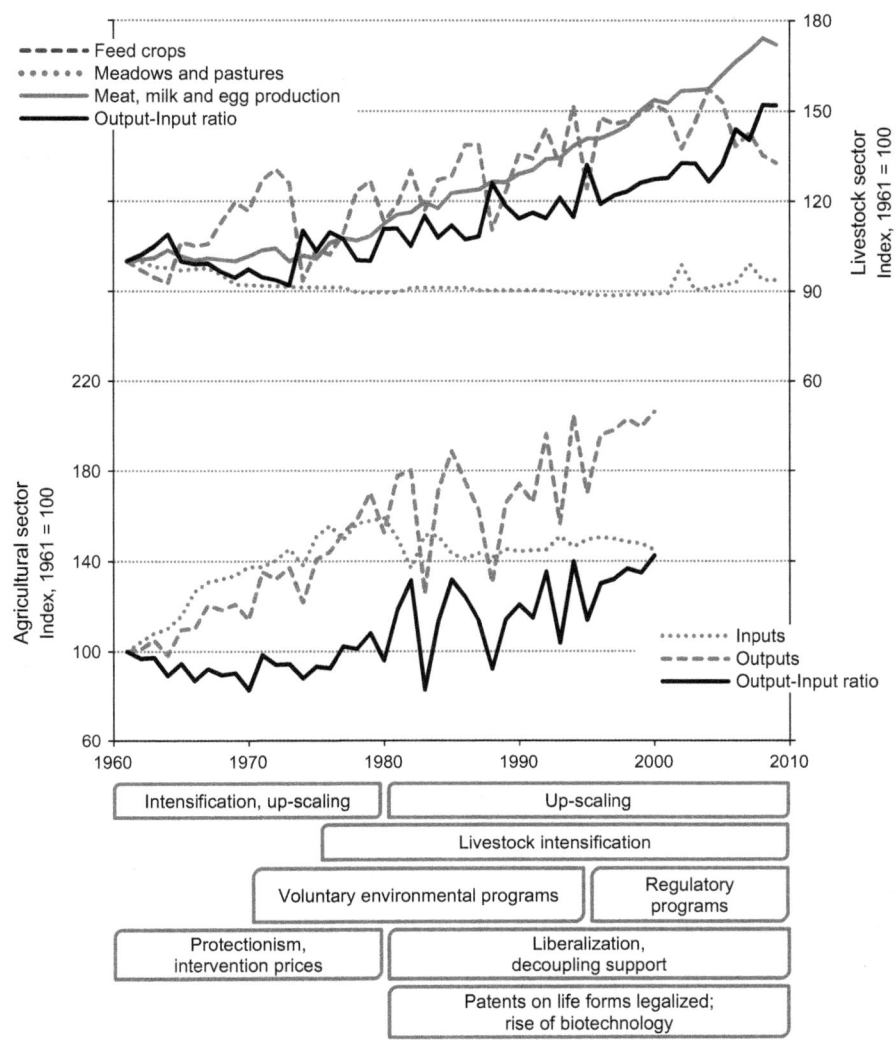

Figure A7. Productivity developments in the American agricultural and livestock sector and institutional, economic and technological/management developments.

Cattle are kept for ploughing, milk, manure, but also for savings (Dixon et al. 2001). Cereal-root crop mixed systems are situated in regions of lower altitude and higher temperatures. The number of livestock per household is higher compared to the maize mixed system. The major sources of income are corn, sorghum, millet, cassava, yams, legumes, and cattle (Dixon et al. 2001).

In Zambia and other South African countries, the major environmental problem related to agriculture is declining soil fertility (Dixon et al. 2001). Soil degradation is caused by inappropriate management practices such as continuous cropping and overgrazing (Dixon et al. 2001). Average agricultural input levels in Zambia are low (Fig. 2).

Economic and institutional developments

Zambia is a former colony of Britain (the region was named Northern Rhodesia) and gained independence in 1964. After independence, the economy heavily depended on copper exports and many people lived in the urban mining areas. To ensure food supply to these areas, the new government aimed to increase national corn production. In the colonial period, however, commercial corn production had mainly relied on large-scale European farmers. The new objective was to enhance the participation of smallholders in the commercial corn market (Howard and Mungoma 1996). The agricultural intervention system of price controls and subsidies, which also dated from the colonial period, was maintained. But, new pricing policies favored smallholders in remote areas over commercial farmers with good market access (Howard and Mungoma 1996; Pletcher 2000). In the early 1970s, agricultural policies were expanded with fertilizer subsidies, which mainly benefitted the corn sector (Howard and Mungoma 1996; Pletcher 2000). Both corn seed and fertilizer were made accessible to smallholder farmers in remote areas through a network of cooperative depots. In addition, farmers could sell their corn to these depots (Howard and Mungoma 1996). In the meantime, the Zambian government had invested in a corn breeding program which resulted in the release of twelve new varieties between 1977 and 1994. The program was started in the early 1960s to reduce the import of crop varieties from Zimbabwe, on which European farmers had relied during the colonial period (Howard and Mungoma 1996).

Between 1973 and 1991, Zambia had been governed by single party rule. This period coincided with an economic crisis in the late 1970s and 1980s due to a collapse of the copper price in 1975 and poorly managed governmental interventions in the market (Pletcher 2000). Although attempts were made to reform (agricultural) policies in the 1980s, economic liberalization only started when a new government came to power in 1991 (Pletcher 2000; Jayne et al. 2003). Through liberalization, the corn market was fully privatized. But, intervention in the input markets for fertilizer and credit remained (Pletcher 2000). Fertilizer price subsidies had been eliminated in 1988, which resulted in high input costs for corn. Therefore, smallholders reduced their use of fertilizer and hybrid corn varieties and returned to the cultivation of traditional corn varieties and subsistence crops like sorghum (Howard and Mungoma 1996; Dixon et al. 2001). The government then decided to continue fertilizer distribution on loan, but this undermined the ability of the private market to distribute fertilizer commercially (Jayne et al. 2003). Also, underinvestment in infrastructure and other public goods had made the purchase of fertilizer unprofitable to many farmers (Jayne et al. 2003). In response to the reduced fertilizer use (Fig. 2) and corn production in the 1990s, a new policy for fertilizer distribution and subsidy (the Fertilizer Support Program) was implemented in 2002 (Djurfeldt et al. 2011).

Efforts to control soil erosion started in the mid-1980s, driven by the spreading problem of land degradation and the economic reforms in late 1980s and early 1990s. At first, commercial farmers adopted conservation farming technologies to improve the profitability of mechanized corn production (Haggblade and Tembo 2003). In 1995, appropriate technologies for smallholders were introduced as well. The development and promotion of the technologies was collectively conducted by farmer organizations, private companies, NGOs and the government (Haggblade and Tembo 2003).

Yield developments

After corn yields declined and fertilizer use increased slowly in the 1960s, the introduction of fertilizer subsidies (Howard and Mungoma 1996; Pletcher 2000) caused these levels to increase significantly in the 1970s (Fig. 2, Table 1). The fact that agricultural policies were mainly focused on corn production (Howard and Mungoma 1996; Pletcher 2000) is clearly reflected in the high yield improvement rate of 4.5% year^{-1} for corn compared to 1.2 % year^{-1} for soybeans and 0.1 % year^{-1} for sugarcane between 1971 and 1990 (relative to 1971, 1973 for soy). After the elimination of fertilizer subsidy in 1988 due to economic liberalization (Howard and Mungoma 1996; Pletcher 2000), fertilizer use and corn yields declined in the 1990s. It appears that commercial farmers benefitted from the economic reforms, as irrigation levels increased considerably and soybean yields improved at a very high rate of almost 20 % year^{-1} in the 1990s[3] (relative to 1991). Sugarcane yields increased at 0.7 % year^{-1} in the same decennium. In addition, Figure A8 shows that the output-input ratio of Zambia's agricultural sector improved in the late 1980s and 1990s. Besides reduced fertilizer use, these advances in productivity may also be the result of the introduction of conservation farming technologies (Haggblade and Tembo 2003). After the adoption of the Fertilizer Support Program in 2002 (Djurfeldt et al. 2011), however, fertilizer use increased again, the area equipped for irrigation stabilized (Fig. 2) and overall agricultural productivity dropped. Corn yields rose again and sugarcane yields stabilized. The effect on soybean yields is unclear; after a steep decline in 2001 due to drought, yields recovered and returned to levels comparable to the 1990s.

The FAO data show constant milk yields from 1961 until 2010. Beef yields have also been relatively stable, except for a decline of 1.8% year^{-1} between 1968 and 1980 (relative to 1970). An explanation may be that the shift in focus of agricultural policies towards smallholders in this period has affected commercial farmers. This theory can, however, not be confirmed by statistics.

Appendix 8

Zimbabwe

Agricultural characteristics

Until 2000, the agricultural sector of Zimbabwe consisted of two major farming systems, which both occupied half of the arable land (Eicher 1995). The commercial farming system was dominated by a relatively small group of European farmers. These large scale farms were located in the higher rainfall areas in North-Eastern Zimbabwe (Whitlow 1985; Matondi, 2012c). Production was mainly focused on crops and input intensive (Whitlow 1985). The smallholder farming system involved a large number of African farmers. These small-scale farms were located

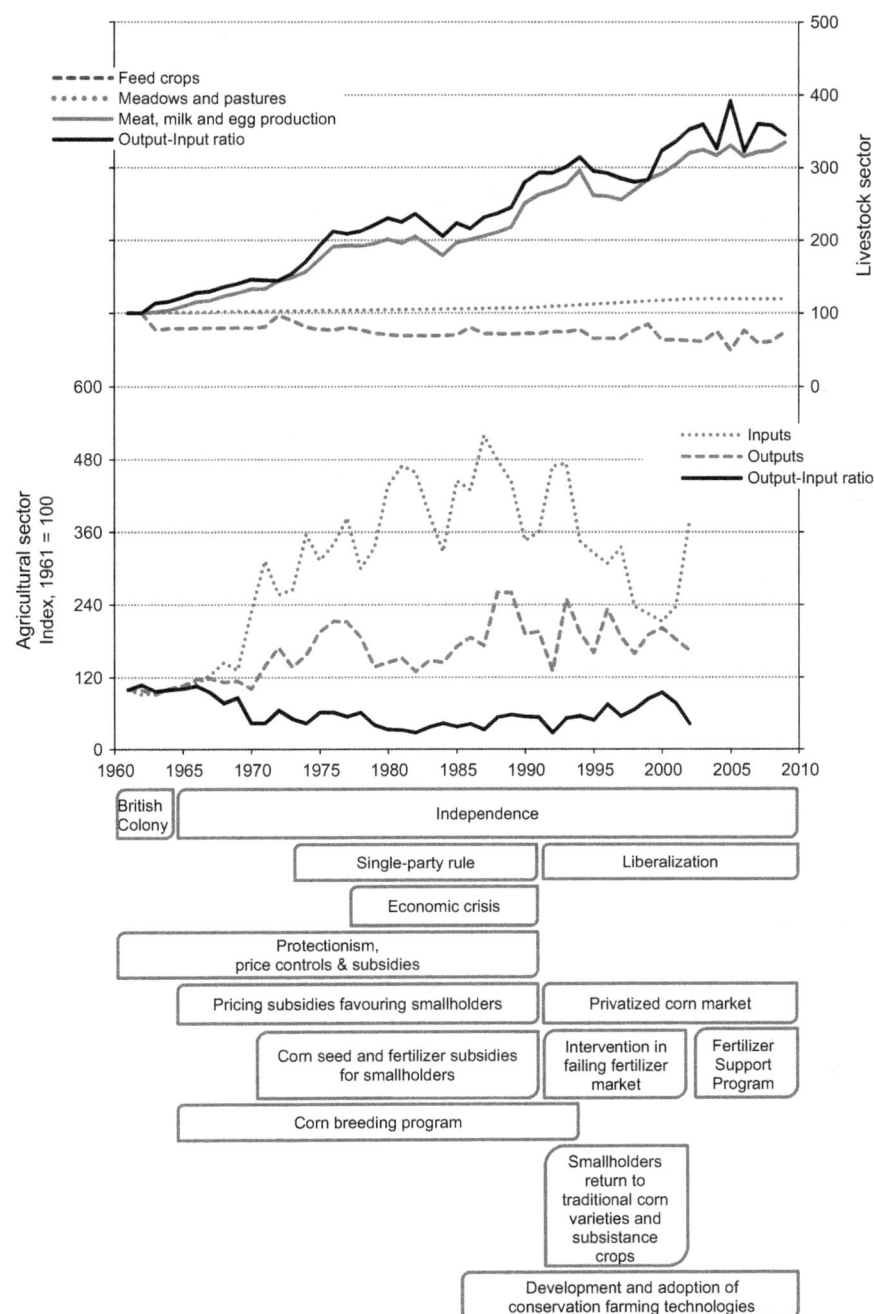

Figure A8. Productivity developments in the Zambian agricultural and livestock sector and institutional, economic and technological/management developments. Because of limited data, agricultural tractors are not included in the inputs and in the output-input ratio for the agricultural sector.

in the drier and more remote areas with poor market access. Farming included crop and livestock production and was mainly subsistence driven (Whitlow 1985).

From 2000, the Zimbabwean government acquired land from the European commercial farmers on a large scale. The land was divided into smallholder farms and commercial farms of varying scales, and redistributed to black farmers (Matondi, 2012a). Because of the limited knowledge and skills of the new farmers and poor access to inputs

and new technologies, the national level of irrigation and fertilizer use declined after 2000, see Figure 2 (Matondi, 2012b). Also, a loss of knowledge about livestock management led to more disease related deaths (Matondi, 2012b). As a result, the number of ruminants and the ruminant density on pastures declined (Fig. A10).

At the start of the 21st century, the major farming system was maize mixed (Dixon et al. 2001), see the description for Zambia. As more than half of the land is not suitable for crop production without irrigation, cattle production plays an important role in Zimbabwe. The relatively high human and livestock populations and densities on marginal suitable lands, however, has resulted in large-scale soil degradation in smallholder farming areas (Whitlow 1988, 1985).

Economic and institutional developments

Together with Zambia, Zimbabwe is a former colony of Britain (the region was named Southern Rhodesia). But, while Zambia was directly administered by the British during its colonization period, Zimbabwe was a self-governing colony. In 1965, The Zimbabwean government declared independence unilaterally, which was only recognized internationally in 1980. During the period of unilateral independence, the United Nations (UN) imposed sanctions on exports (Eicher 1995). To face these embargos, the government adopted a policy of import substitution (Marquette 1997). Policies aimed at agricultural diversification and commercial production of export oriented tobacco was replaced by cultivation of previously imported crops like corn, wheat and soybeans (Whitlow 1988; Eicher 1995). The period between 1965 and 1980 was also accompanied by a civil war, which was partly concerned with the uneven distribution of land between commercial farmers and smallholders (Whitlow 1985). Intensification of this guerilla in the late 1970s led to the abandonment of commercial farms in more remote areas and occupation by peasants (Whitlow 1988).

After independence in 1980, the government aimed to support the development of smallholders. A new land reform policy allowed the sale of commercial farmland on a 'willing buyer, willing seller' basis (Eicher 1995). In addition, smallholders were enabled to obtain credit to purchase seed and fertilizer and to make use of subsidized marketing services. This led to a rapid adoption of hybrid corn varieties (Eicher 1995). According to Eicher (1995), this successful smallholder green revolution in the first half of the 1980s could be realized because of good political, institutional, technological and economic conditions. An important factor to success was the investment in research, education and farmer support in previous decennia, which had already led to a green revolution by white commercial farmers in the 1960s (Eicher 1995; Langyintuo and Setimela 2009). Government financed research on high yielding crops in the 1970s and 1980s led to the release of more than 30 new hybrid corn varieties by 1990 (Langyintuo and Setimela 2009; Tauger 2011).

The success of the smallholder support system, however, resulted in high expenses for subsidies. In the late 1980s and early 1990s, the government lowered subsidies and encouraged farmers to diversify crop production (Eicher 1995). These reductions in public spending were part of an economic structural adjustment program (ESAP), which also included other measures to liberalize the economy (Marquette 1997). In addition, Zimbabwe's public R&D system slowly started to deteriorate; many European agricultural experts left Zimbabwe in the years after independence, while the shifted focus of agricultural research programs from commercial farmers to smallholders required experienced researchers (Eicher 1995).

A series of events in the 1990s led to hyper-inflation and a collapse of the economy in the 2000s (Matondi, 2012b). The ESAP had seriously affected Zimbabwe's economy and the fast-track land reform program in 2000 disrupted commercial agricultural production. Also, beef exports to the EU were suspended because of foot-and-mouth disease (Marquette 1997; Matondi, 2012b). Due to this hyper-inflation, farmers' incomes dropped dramatically and inputs became unaffordable to many farmers (Langyintuo and Setimela 2009). While Zimbabwe was once called the bread-basket of South-Africa, now international support programs are needed to improve food security among impoverished rural households (Langyintuo and Setimela 2009). The program initiated by the British government in 2003 also aimed at promoting conservation farming practices (Langyintuo and Setimela 2009; Marongwe et al. 2012). The first agri-environmental policies, however, were already introduced in the early 20th century. The government provided significant financial support to apply conservation farming practices on commercial farms, which was very successful in the 1960s and 1970s (Whitlow 1988). Policies to address soil degradation in peasant farming had some success in the 1960s, but ceased in 1970s due to the political situation, increasing human and livestock populations, and a lack of (financial) support (Whitlow 1988). In the 1980s and 1990s, several research activities on conservation agriculture were initiated, but did not lead to significant uptake of conservation practices by smallholders (Marongwe et al. 2012).

Appendix 9

Additional figures and tables

See Figures A9–A12 and Tables A1 and A2.

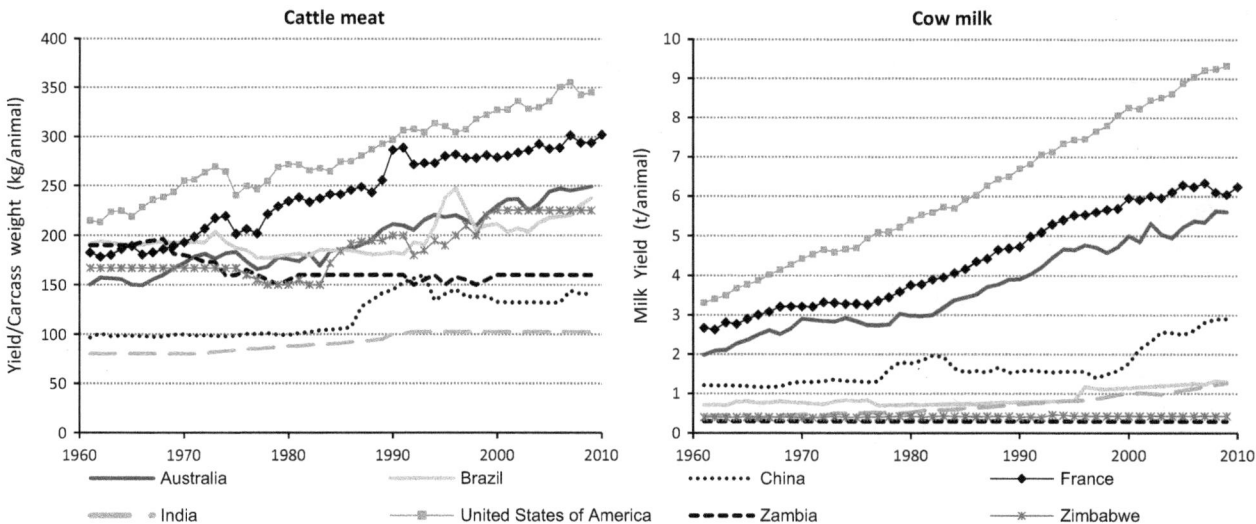

Figure A9. Historical yield developments (1961–2010) for the production of cattle meat and cow milk (FAOSTAT).

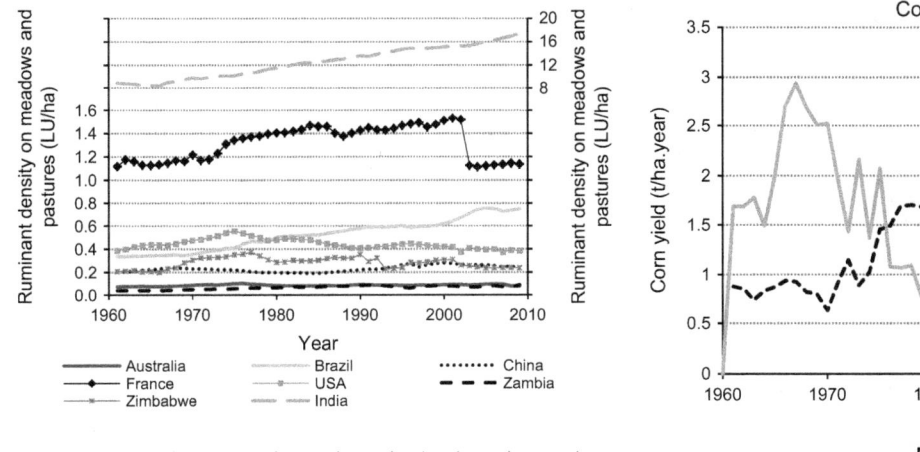

Figure A10. Development in livestock production intensity: ruminant density on pastures and meadows in livestock units per hectare (LU ha⁻¹), derived from FAOSTAT data (FAOSTAT). Ruminants included: buffaloes, camel, cattle, goats, and sheep. Note the different scale for India compared to the other countries.

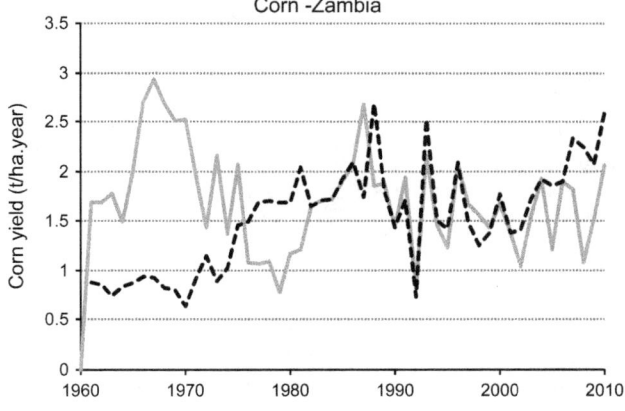

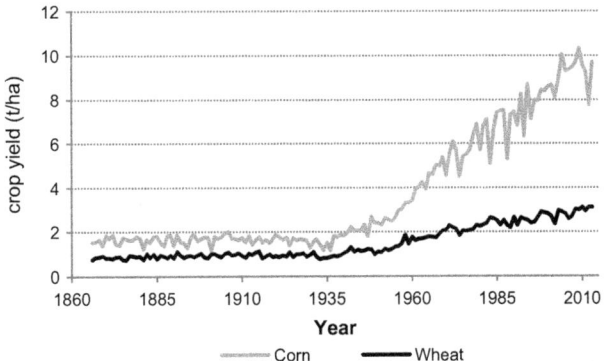

Figure A11. Long term historical yield trends for corn and wheat in the USA (Quick Stats).

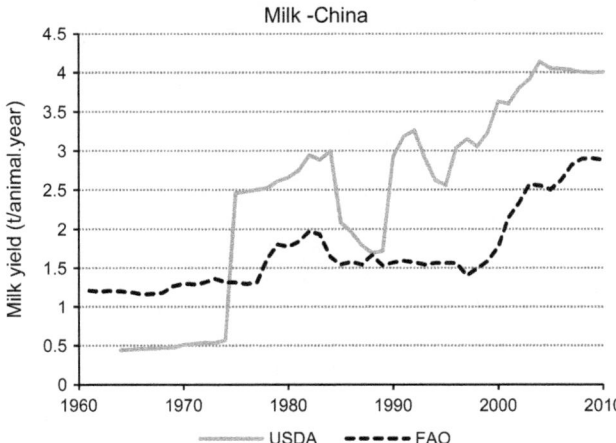

Figure A12. Examples of diverging statistical yield data between FAO (FAOSTAT) and USDA (PSD Online).

Table A1. Comparison of average annual crop yield growth rates (% year[-1]) derived from (a) extrapolation of linear regression (2010–2050); (b) IMAGE projections (2010–2050); (c) MIRAGE projections (2008–2020); (d) projections from Jaggard et al. (2007–2050).

		Australia	Brazil	China	India	Zambia	Zimbabwe	USA	World
Wheat									
Linear extrapolation[1]		0.7%		1.8%	1.5%			0.8%	1.3%
IMAGE projection[1,4]		1.2%[6]		2.0%	1.4%			0.5%	1.2%
MIRAGE projection[2]	bau			1.4%				0.8%	1.0%
Yield projections in Jaggard et al.[3]	bau	1.1%		2.4%	2.0%			1.2%	
	min	0.8%		1.7%	1.4%			0.9%	
	max	1.9%		4.6%	3.9%			2.2%	
Corn									
Linear extrapolation[1]			1.6%	1.6%		1.2%	−1.6%	1.2%	1.3%
IMAGE projection[1]			2.4%	1.8%		1.6%[7]		1.2%	0.9%
MIRAGE projection[2]	bau		4.9%	1.8%		1.6%[8]		1.2%	1.5%
Yield projections in Jaggard et al.[3]	bau		1.7%	1.9%				1.3%	
	min		1.2%	1.3%				0.8%	
	max		3.5%	3.8%				2.7%	
Rice									
Linear extrapolation[1]		0.7%		1.3%	1.3%			1.0%	1.2%
IMAGE projection[1]		0.5%[6]		0.4%	1.8%			1.1%	1.1%
MIRAGE projection[2]	bau			1.6%				0.9%	2.2%
Yield projections in Jaggard et al.[3]	bau	0.9%		1.5%	1.4%			1.2%	
	min	0.6%		1.0%	0.9%			0.9%	
	max	1.7%		3.1%	3.0%			2.4%	
Soybean									
Linear extrapolation[1]			1.4%		1.2%	1.2%	0.6%	1.0%	1.1%
IMAGE projection[1,5]			1.0%		1.5%	1.7%[7]		1.4%	1.0%
MIRAGE projection[2]	bau		3.1%			2.0%[8]		1.1%	2.0%
Yield projections in Jaggard et al.[3]	bau		1.6%		1.4%			1.1%	
	min		1.1%		1.0%			0.8%	
	max		3.2%		2.9%			2.2%	
Sugarcane									
Linear extrapolation[1]		0.3%	1.0%		0.8%	−0.1%	−0.3%		0.6%
IMAGE projection[1]									
MIRAGE projection[2]	bau								
Yield projections in Jaggard et al.[3]	bau	0.4%	1.2%		1.1%				
	min	0.3%	0.9%		0.8%				
	max	0.7%	2.4%		2.1%				

[1]Relative to 2010 yields.
[2]Relative to 2008 yields.
[3]Relative to 2007 yields.
[4]In IMAGE, wheat is aggregated into the product group temperate cereals.
[5]In IMAGE, soybean is aggregated into the product group oil crops.
[6]In IMAGE, Australia is aggregated into the region Oceania.
[7]In IMAGE, Zambia and Zimbabwe are aggregated into the region Southern Africa.
[8]In MIRAGE, Zambia and Zimbabwe are aggregated into the region sub-Saharan Africa.
The IMAGE model adopts yield projections from the FAO and combines these with endogenous assumptions on yield changes (Bruinsma 2003; OECD 2012). The MIRAGE model uses a baseline scenario from Aglink-Cosimo, which is also complemented by endogenous assumptions on yield developments (Laborde 2011). Jaggard et al. (2010) assume a continuation of current yield trends, but also take into account relative changes owing to increasing carbon dioxide (CO_2) and ozone (O_3) concentrations, climate change and technological developments.

Table A2. Recent yields (FAOSTAT), maximum attainable yields and yield gaps (GAEZ Global Agri-Ecological Zones) (t ha⁻¹ year⁻¹).

		AU	BR	CN	FR	IN	ZM	ZW	US
Wheat									
Average yield 2008–2010	t ha⁻¹	1.6		4.7	7.1	2.8			3.0
Max attainable yield[1,3]	t ha⁻¹	3.8		6.4	8.2	3.9			6.1
Yield gap	t ha⁻¹	2.2		1.7	1.1	1.1			3.1
Current yield as % of max	%	42		74	87	72			50
Corn									
Average yield 2008–2010	t ha⁻¹		4.1	5.4	9.1		2.3	0.5	9.9
Max attainable yield[1,4]	t ha⁻¹		6.6	8.4	8.8		10.9	9.8	12.6
Yield gap	t ha⁻¹		2.5	2.9	−0.3		8.6	9.2	2.7
Current yield as % of max	%		62	65	103		21	6	78
Rice									
Average yield 2008–2010	t ha⁻¹	8.9		6.6		3.3			
Max attainable yield[2,5,6]	t ha⁻¹	10.7		9.5		9.2			
Yield gap	t ha⁻¹	1.9		3.0		5.9			
Current yield as % of max	%	82		69		36			
Soybean									
Average yield 2008–2010	t ha⁻¹		2.8			1.0	1.7	1.5	2.9
Max attainable yield[1,7]	t ha⁻¹		3.4			3.1	4.4	4.1	3.7
Yield gap	t ha⁻¹		0.6			2.0	2.8	2.6	0.8
Current yield as % of max	%		82			34	38	36	78
Sugarcane									
Average yield 2008–2010	t ha⁻¹	80.3	79.1			66.5	105.4	79.5	
Max attainable yield[2,5]	t ha⁻¹	135.3	106.8			123.1	144.1	142.1	
Yield gap	t ha⁻¹	55.1	27.6			56.6	38.8	62.6	
Current yield as % of max	%	59	74			54	73	56	

AU, Australia; BR, Brazil; CN, China; FR, France; IN, India; ZM, Zambia, ZW, Zimbabwe; US, United States.

[1]Maximum attainable yield in 2020 as calculated for the IPCC SRES B1 Scenario from the Australian Commonwealth Scientific and Research Organization (CSIRO) Mark 2 Model (GAEZ Global Agri-Ecological Zones).

[2]Maximum attainable yield based on the average climatic conditions for the period 1961–1990, applied in case no projection for 2020 was available (GAEZ Global Agri-Ecological Zones).

[3]Australia, France, and the United States: high input level, rain-fed conditions; China and India: high input level, irrigated conditions.

[4]All countries except France and the United States: high input level, rain fed conditions; France and the United States: high input level, irrigated conditions.

[5]High input level, irrigated conditions.

[6]Maximum agriecological attainable yield for Indica wetland rice (150 days).

[7]High input level, rain-fed conditions.

High input level: "Under a high level of input (advanced management assumption), the farming system is mainly market oriented. Commercial production is a management objective. Production is based on improved or high yielding varieties, is fully mechanized with low labor intensity and uses optimum applications of nutrients and chemical pest, disease and weed control." (GAEZ Global Agri-Ecological Zones).

10

Alternate wetting and moderate drying increases rice yield and reduces methane emission in paddy field with wheat straw residue incorporation

Guang Chu[1], Zhiqin Wang[1], Hao Zhang[1], Lijun Liu[1], Jianchang Yang[1] & Jianhua Zhang[2]

[1]Jiangsu Key Laboratory of Crop Genetics and Physiology/Co-Innovation Center for Modern Production Technology of Grain Crops, Yangzhou University, Yangzhou, Jiangsu, China
[2]School of Life Sciences and State Key Laboratory of Agrobiotechnology, The Chinese University of Hong Kong, Hong Kong, China

Keywords

Alternate wetting and drying, grain yield, methane, nitrous oxide, rice (Oryza sativa), wheat straw

Correspondence

Jianchang Yang, Jiangsu Key Laboratory of Crop Genetics and Physiology/Co-Innovation Center for Modern Production Technology of Grain Crops, Yangzhou University, Yangzhou, Jiangsu, China.
E-mail: jcyang@yzu.edu.cn

Funding Information

We are grateful for grants from the National Natural Science Foundation of China (31461143015; 31271641, 31471438, 91317307), the National Key Technology Support Program of China (2011BAD16B14; 2012BAD04B08; 2014AA10A605), the Priority Academic Program Development of Jiangsu Higher Education Institutions (PAPD), Hong Kong Research Grant Council (AoE/M-05/12) and Shenzhen Overseas Talents Innovation & Entrepreneurship Funding Scheme (The Peacock Scheme).

Abstract

Wheat residue incorporation into the rice paddy field is becoming a popular practice in rice production in China's main rice-growing area but risks an increased emission of greenhouse gases. This study investigated if an alternate wetting and moderate drying (AWMD) irrigation regime in rice production reduces CH_4 emission and increases grain yield when wheat straw residues are incorporated into rice paddy field. One super rice variety was field-grown in 2012 and 2013 and subjected to four irrigation and straw incorporation treatments: continuously flooded (CF) without straw incorporation (−S), AWMD without straw incorporation (AWMD−S), then CF with straw incorporation (CF + S) and AWMD + S. When compared with the CF, the AWMD regime increased grain yield and water use efficiency (WUE, grain yield over the amount of water used) by 2.7% and 27.6%, respectively, under −S, and by 18.0 and 50.0%, respectively under +S. The AWMD + S treatment also significantly increased nitrogen use efficiency (NUE) compared with the CF + S treatment. The increase in grain yield, WUE and NUE in the AWMD regime, especially under +S, were attributed mainly to a greater root oxidation activity, deeper root distribution and increases in productive tillers, crop growth rate and non-structural carbohydrate remobilization during grain filling. There was a total of 0.49 kg N_2O-N ha^{-1} more loss in the AWMD than in the CF regime. However, the AWMD regime substantially decreased seasonal CH_4 emissions, global warming potential (GWP, including both CH_4 and N_2O) and greenhouse gas intensity (grain yield over GWP) by 49.8%, 45.2% and 46.7%, respectively, under −S, and by 57.5, 55.9% and 62.6%, respectively, under +S, when compared with the CF regime. The results demonstrate that the AWMD is an effective practice to increase grain yield and resource-use efficiency and reduce environmental risks especially, when wheat straw is incorporated into paddy field.

Introduction

Global agriculture in the 21st century faces the tremendous challenge of providing enough food for a growing population under increasing scarcity of water resources, while minimizing environmental consequences (Bouman 2007; Linquist et al. 2015). Rice (Oryza sativa L.) is one of the most important food crops in the world and consumed by more than 3 billion people (Fageria 2007). It is estimated that, by the year 2025, it will be necessary to produce about 60% more rice than what is currently produced to meet the food needs (Fageria 2007; GRiSP 2013). About 75% of total rice production comes from irrigated lowlands (Maclean et al. 2002; GRiSP 2013).

Irrigated rice accounts for about 80% of the total fresh water resources used for irrigation in Asia (Bouman and Tuong 2001). Fresh water for irrigation, however, is becoming increasingly scarce because of population growth, increasing urban and industrial development, and the decreasing availability resulting from pollution and resource depletion (Belder et al. 2005; Bouman 2007). On the other hand, rice fields have been identified as an important source of atmospheric methane (CH_4), one of the major potent greenhouse gases (GHG), and contribute approximately 15–20% of the global total anthropogenic CH_4 emission (Aulakh et al. 2001; Yan et al. 2005). Several recent experiments have shown that nitrous oxide (N_2O), another potent GHG, could be emitted from rice fields, which may be attributed to the combined effect of nitrogen (N) fertilization and water management (Zou et al. 2007; Shan and Yan 2013; Li et al. 2014). It would have great significance to develop technologies and practices for reducing GHG emissions and water use while increasing grain yield in rice.

China is one of the largest rice producing countries, accounting for 18.6% of the world rice harvested area and 30% of total world production of rice grain (GRiSP 2013). The rice-wheat rotations, with the acreage of 13 million hectares, are a major cropping system in the Yangtze River Valley in this country (Ma et al. 2009). Crop production inevitably results in large amounts of straw residues. Farmers usually burn wheat residues particularly when they want to establish rice crop rapidly while labor is limited. This leads to loss of most organic carbon and large losses (up to 80%) of N (Raison 1979), 25% of phosphorus (P), and 21% of potassium (K) (Ponnamperuma 1984) as well as significant air pollution and death of beneficial soil fauna and microorganisms. Therefore, incorporating crop residues into the field has currently been highly recommended in China as a measure to promote organic matter recycling and environmental friendly, sustainable agricultural production (Yao et al. 2013). However, this measure undoubtedly provides the readily available carbon and N substrate, inducing greater CH_4 release from rice paddy fields and also influencing N_2O emissions (Ma et al. 2009; Wang et al. 2010; Shan and Yan 2013; Yao et al. 2013; Li et al. 2014). Ways must be sought to reduce CH_4 emissions when the crop residues are incorporated into rice fields.

To counter water shortage and increase water use efficiency (WUE), alternate wetting and drying (AWD) irrigation in rice has been developed as a novel water-saving technique and adopted in many countries such as China, Bangladesh, India, and Vietnam (Bouman and Tuong 2001; Belder et al. 2004; Yang et al. 2007; Zhang et al. 2008; Yao et al. 2012; Liu et al. 2013). This technique could substantially reduce irrigation water by introducing alternation of periods of soil submergence with periods of nonsubmergence during the growing season (Belder et al. 2004). It is reported that AWD practices could reduce both water use and GHG emissions without seriously sacrificing grain yield in rice systems (Linquist et al. 2015). Our earlier work (Yang et al. 2007; Zhang et al. 2009, 2010, 2012; Liu et al. 2013) has shown that an alternate wetting and moderate drying (AWMD) regime could not only save water but also maintain or even increase rice yield. However, little is known whether an AWMD regime could substantially reduce GHG emissions meanwhile markedly increase grain yield and WUE under the condition of wheat straw incorporation into paddy fields.

The objective of this study was to test the hypothesis that an AWMD regime may decrease global warming potentials (GWP) through reducing CH_4 emissions and increase grain yield through enhancing shoot and root growth when wheat straw is incorporated into paddy fields. Both CH_4 and N_2O emissions, WUE and N use efficiency (NUE) were determined. Tiller number, root and shoot biomasses, root oxidation activity (ROA), leaf photosynthetic rate, crop growth rate (CGR), and nonstructural carbohydrate (NSC) remobilization were investigated to understand the biological process in which water and straw management affects rice growth.

Materials and Methods

Plant materials and growth conditions

The experiment was conducted at a research farm of Yangzhou University, Jiangsu Province, China (32°30′N, 119°25′E, 21 m altitude) during the rice growing season (May to October) of 2012, and repeated in 2013. The soil was a sandy loam (Typic fluvaquents, Etisols [U.S. taxonomy]) with 24.3 g kg^{-1} organic matter, 101 mg kg^{-1} alkali hydrolysable N, 34.5 mg kg^{-1} Olsen-P, and 65.6 mg kg^{-1} exchangeable K. The field capacity soil moisture content was 0.188 g g^{-1}, and bulk density of the soil was 1.33 g cm^{-3}. The average air temperature, precipitation, and sunshine hours during the rice growing season across the two study years measured at a weather station close to the experimental site are shown in Fig. S1.

A "super" rice (Oryza sativa. L) variety Yangjing 4038 (japonica), currently used in local production, was grown in the field. Seedlings were raised in the field with sowing date on 15 May and transplanted on 10 June at a hill spacing of 0.16 m × 0.25 m with two seedlings per hill. In both years, N (60 kg ha^{-1} as urea), P (30 kg ha^{-1} as single superphosphate) and K (40 kg ha^{-1} as KCl) were applied and incorporated just before transplanting. Nitrogen as urea was also applied at early tillering (8 days after transplanting (DAT) (40 kg ha^{-1}), panicle initiation (45 DAT, 50 kg ha^{-1}) and the initial of spikelet differentiation (62 DAT, 50 kg ha^{-1}). The total N application was 200 kg ha^{-1} which is within the recommended range. The variety (50% of plants) headed on 27–28 August, and was harvested on 15–16 October.

Treatments

The experiment was laid out in a complete randomized block design with four replicates. Plot dimensions were 6.4 m × 5 m and plots were separated by an alley 1 m wide with plastic film inserted into the soil to a depth of 50 cm to form a barrier. Treatments consisted of four water and straw management combinations including continuously flooded (CF) without wheat straw incorporation (−S) (CF−S), alternate wetting and moderate drying (AWMD) without wheat straw incorporation (AWMD−S), CF with wheat straw incorporation (+S) (CF + S), and AWMD with wheat straw incorporation (AWMD + S). In the +S plots, wheat straw from the preceding season was chopped to approximately 10 cm in length and incorporated freshly to a soil depth of 0–15 cm during the tillage, prior to rice transplantation. The amount of the wheat straw incorporated was approximately 6500 kg ha^{-1} in dry weight containing 36 kg N ha^{-1}, 6.1 kg P ha^{-1} and 72 kg K ha^{-1}, on average. Wheat straw was removed from the −S plots. Except drainage at the mid season, the field was continuously flooded with 2–3 cm water level until 1 week before harvest in CF regimes. In AWMD regimes, plots were kept a 2–3 cm water level during the first 12 days after transplanting (DAT), anthesis, and the timing for N top dressing. At other growth stages, fields were not irrigated until the soil water potential reached −15 kPa (soil moisture content 0.167 g g^{-1}) at 15–20 cm depth. Soil water potential of −15 kPa in the AWMD regime was chosen as our earlier work (Yang et al. 2007) has shown that a mild soil drying (soil water potential −15 kPa at 15–20 cm depth) during the growing season could not reduce grain yield when compared the CF regime. Soil water potential was monitored at 15–20 cm soil depth with a tensiometer consisting of a sensor of 5 cm length (Soil Science Research Institute, Nanjing, China). Four tensiometers were installed in each plot, and readings were recorded at 1200 h each day. When soil water potential reached the threshold, a flood with 2.0–3.0 cm water depth was applied to the plots. The amount of irrigation water was monitored with a flow meter (LXSG-50 Flow meter, Shanghai Water Meter Manufacturing Factory, Shanghai, China) installed in the irrigation pipelines. Both irrigation and drainage systems were built between blocks. Each plot was irrigated or drained independently.

Soil redox potential, leaf water potential, and photosynthetic rate measurement

Soil redox potential (Eh) was measured at 10, 19, 39, 55, 78, 90, 104, and 127 DAT in 2012 and at 10, 20, 40, 56, 79, 91, 105, and 128 DAT in 2013. The growth stages corresponding above dates were early tillering, mid tillering, late tillering, the initial of spikelet differentiation, heading time, early grain filling, mid grain filling, and maturity, respectively. Soil Eh was monitored at 10 cm soil depth by using Pt-tipped electrodes (Hirose Rika Co. Ltd. Niwa-Gun, Aichi, Japan) and an oxidation–reduction potential meter with a reference electrode (Toa PRN-41), with four replications for each treatment.

Leaf water potentials of flag leaves were measured at 2-h intervals from 0600 h to 1800 h at 72 and 73 DAT when days were clear and soil water potential was approximately −15 kPa in the AWMD regime (Fig. S2). A pressure chamber (Model 3000, Soil Moisture Equipment Corp., Santa Barbara, CA) was used for leaf water potential measurement, with eight leaves for each treatment.

The photosynthetic rate of the flag leaves were determined at 72 and 73 DAT (D1) in both years and 93 and 98 DAT (D2), respectively, in 2012 and 2013 when soil water potential was approximately −15 kPa in the AWMD regime and at 74 and 75 DAT (W1) in both years and at 95 and 100 DAT (W2), respectively, in 2012 and 2013 when plants were rewatered. A gas exchange analyzer (Li-Cor 6400 portable photosynthesis measurement system, Li-Cor, Lincoln, NE) was used for measurement of the photosynthetic rate during 0900 to 1100 h when photosynthetic active radiation above the canopy was 1300–1500 μmol m^{-2} s^{-1}. The measurement was made on the upper surface of the flag leaf using eight leaves from each treatment.

Measurements of tiller number, root and shoot biomass and root activity

Twenty plants in each plot were tagged for observation of tiller number. The observation was made at transplanting, jointing stage, heading time, and maturity. The percentage of productive tillers was defined as the number of panicles developed from tillers as a percentage of the number of tillers at the jointing stage.

Root and shoot biomass and ROA were determined at 20–21, 43–44 (panicle initiation), 78–79 and 127–128 DAT. Shoot biomass was also measured at transplanting, and ROA was determined at mid grain filling (104–105 DAT) instead of maturity. When root biomass and ROA were measured, soil water potential was about −15 kPa at 20–21 DAT, 104–105 DAT, and 127–128 DAT and was 0 kPa at 43–44 DAT and 78–79 DAT in the AWMD regime.

To maintain canopy conditions, the vacant spaces left after sampling for measurements of root and shoot biomasses were immediately replaced with hills taken from the borders and these replanted hills were not subjected to sampling any more.

For each root sampling, a cube of soil (25 cm in length × 16 cm in width × 20 cm in depth) around each individual hill was removed up using a sampling core. Such a cube contains approximately 95% of total root biomass (Kukal and Aggarwal 2003; Yang et al. 2008). Plants of four hills from each plot formed a sample at each measurement. The cube of soil was cut into two parts, with 10 cm depth for each part. The roots in each cube of soil were carefully rinsed with hydropneumatic elutriation device (Gillison's Variety Fabrications, Benzonia, MI). After combining roots of four hills and recording fresh weight, portions of each root sample were used for measurements of root oxidation activity (ROA). The rest of the roots were dried in an oven at 70°C to constant weight and were weighed. The ROA was determined by measuring oxidation of alpha-naphthylamine (α-NA) according to the method of Ramasamy et al. (1997), and was expressed as μg α-NA per gram DW per hour (μg α-NA g^{-1} DW h^{-1}). Before root sampling, aboveground plants were sampled and separated into leaves, stems, panicles (at heading time and maturity) and dead shoot parts, and were dried in an oven at 70°C to constant weight for determining shoot biomass. The amount of nonstructural carbohydrate (NSC) in the stem (culm + sheath) was determined at heading time and maturity according to the method described by Yoshida et al. (1976). Crop growth rate (CGR) was calculated using the following formula:

$$\text{CGR} (g\,m^{-2}d^{-1}) = (W_2 - W_1)/(t_2 - t_1)$$

where W_1 and W_2 are the first and second measurement of shoot biomass (g m^{-2}), respectively, and t_1 and t_2 present the first and second time (d), respectively, of the measurement.

Greenhouse gas flux measurements

Fluxes of CH_4 and N_2O were measured using static vented flux chamber technique (Hutchinson and Livingston 1993). The chamber included a permanent base that was inserted into the soil (with rice plants growing inside); extensions of varying length to accommodate the growing plants; and a lid which was equipped with vent tube, fan and thermocouple wire. The base was made of PVC frame (0.5 m × 0.5 m) and inserted to a depth of 15 cm which left about 10 cm above the soil line. Holes drilled in the base above and below the soil line allowed for relatively free root and water movement. During sampling, holes above the water line were plugged with rubber stoppers when the water level was below the holes to ensure chambers were airtight. One chamber was employed in each plot and positioned at least 1 m inside the plots and sampling locations were connected using board walks to prevent soil disturbance when sampling. The chamber was wrapped with a layer of sponge and aluminum foil to minimize the air temperature changes inside the chamber during the period of sampling.

Gas flux measurements were conducted at daily during the entire growing season. For each flux measurement, gas samples were collected from 0900 to 1100 h by a 20-mL syringe at 0, 10, 20, and 30 min after the chamber closure. Gas samples were analyzed for CH_4, N_2O, and CO_2 concentrations by a gas chromatograph (Agilent 7890A, Agilent Technologies, Palo Alto, CA) equipped with two detectors. N_2O was detected by an electron capture detector (ECD), and CH_4 was detected by a hydrogen flame ionization detector (FID). CO_2 was reduced with hydrogen to CH_4 in a nickel catalytic converter at 375°C and then detected by the FID. The carrier gas was argon-methane (5%) at a flow rate of 40 mL min^{-1}. The temperatures for the column and ECD detector were maintained at 40°C and 300°C, respectively. The oven and FID were operated at 50°C and 300°C, respectively. Sample sets were rejected unless they yielded a linear regression value of r^2 greater than 0.90. The average fluxes and standard deviations (SDs) of CH_4 and N_2O were calculated from four replicates. The seasonal amounts of CH_4 and N_2O emissions were calculated from the daily measurement.

The GWP of N_2O and CH_4 was calculated in mass of CO_2 equivalents (kg CO_2 eq ha^{-1}) over a 100-year time horizon. A radiative forcing potential relative to CO_2 of 298 was used for N_2O and 25 for CH_4 (Ma et al. 2013).

Final harvesting

Plants were hand-harvested on 15 October in 2012 and 16 October in 2013. The measurement of grain yield and yield components was followed the procedure as described by Yoshida et al. (1976). Plants in two rows on each side of the plot were discarded to avoid border effects. Grain yield was determined from a harvest area of 6.0 m^2 in each plot (not including plants in the chamber) and adjusted to 14% moisture. Aboveground biomass and yield components, i.e., the number of panicles per square meter, number of spikelets per panicle, percentage of filled grains, and grain weight, were determined from plants of 0.6 m^2 (excluding the border ones and those in the chamber) sampled randomly from each plot. The percentage of filled grains was defined as the filled grains (specific gravity $\geq$ 1.06 g cm^{-3}) as a percentage of total number of spikelets.

Aboveground plants sampled at maturity were separated into straw, filled and unfilled grains, and rachis. Dry weight of each component was determined by oven-drying at 70°C to constant weight and weighed. Tissue N content was determined by micro Kjeldahl digestion, distillation and titration to calculate aboveground N uptake (Yoshida et al. 1976). The methods for calculating NUE were according to Xue et al. (2013), i.e., the internal N use efficiency (IE_N) = grain yield/the total amount of N uptake in plants at maturity, and N partial factor productivity (PFP_N) = grain yield/the amount of N applied. The WUE was calculated from and grain yield and amount of irrigation water and precipitation (grain yield over the amount of irrigation water and precipitation, kg m^{-3}), and GHG intensity (GHGI) was expressed as GWP per unit mass of rice grain (kg CO_2 eq kg^{-1} grain) (Mosier et al. 2006).

Statistical analysis

Analysis of variance was performed using SAS/STAT statistical analysis package (version 6.12, SAS Institute, Cary, NC). The statistical model included sources of variation due to replication, year (Y), irrigation regime (I), straw incorporation (S) and the interaction of Y × I, Y × I, Y × S, and I × S. Data from each sampling date were analyzed separately. Means were tested by least significant difference at $P < 0.05$ ($LSD_{0.05}$).

Results

Soil and leaf water potentials and soil Eh

The difference in total rainfall during the growing season was rather small between the two study years (561.6 mm in 2012 and 516.6 mm in 2013), especially during the mid and late growing stages (40–128 DAT) (Fig. S1c). Changes in soil water potentials were similar in both years (Fig. S2a and b). It took 6–10 days to reach soil water potential of −15 kPa for the AWMD regime under either without (−S) or with (+S) wheat straw incorporation. The CF regimes received 18–20 times of irrigation, whereas AWMD regimes were applied 11–13 times of irrigation, from transplanting to maturity. The differences in soil water potentials were not significant between −S and +S when the irrigation regime was the same (Fig. S2).

The amount of irrigation water from land preparation to rice harvest was 371 mm for the AWMD−S treatment and 417 mm for the AWMD + S treatment, which was 62.5% and 61.9% of that for the CF−S (594 mm) and CF + S (674 mm) treatments, respectively (Fig. 1). More water use under +S than under −S was mainly resulted from more irrigation water for land preparation for the

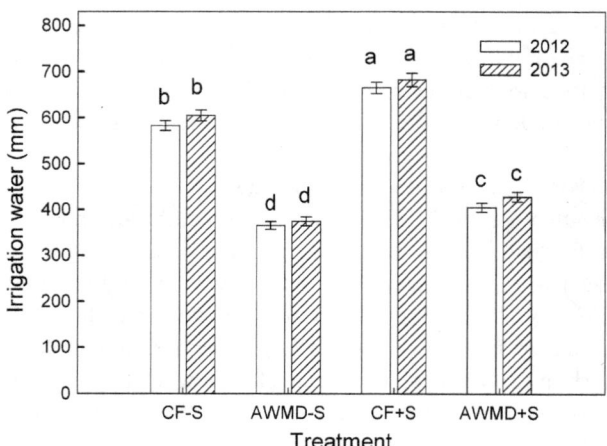

Figure 1. Irrigation water under various irrigation and wheat straw incorporation treatments in 2012 and 2013. CF, AWMD, −S and +S represent continuously flooded, alternate wetting, and moderate drying, without straw incorporation and with straw incorporation, respectively. Vertical bars represent ± standard error of the mean ($n = 4$), where these exceed the size of the symbol. Different letters above the column indicate statistical significance ($P < 0.05$).

former, such as soaking the field and puddling before rice transplanting, due partly to more water to rehydrate the dry straw and partly to more water to flood the increased field level, when wheat straw was incorporated into soil.

Figure 2 shows diurnal changes of leaf water potentials when soil water potentials were approximately −15 kPa in AWMD regimes. The leaf water potential ranged from −0.28 to −0.32 MPa at predawn (0600 h) to −0.53 to −0.56 MPa at mid-day (1200 h) for the plants in CF regimes. It was greatly reduced for the plants in AWMD regimes during the day, and reached −0.87 to −0.91 MPa at mid-day (Fig. 2A and B). However, the differences in leaf water potentials in the morning (0600 h and 0800 h) were very small between the plants in CF and AWMD regimes, indicating that plants in AWMD regimes could rehydrate overnight. No significant difference was observed in leaf water potentials between −S and +S in the same irrigation regime (Fig. 2).

As shown in Figure 3, soil Eh was very low at 10 DAT (before the start of AWMD treatments) and was −75 to −98 mV under −S and −174 to −185 mV under +S. It was substantially increased by the AWMD treatment, and ranged from 121 to 233 mV during the soil drying period and from −4.5 to −32.5 mV during the wetting period, and showed no significant difference between AWMD−S and AWMD + S treatments (Fig. 3A and B). When compared with CF−S treatment, the CF + S treatment showed much lower soil Eh during the flooding period. Soil Eh in CF regimes was markedly increased during mid season drainage and the drainage before the final harvesting (Fig. 3A and B).

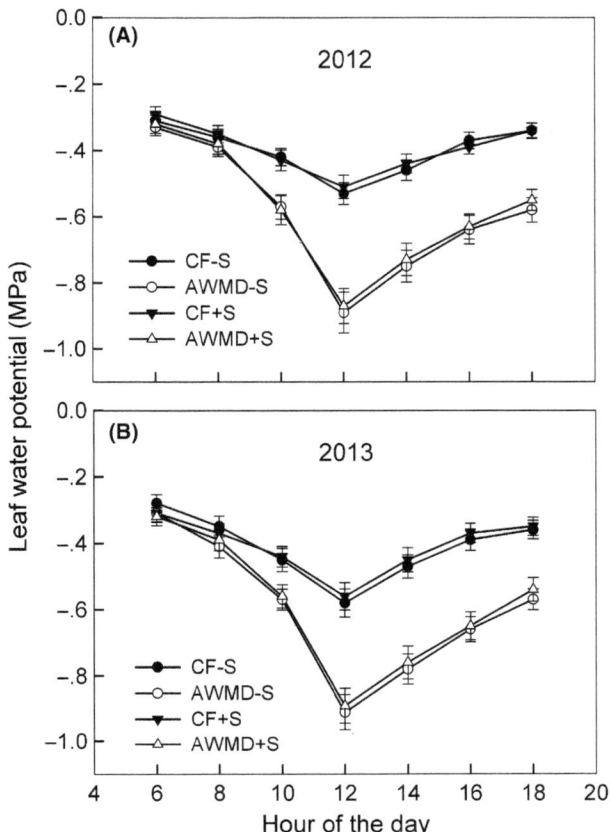

Figure 2. Diurnal changes in leaf water potentials of rice under various irrigation and wheat straw incorporation treatments in 2012 (A) and 2013 (B). CF, AWMD, −S and +S represent continuously flooded, alternate wetting and moderate drying, without straw incorporation and with straw incorporation, respectively. Measurements were made on the flag leaves at 2-h intervals from 0600 h to 1800 h on the 72 and 73 day after transplanting when soil water potentials were about −15 kPa in AWMD regimes. Vertical bars represent ± standard error of the mean (*n* = 8), where these exceed the size of the symbol.

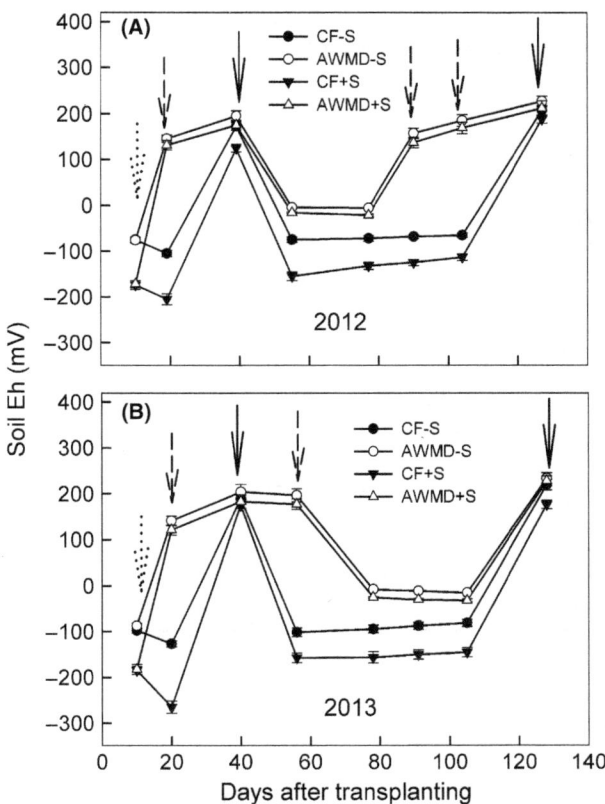

Figure 3. Soil redox potential (Eh) under various irrigation and wheat straw incorporation treatments in 2012 (A) and 2013 (B). CF, AWMD, −S and +S represent continuously flooded, alternate wetting and moderate drying, without straw incorporation and with straw incorporation, respectively. Vertical bars represent ± standard error of the mean (*n* = 4), where these exceed the size of the symbol. Dotted and dash arrows indicate the start of soil drying and during the soil drying period, respectively, in the AWMD treatments. Solid arrows represent mid season drainage and the drainage before the final harvesting for all the treatments.

Tiller number, leaf photosynthesis, and crop growth rate

The number of tillers varied with treatments (Table 1). The CF−S treatment always showed more, whereas the CF + S treatment exhibited less, tiller number than any other treatment. The percentage of productive tillers under both AWMD−S and AWMD + S treatments were greater than that under the CF−S or CF + S treatment, and showed no significant difference between the treatments of AWMD−S and AWMD + S or between CF−S and CF + S (Table 1).

Similar to the tiller number, the photosynthetic rate of leaves was significantly smaller for the CF + S treatment than for any other treatment at all the measurement times (Fig. 4A and B). It showed no significant difference among the CF−S, AWMD−S and AWMD + S treatments

during the soil drying period in AWMD regimes (D1 and D2), but was significant greater for both AWMD−S and AWMD + S treatments than for the CF−S treatment when plants in AWMD regimes were rewatered (W1 and W2), indicating a beneficial effect of the AWMD irrigation.

Consistent with the number of tillers and leaf photosynthetic rate, the crop growth rate (CGR) of the CF + S treatment was the smallest among the four treatments during the whole growing season (Fig. 5A and B). It was significantly greater under the CF−S treatment than the AWMD−S or AWMD + S treatment from mid tillering to panicle intimation, and showed no significant difference among the three treatments from transplanting to mid tillering and from panicle initiation to heading time. When compared with the CF regime, the AWMD regime significantly increased CGR from heading to maturity under either

Table 1. Number of tillers and the percentage of productive tillers of rice under various irrigation and wheat straw incorporation treatments[1].

Year/Treatment	Number of tillers per m[2]				Productive tillers (%)[2]
	Mid tillering	Jointing	Heading	Maturity	
2012					
CF–S	219a[3]	338a	278a	248a	73.4b
A WMD–S	218a	311b	256b	245a	78.8a
C F+S	205b	300c	243c	224b	74.6b
A WMD+S	216a	310b	257b	243a	78.5a
2013					
C F–S	226a	349a	291a	259a	74.3b
A WMD–S	224a	316b	265b	254a	80.5a
C F+S	195b	297c	251c	224b	75.3b
AWMD + S	223a	320b	268b	255a	79.6a
Analysis of variance					
Year (Y)	NS[4]	NS	NS	NS	NS[3]
Irrigation (I)	9.8**	4.8*	6.1*	8.7**	23.9**
Straw incorp. (S)	17.4**	49.6**	42.4**	12.5**	NS
Y × I	NS	NS	NS	NS	NS
Y × S	NS	NS	NS	NS	NS
I × S	13.3**	56.7**	47.5**	10.6**	NS

[1]CF, AWMD, –S and +S represent continuously flooded, alternate wetting and moderate drying, without straw incorporation and with straw incorporation, respectively.
[2]The number of panicles developed from tillers (tillers at maturity)/the maximum number of tillers at the jointing stage.
[3]Different letters indicate statistical significance at the $P < 0.05$ level within the same column and the same year.
[4]NS, not significant ($P > 0.05$).
*Significant at the $P < 0.05$ level; **Significant at the $P < 0.01$ level.

–S or +S, suggesting that the AWMD regime is more favorable to plant growth during the grain filling period.

When compared with those in the CF regime, NSC accumulation in the stem at the heading time and NSC remobilization during the grain filling period were greater in the AWMD regime under either –S or +S (Table 2). The AWMD–S treatment showed the greatest, while the CF + S treatments exhibited the smallest, NSC accumulation and remobilization among the four treatments, indicating an interaction between the irrigation regime and straw incorporation on NSC accumulation and remobilization.

Shoot and root biomass and root oxidation activity

At each growth stage, both shoot and root biomasses showed no significant difference among the CF–S, AWMD–S, and AWMD + S treatments (Fig. 6A–D). The CF + S treatment exhibited the smallest shoot and root biomasses among the four treatments, in good agreement with the tiller number and CGR (Table 1 and Fig. 5). The root/root ratio had no significant difference among the four treatments at the same growth state (Fig. 6E and F).

Although the difference in total root biomass in 0–20 cm soil layer was not significant among the CF–S, AWMD–S, and AWMD + S treatments, both AWMD–S and

AWMD + S treatments exhibited significantly greater root dry weight in 10–20 cm soil layer than the CF–S or CF + S treatment at all growth stages (Table 3). A similar observation was made on ROA (Fig. 7A and B). The ROA was the greatest under the AWMD–S or AWMD + S among the four treatments, followed by the CF–S, and the smallest under the CF + S treatment (Fig. 7A and B).

Grain yield and N and water use efficiencies

There was a very significant interaction between the irrigation regime and the straw incorporation on grain yield (Table 4). When compared with that in CF regimes, grain yield in AWMD regimes was increased by 2.7% under –S and by 18.0% under +S (Table 4). The significant increase in grain yield under the AWMD + S than under the CF + S was due mainly to increases in the panicle number per m[2], percentage of filled grains and grain weight. A low grain yield under the CF + S treatment was mainly attributed to a decreased panicle number per m[2], which was closely associated with the decreased tillering ability resulted from a strong soil reduction condition when wheat straw was incorporated into soil and the field was continuously flooded (Table 1 and Fig. 3).

As shown in Table 4, both AWMD–S and AWMD + S treatments showed a higher harvest index than the CF–S or CF + S treatment. A higher harvest index in AWMD

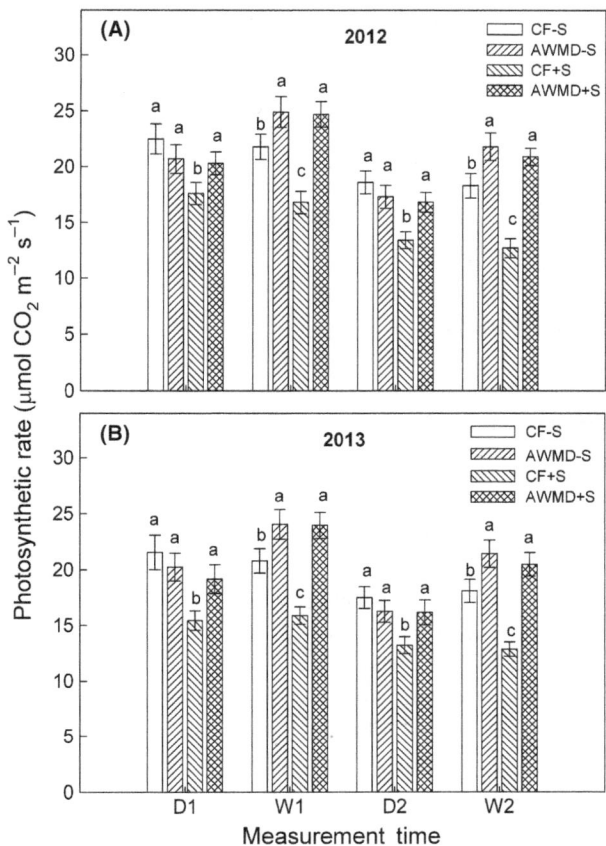

Figure 4. Photosynthetic rate of the flag leaf of rice under various irrigation and wheat straw incorporation treatments in 2012 (A) and 2013 (B). CF, AWMD, −S and +S represent continuously flooded, alternate wetting and moderate drying, without straw incorporation and with straw incorporation, respectively. D1 and D2 are measurement times at 72–73 and 93–98 days after transplanting (DAT), respectively, when soil water potential was about −15 kPa in the AWMD plot, and W1 and W2 are measurement times at 74–75 and 95–100 DAT, respectively, when plants were rewatered. Vertical bars represent ± standard error of the mean (n = 8), where these exceed the size of the symbol. Different letters above the column indicate statistical significance (P < 0.05) within the same measurement date.

regimes may be attributed partly to more NSC remobilization during the grain filling period (Table 2) and partly to a greater CGR during maturity (Fig. 5).

Although N content in plants at maturity showed no significant difference among the four treatments, NUE varied with treatments (Table 5). When wheat straw was incorporated into the field, the total N uptake in plants at maturity, IE_N and PEP_N were significantly higher in the AWMD regime than in the CF regime. They showed no significant difference between CF and AWMD regimes when wheat straw was not incorporated. Both AWMD−S and AWMD + S exhibited the highest, whereas the CF + S treatment showed the smallest, WUE among the four treatments (Table 5), in good agreement with harvest index (Table 4).

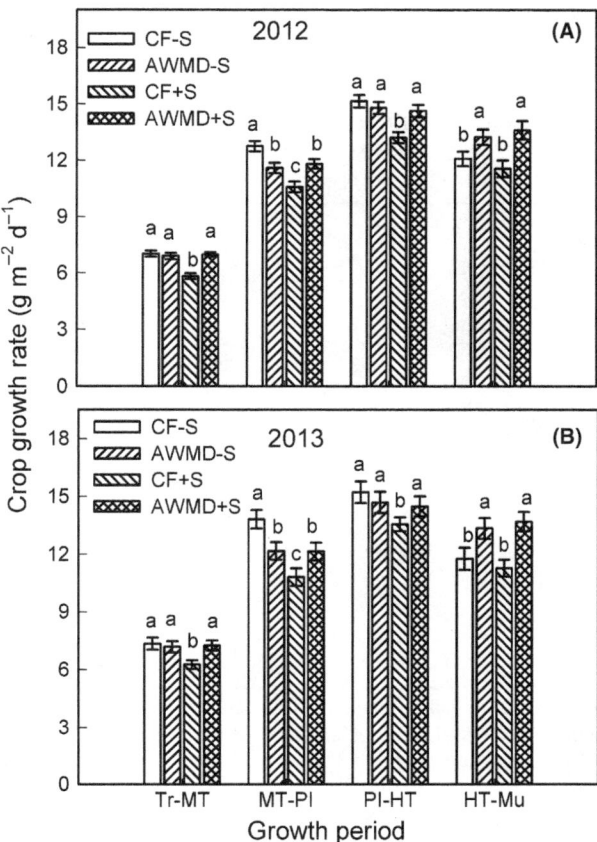

Figure 5. Crop growth rate of rice in 2012 (A) and 2013 (B). CF, AWMD, −S and +S represent continuously flooded, alternate wetting and moderate drying, without straw incorporation and with straw incorporation, respectively. Tr-MT, from transplanting to mid tillering; MT-PI, from mid tillering to panicle initiation; PI-HT, from panicle initiation to heading time; HT-Mu, from heading time to maturity. Vertical bars represent ± standard error of the mean (n = 4), where these exceed the size of the symbol. Different letters above the column indicate statistical significance (P < 0.05) within the same growth period.

GHG emissions and GHG intensity

Fluxes of CH_4 were highly dependent on water management and straw incorporation conditions (Fig. 8A and B). In all treatments CH_4 emissions were generally higher during the continuous flooding period at the early stage of rice growth and were fluctuated in AWMD regimes and peaked 24–25 DAT in CF regimes. They were markedly decreased during the mid season drainage in CF regimes and during the soil drying period in AWMD regimes. The peak of CH_4 fluxes also appeared 47–48 DAT under the all treatments which may be attributed to higher temperature (Fig. S1a) and vigorous rice growth during this period, and then CH_4 fluxes maintained at a very low level. Wheat straw incorporation substantially increased, while the AWMD regime greatly reduced, CH_4

Table 2. Nonstructural carbohydrate (NSC) accumulation and remobilization of rice under various irrigation and wheat straw incorporation treatments[1].

Year/Treatment	NSC at heading (g m^{-2})	NSC at maturity (g m^{-2})	NSC remobilization during grain filling (%)[2]
2012			
CF−S	268b[3]	125a	53.4b
AWMD−S	297a	124a	58.2a
CF+S	213c	119a	44.1c
AWMD+S	292a	126a	56.8a
2013			
CF−S	274b	131a	52.2b
AWMD−S	302a	126a	58.3a
CF+S	217c	121a	44.2c
AWMD+S	298a	130a	56.4a
Analysis of variance			
Year (Y)	NS[4]	NS	NS
Irrigation (I)	445**	NS	56.6**
Straw incorp. (S)	139**	NS	18.7**
Y × I	NS	NS	NS
Y × S	NS	NS	NS
I × S	101**	NS	8.7**

[1]CF, AWMD, −S and +S represent continuously flooded, alternate wetting and moderate drying, without straw incorporation and with straw incorporation, respectively.

[2](NSC in stems at heading time – NSC in stems at maturity)/NSC in stems at heading time ×100.

[3]Different letters indicate statistical significance at the $P < 0.05$ level within the same column and the same year.

[4]NS, not significant ($P > 0.05$).

**Significant at the $P < 0.01$ level.

emissions. During the entire rice growing season, the average CH$_4$ flux was 5.7, 2.8, 29.2, and 6.2 mg CH$_4$ m^{-2} h^{-1}, respectively, under CF−S, AWMD−S, CF + S and AWMD + S treatments.

Nitrous oxide emissions were only observed when soil water potential ≤−10 kPa, and exhibited a pulse-like pattern which was coincided with the mid season drainage in CF regimes and the soil drying period in AWMD regimes (Fig. 8C and D). The AWMD regime increased, whereas wheat straw incorporation showed some reduction in, N$_2$O emissions. The average N$_2$O flux was 10.0, 26.6, 8.1, and 23.2 μg N$_2$O m^{-2} h^{-1}, respectively, under CF−S, AWMD−S, CF + S and AWMD + S treatments during the whole growing season.

During the experimental period, the GWP of CH$_4$ and N$_2$O under various treatments varied between 2.3 and 11.2 t CO$_2$-eq ha^{-1} in 2012 and between 2.5 and 11.4 t CO$_2$-eq ha^{-1} in 2013 (Table 6). Compared with CF regimes, AWMD regimes decreased the GWP by 45.2% and 55.9%, respectively, under −S and +S. Similarly, AWMD regimes decreased the GHGI, i.e., yield-scaled GWP, by 46.7%

under −S and by 62.6% under +S, when compared with CF regimes (Table 6). As shown in Table 6, the GWP of N$_2$O accounted for 2.1%, 10.1%, 0.7%, and 4.3% of the total GWP of CH$_4$, and N$_2$O, respectively, under the treatments of CF−S, AWMD−S, CF + S, and AWMD + S, indicating that CH$_4$ emissions are the dominant in the total GWP from rice fields in either CF or AWMD regimes.

Discussion

Integrative effect of AWMD and straw incorporation on grain yield and WUE

Although the effect of AWD technology or straw application on rice yield, WUE or GHG emissions has been studied previously (Ma et al. 2009; Yao et al. 2012; Zhang et al. 2012; Li et al. 2014; Linquist et al. 2015), information on the integrative effect of AWMD and straw incorporation on grain yield, WUE, NUE, and GHG emissions is unavailable. Our results showed that, when compared with the CF regime, the AWMD regime increased grain yield by 2.7% and 18.0%, reduced irrigation water by 37.7% and 38.2% and increased WUE by 27.6% and by 50.0%, respectively, under −S and +S (Tables 4 and 5, Fig. 1). The results suggest that there is an interaction between AWMD and straw incorporation on rice yield and WUE, and adoption of the AWMD technology is more effective to increase grain yield and resource-use efficiency if wheat straw is incorporated into paddy fields.

How could an AWMD regime lead to higher rice yield and WUE, especially under wheat straw incorporation? The physiological mechanism is not understood. It has been observed that a long period of flooding in the paddy field could produce high concentrations of toxic reduction products such as Fe^{2+}, H$_2$S, and organic compounds (Ramasamy et al. 1997). These toxic reduction products are more aggravated when wheat straw is incorporated into soil, and thereby seriously inhibit root growth (Yao et al. 2013; Li et al. 2014). In this study, we observed that soil Eh was much lower under the CF + S treatment than under the CF−S treatment (Fig. 3), indicating a strong soil reduction condition when wheat straw was incorporated into soil and the field was continuously flooded. A strong soil reduction condition could inhibit root growth, which was evidenced by much smaller root biomass and much lower ROA under the CF + S treatment than those under the CF−S treatment (Figs. 6, 7). Drainage during mid season, especially an AWD regime, could greatly improve soil redox conditions and remove toxic reduction substances (Ramasamy et al. 1997; Yang et al. 2007; Liu et al. 2013), and therefore benefit root growth, which was evidenced by our observations that

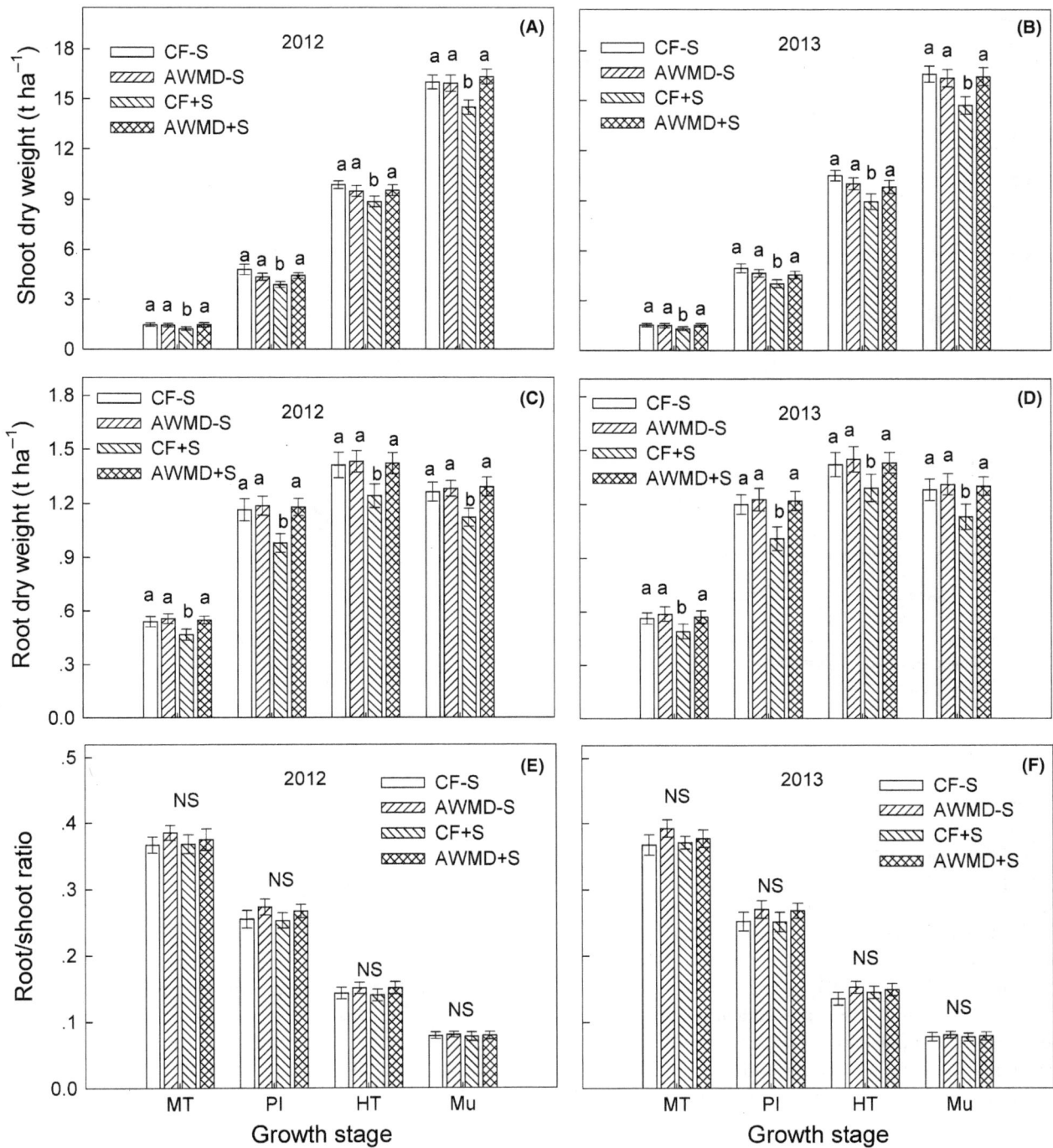

Figure 6. Shoot dry weight (A and B), root dry weight (C and D) and root/shoot ratio (E and F) of rice in 2012 (A, C, and E) and 2013 (B, D, and F). CF, AWMD, −S and +S represent continuously flooded, alternate wetting, and moderate drying, without straw incorporation and with straw incorporation, respectively. MT, PI, HT and Mu represent mid tillering, panicle initiation, heading time and maturity, respectively. Vertical bars represent ± standard error of the mean ($n = 4$), where these exceed the size of the symbol. Different letters above the column indicate statistical significance ($P < 0.05$) within the same measurement stage, and NS means not significant ($P > 0.05$).

soil Eh and ROA were much higher under both AWMD−S and AWMD + S treatments than under either CF−S or CF + S treatments (Figs. 3, 7), and that root biomass was greater in AWMD regimes than in CF regimes when

wheat straw was applied (Fig. 6). Furthermore, there was a greater root biomass in 10–20 cm soil depth throughout the growing season in AWMD than in CF regimes irrespective of −S and +S (Table 3), suggesting a deeper

Table 3. Root dry weight of rice in 10–20 cm soil layer under various irrigation and wheat straw incorporation treatments[1].

Year/Treatment	Mid tillering (g m^{-2})	Panicle initiation (g m^{-2})	Heading time (g m^{-2})	Maturity (g m^{-2})[3]
2012				
CF–S	3.58b[2]	25.4b	42.7b	17.2b
AWMD–S	4.85a	30.9a	48.5a	20.3a
CF+S	2.49c	20.5c	35.2c	13.9c
AWMD+S	4.61a	28.8a	46.4a	19.6a
2013				
CF–S	3.67b	26.3b	43.5b	18.1b
AWMD–S	4.96a	31.5a	48.8a	21.5a
CF+S	2.53c	21.8c	36.4c	14.2c
AWMD+S	4.74a	29.1a	47.3a	20.0a
Analysis of variance				
Year (Y)	NS[3]	NS	NS	NS
Irrigation (I)	386**	204**	89.5**	57.6**
Straw incorp. (S)	58.9**	43.7**	45.6**	26.8**
Y × I	NS	NS	NS	NS
Y × S	NS	NS	NS	NS
I × S	25.5**	21.8**	16.4**	14.5**

[1]CF, AWMD, −S and +S represent continuously flooded, alternate wetting and moderate drying, without straw incorporation and with straw incorporation, respectively.
[2]Different letters indicate statistical significance at the $P < 0.05$ level within the same column and the same year.
[3]NS, not significant ($P > 0.05$).
**Significant at the $P < 0.01$ level.

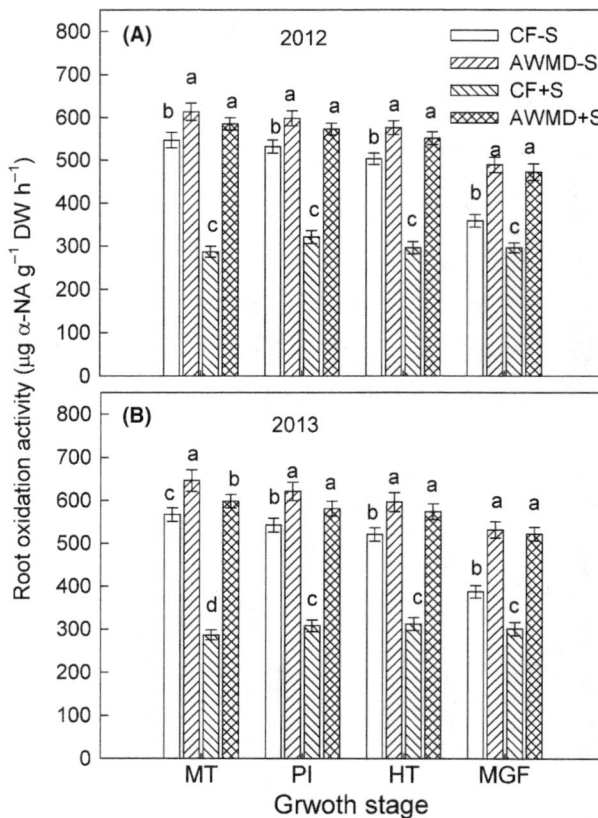

Figure 7. Root oxidation activity of rice in 2012 (A) and 2013 (B). CF, AWMD, −S and +S represent continuously flooded, alternate wetting and moderate drying, without straw incorporation and with straw incorporation, respectively. MT, PI, HT, and MGF represent mid tillering, panicle initiation, heading time and mid grain filling, respectively. Vertical bars represent ± standard error of the mean ($n = 4$), where these exceed the size of the symbol. Different letters above the column indicate statistical significance ($P < 0.05$) within the same measurement stage.

root distribution in the soil. There is a proposal that plants that a deeper root distribution in the soil could maximize soil moisture capture and thereby maintain a high plant water status under drought conditions, and consequently contributes to a higher crop yield and WUE (Garnett et al. 2009; Luo et al. 2011; Lynch 2013; Chu et al. 2014). Therefore, we conclude that a deeper root distribution contributes, at least partly, to higher grain yield and WUE in AWMD regimes.

It is proposed that an interdependent relationship exists between the root and shoot (Osaki et al. 1997; Yang et al. 2004), that is, active shoots can ensure a sufficient supply of carbohydrates to roots and maintain active root functions; the activation of root functions can improve shoot growth by supplying a sufficient amount of nutrients, water and phytohormones to shoots, thus ensures an increase in crop productivity (Osaki et al. 1997; Yang et al. 2004; Zhang et al. 2009). Our earlier work (Yang et al. 2002, 2003) has shown that a moderate soil drying during grain filling of rice and wheat (*Triticum asetivum* L.) could increase abscisic acid levels in root exudates, and consequently, enhance NSC remobilization from vegetative tissues to grains and increase grain filling rate. The present results showed that, when compared with the CF + S treatment, the AWMD + S treatment markedly increased

tiller number, the percentage of productive tillers, leaf photosynthetic rate, crop growth rate, NSC accumulation in stems before heading, NSC remobilization during grain filling and shoot biomass during the whole growing season (Tables 1 and 2, Figs. 4, 5 and 6). We speculate that the AWMD regime, especially under +S, improves root growth which benefits other physiological processes, leading to higher grain yield and better WUE.

There are reports showing that an increase in WUE usually accompanies a yield penalty under an AWD compared with that under the CF (Bouman and Tuong 2001; Belder et al. 2004; Yao et al. 2012; Linquist et al. 2015). The results herein demonstrated that the AWMD regime could not only increase WUE but also increase grain yield when compared the CF regime under either −S or +S (Tables 4 and 5). The discrepancies between previous studies and our work are probably attributed to many reasons, such as variations in soil hydrological conditions, climate

Table 4. Effect of water and straw management on grain yield, yield components and harvest index of rice under various irrigation and wheat straw incorporation treatments[1].

Year/Treatment	Grain yield (t ha^{-1})	Panicles per m^2	Spikelets per panicle	Filled grains (%)	Grain weight (mg)	Harvest index[2]
2012						
CF−S	9.01a[3]	298a	136a	85.9b	26.2b	0.484b
AWMD−S	9.32a	295a	133a	88.7a	26.9a	0.503a
CF+S	8.05b	274b	137a	84.5b	26.2b	0.478c
AWMD+S	9.48a	293a	135a	89.2a	27.1a	0.499a
2013						
CF−S	9.36a	309a	134a	88.9b	26.5b	0.486b
AWMD−S	9.54a	304a	132a	91.3a	27.3a	0.502a
CF+S	8.19b	274b	135a	85.5c	26.4b	0.479c
AWMD+S	9.67a	305a	133a	91.6a	27.1a	0.501a
Analysis of variance						
Year (Y)	NS[4]	NS	NS	7.4*	NS	NS
Irrigation (I)	53.8**	7.1*	NS	23.4**	26.4**	162**
Straw incorp. (S)	15.7**	14.5**	NS	NS	NS	8.6**
Y × I	NS	NS	NS	NS	NS	NS
Y × S	NS	NS	NS	NS	NS	NS
I × S	27.2**	13.6**	NS	NS	NS	NS

[1]CF, AWMD, −S and +S represent continuously flooded, alternate wetting and moderate drying, without straw incorporation and with straw incorporation, respectively. Values of grain yield are means of plants of 6 m^2 harvested from each plot. Values of panicles per m^2, spikelets per panicle, filled-grain percentage, and 1000-grain weight are means of plants harvested from 0.6 m^2 from each plot in each treatment.
[2]Total grain weight (dry weight)/total aboveground biomass (dry weight).
[3]Different letters indicate statistical significance at the $P < 0.05$ level within the same column and the same year.
[4]NS, not significant ($P > 0.05$).
*Significant at the $P < 0.05$ level; **Significant at the $P < 0.01$ level.

Table 5. Nitrogen content and uptake at maturity, nitrogen use efficiency and water use efficiency (WUE) of rice under various irrigation and wheat straw incorporation treatments[1].

Year/Treatment	N content (%)	N uptake (kg ha^{-1})	IE$_N$[2] (kg kg^{-1})	PFP$_N$[3] (kg kg^{-1})	WUE[4] (kg m^{-3})
2012					
CF−S	1.05a[5]	168a	53.6ab	45.1a	0.79b
AWMD−S	1.04a	166a	56.1a	46.6a	1.01a
CF+S	1.06a	154b	52.4b	40.3b	0.66c
AWMD+S	1.05a	172a	55.2a	47.4a	0.98a
2013					
CF−S	1.03a	171a	54.6ab	46.8a	0.84b
AWMD−S	1.02a	167a	57.2a	47.7a	1.07a
CF+S	1.05a	154b	53.0b	41.0b	0.68c
AWMD+S	1.04a	172a	56.1a	48.4a	1.03a
Analysis of variance					
Year (Y)	NS[6]	NS	NS	NS	5.4*
Irrigation (I)	NS	45.4**	10.9**	52.5**	209**
Straw incorp. (S)	NS	20.2**	NS	14.8**	21.6**
Y × I	NS	NS	NS	NS	NS
Y × S	NS	NS	NS	NS	NS
I × S	NS	88.9**	NS	25.6**	8.1**

[1]CF, AWMD, −S and +S represent continuously flooded, alternate wetting and moderate drying, without straw incorporation and with straw incorporation, respectively.
[2]IE$_N$, internal N use efficiency: grain yield (kg)/N uptake of plants (kg).
[3]PFP$_N$, N partial factor productivity: grain yield (kg)/N rate (200 kg ha^{-1}).
[4]WUE, water use efficiency: grain yield (kg)/(amount of irrigation water + precipitation) (m^3).
[5]Different letters indicate statistical significance at the $P < 0.05$ level within the same column and the same year.
[6]NS, not significant ($P > 0.05$).
*Significant at the $P < 0.05$ level; **Significant at the $P < 0.01$ level.

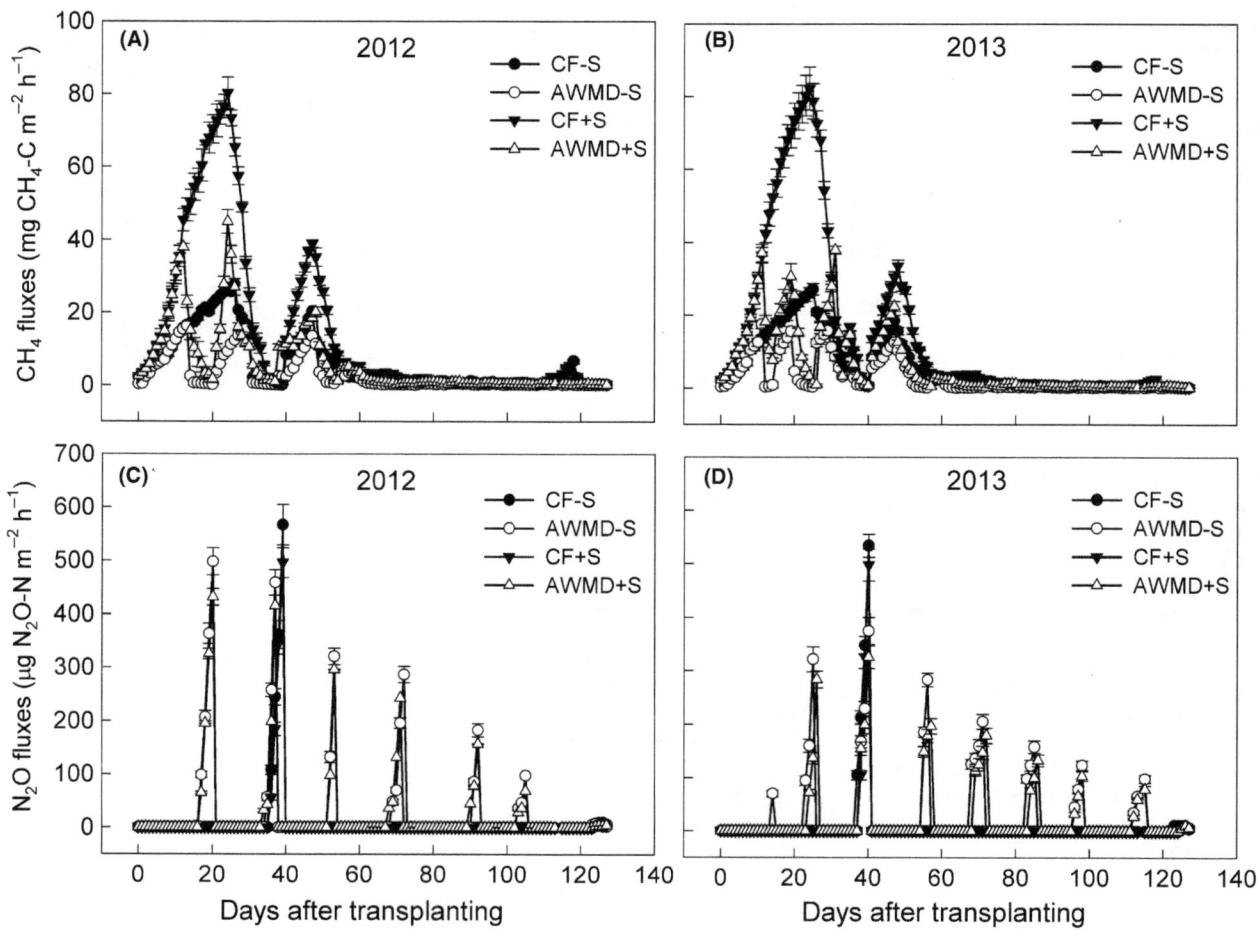

Figure 8. Fluxes of CH_4 (A and B) and N_2O (C and D) during the rice growing season in 2012 (A and C) and 2013 (B and D). CF, AWMD, −S and +S represent continuously flooded, alternate wetting and moderate drying, without straw incorporation and with straw incorporation, respectively. Vertical bars represent ± standard error of the mean ($n = 4$), where these exceed the size of the symbol.

during rice growing season, rice varieties used, N nutrition and the timing of the irrigation method applied (Belder et al. 2004; Yang et al. 2004; Zhang et al. 2008, 2009). In this study, the soil drying in the AWMD regime was very mild which was evidenced that the difference in leaf water potentials in the morning (0600–0800 h) was not significant between CF and AWMD regimes, and the mid-day leaf water potential was not lower than −0.91 MPa under the AWMD regime (Fig. 4). There is a report showing that mid day leaf water potential > −1.2 MPa could not seriously affect leaf photosynthesis (Yang et al. 2007). The results herein also demonstrated that the leaf photosynthetic rate was not inhibited by the soil drying, but it was greatly enhanced when plants were rewatered in AWMD regimes (Fig. 4). We suggest that the diagnosis that mid-day leaf water potential is not lower than −1.0 MPa and plants could rehydrate overnight could be used as an index for soil drying in a safe AWD to increase both grain yield and WUE.

Integrative effect of AWMD and straw incorporation on GHG emissions and NUE

It is generally believed that rice systems produce more GWP of GHG (mainly CH_4) than other cereal crops, because flooded paddy soils create an anaerobic environment favorable for methanogenesis (Yan et al. 2005; Linquist et al. 2012). Straw incorporation in the paddy fields could provide a source of readily available C and a predominant source of methanogenic substrates, and straw decomposition in anaerobic flooded soils could not only result in accumulation of acetate, one of the most important substrates for methanogens in flooded soils but also acted as an electron donor, helping to stimulate soil reduction and create strict reductive conditions for CH_4 production (Gao et al. 2004; Minamikawa and Sakai 2006; Lee et al. 2010; Li et al. 2014). We observed that CH_4 emissions in CF regimes were very high during the early and mid growing periods, and were greatly aggravated

Table 6. Seasonal CH_4 and N_2O emissions, global warming potential (GWP) and greenhouse gas intensity (GHGI) under various irrigation and wheat straw incorporation treatments[1].

Year/Treatment	CH_4 (kg CH_4-C ha^{-1})	N_2O (kg N_2O-N ha^{-1})	GWP[2] (kg CO_2 eq ha^{-1})	GHGI[3] (kg CO_2 eq kg^{-1} grain)
2012				
CF−S	171.4c[4]	0.316c	4379c	0.486b
AWMD−S	83.6d	0.833a	2338d	0.251c
CF + S	442.8a	0.265d	11 149a	1.385a
AWMD + S	186.4b	0.718b	4874b	0.514b
2013				
CF−S	177.2c	0.298c	4519d	0.483b
AWMD−S	91.7d	0.807a	2533b	0.266c
CF + S	452.9a	0.231d	11 391a	1.391a
AWMD + S	194.6b	0.710b	5077c	0.525b
Analysis of variance				
Year (Y)	NS[5]	NS	NS	NS
Irrigation (I)	1039**	2120**	955**	1818**
Straw incorp. (S)	1243**	60.3**	1213**	2058**
Y × I	NS	NS	NS	NS
Y × S	NS	NS	NS	NS
I × S	256**	4.9*	254**	625**

[1]CF, AWMD, −S and +S represent continuously flooded, alternate wetting and moderate drying, without straw incorporation and with straw incorporation, respectively.
[2]GWP = 25 × CH_4 + 298 × N_2O.
[3]GHGI = GWP/grain yield.
[4]Different letters indicate statistical significance at the $P < 0.05$ level within the same column and the same year.
[5]NS, not significant ($P > 0.05$).
*Significant at the $P < 0.05$ level; **Significant at the $P < 0.01$ level.

when wheat straw was incorporated in to the soil (Fig. 8A and B), consistent with the previous reports (Shan and Yan 2013; Yao et al. 2013; Li et al. 2014). The present results showed that the AWMD regime could substantially decrease CH_4 fluxes, and the seasonal CH_4 emissions in such a regime were reduced by 86.7 kg CH_4-C ha^{-1} under −S and by 258 kg CH_4-C ha^{-1} under +S when compared with those in the CF regime (Table 6). The mechanism that the AWMD regime reduces CH_4 emissions is not clear. A probably explanation is that an alternate wetting and drying regime could greatly improve soil redox conditions (Fig. 3), which prevent CH_4 formation by inhibiting methanogenic bacteria and hence reduce CH_4 emissions (Li et al. 2014). The results suggest that the AWMD irrigation is a good practice to increase grain yield meanwhile reduce CH_4 emissions in rice production, especially under wheat straw incorporation.

Flooded rice systems generally emit less N_2O than dryland crops because flooding results in most N being lost as N_2 rather than N_2O (Firestone and Davidson 1989). Mid season drainage or an AWD could increase N_2O emissions because of the introduction of aerobic periods (Akiyama et al. 2005; Zou et al. 2007; Li et al. 2014). We also observed that AWMD regimes markedly increased N_2O emissions when compared with CF regimes under either −S or +S (Table 6, Fig. 8C and D). However, the N_2O emission was only detected when soil water

potentials $\leq$ −10 kPa (Figs. S2, 8), implying that N_2O emissions would not be increased if the threshold of soil water potential for rewatering is set at > -10 kPa in AWD. Moreover, the proportion of GWP of N_2O was rather small in the total GWP of CH_4 and N_2O, and was 1.4% and 7.4% under CF and AWMD regimes, respectively (Table 6). Although N_2O emissions were increased in AWMD regimes, the total GWP of CH_4 and N_2O was decreased by 50.1% and the GHGI decreased by 52.9% in these regimes when compared with those in CF regimes (Table 6). The results suggest that reducing CH_4 emissions in rice systems is a major approach to reduce GHG emissions and GHGI.

We observed that wheat straw incorporation reduced N_2O emissions in both CF and AWMD regimes when compared with no straw incorporation (Table 6 and Fig. 8). Similar observations were also made in previous studies (Shan and Yan 2013; Yao et al. 2013; Li et al. 2014). The mechanism in which straw incorporation reduces N_2O emissions is not well known. One possible reason is that straw incorporation could substantially increase C:N ratio in the soil (Yao et al. 2013), and the increase in C:N ratio decreases mineral N availability (Millar et al. 2004). Thus, incorporating wheat straw into rice fields stimulates a net immobilization of plant-available N, thereby lowering the N substrate availability for N_2O production (Yao et al. 2013).

There are reports showing that the adoption of AWD-based technologies could reduce total cumulative plant N and NUE by stimulating N losses through increases in nitrification and denitrification (Sah and Mikkelsen 1983; Eriksen et al. 1985). In present study, the sum of 0.49 kg N_2O-N ha^{-1} was more lost in the AWMD than in the CF regime (Table 6), which accounted for 0.25% of the total amount N application, indicating that N loss due to N_2O emissions is negligible in the AWMD regime. We observed that the total N uptake, internal N use efficiency (IE_N) and N partial factor productivity (PFP_N) showed no significant differences between CF and AWMD regimes when wheat straw was not incorporated, whereas they were significantly higher in the AWMD than in the CF regimes when wheat straw was applied (Table 5). The results demonstrate that the AWMD regime could increase, rather than decrease, the total cumulative plant N and NUE when wheat straw is incorporated in the paddy field. Further research is needed to determine nitrification and denitrification to exactly estimate N loss in the AWMD regime.

It should be noted that we only presented CH_4 and N_2O fluxes, and did not consider CO_2, another source of GHG, although we measured CO_2 fluxes. The first reason for this is that CO_2 emissions are estimated to contribute less than 1% to the GWP of agriculture (Linquist et al. 2015). The second reason is that the net balance between C respiration and fixation in a cropping system is reflected by changes in soil organic C over time (West and Post 2002; Stewart et al. 2007) which is difficult to detect in short-term experiments due to the relatively small change and high degree of spatial variability in soil organic C (Post et al. 2001; Conant et al. 2011; Linquist et al. 2012). In addition, in the rice-dryland crops rotation systems, such as rice-soybean and rice-wheat rotations, an AWD in rice season may not alter soil organic C as soil C stocks are already degraded in these systems (Ma et al. 2013; Linquist et al. 2015).

Conclusion

An AWMD regime, that is, mid-day leaf water potential was not lower than −1.0 MPa and plants could rehydrate overnight during the soil drying period, could increase grain yield, WUE and NUE and decrease GWP and GHGI when compared with the CF regime. The increases or decreases were more remarkable when wheat straw was incorporated in the paddy field. Increases in grain yield, WUE and NUE in the AWMD regime, especially under straw incorporation, were attributed mainly to a greater ROA, deeper root distribution, and increases in the percentage of productive tillers, leaf photosynthetic rate, crop growth rate and NSC remobilization during grain filling and harvest index. While the reduction in GWP and GHGI in the AWMD regime was due primarily to the substantial decrease in CH_4 emissions. The loss of N due to N_2O emissions was negligible in the AWMD regime. There was an interaction between the AWMD and straw incorporation on rice yield, WUE, NUE, and GHG emissions. Adoption of the AWMD technology was more effective to increase grain yield and resource-use efficiency and reduce environmental risks when wheat straw was incorporated into paddy fields.

Acknowledgments

We are grateful for grants from the National Natural Science Foundation of China (31461143015; 31271641, 31471438, 91317307), the National Key Technology Support Program of China (2011BAD16B14; 2012BAD04B08; 2014AA10A605), the Priority Academic Program Development of Jiangsu Higher Education Institutions (PAPD), Hong Kong Research Grant Council (AoE/ M-05/12), and Shenzhen Overseas Talents Innovation & Entrepreneurship Funding Scheme (The Peacock Scheme).

Conflict of Interest

None declared.

References

Akiyama, H., K. Yagi, and X. Yan. 2005. Direct N_2O emissions from rice paddy fields: summary of available data. Global Biogeochem. Cycl. 19:GB1005.

Aulakh, M. S., R. Wassmann, and H. Rennenberg. 2001. Methane emissions from rice fields-quantification, mechanisms, role of management, and mitigation options. Adv. Agron. 70:193–260.

Belder, P., B. A. M. Bouman, R. Cabangon, L. Guoan, E. J. P. Quilang, Y. Li, et al. 2004. Effect of water-saving irrigation on rice yield and water use in typical lowland conditions in Asia. Agric. Water Manage. 65:193–210.

Belder, P., J. H. J. Spiertz, B. A. M. Bouman, G. Lu, and T. P. Tuong. 2005. Nitrogen economy and water productivity of lowland rice under water-saving irrigation. Field Crops Res. 93:169–185.

Bouman, B. A. M. 2007. A conceptual framework for the improvement of crop water productivity at different spatial scales. Agric. Syst. 93:43–60.

Bouman, B. A. M., and T. P. Tuong. 2001. Field water management to save water and increase its productivity in irrigated lowland rice. Agric. Water Manage. 49:11–30.

Chu, G., T. Chen, Z. Wang, J. Yang, and J. Zhan. 2014. Morphological and physiological traits of roots and their relationships with water productivity in water-saving and drought-resistant rice. Field Crops Res. 162:108–119.

Conant, R. T., S. M. Ogle, E. A. Paul, and K. Paustian. 2011. Measuring and monitoring soil organic carbon stocks in agricultural lands for climate mitigation. Front. Ecol. Environ. 9:169–173.

Eriksen, A. B., M. Kjeldby, and S. Nilsen. 1985. The effect of intermittent flooding on the growth and yield of wetland rice and nitrogen-loss mechanism with surface-applied and deep-placed urea. Plant Soil 84:387–401.

Fageria, N. K. 2007. Yield physiology of rice. J. Plant Nutr. 30:843–879.

Firestone, M. K., and E. A. Davidson. 1989. Microbial basis of NO and N_2O production and consumption in soils. Pp. 7–21 in M. O. Andreae and D. S. Schimel, eds. Exchange of trace gases between terrestrial ecosystems and the atmosphere. John Wiley & Sons, New York, NY.

Gao, S., K. K. Tanji, and S. C. Scardaci. 2004. Impact of rice straw incorporation on soil redox status and sulfide toxicity. Agron. J. 96:70–76.

Garnett, T., V. Conn, and B. N. Kaiser. 2009. Root based approaches to improving nitrogen use efficiency in plants. Plant Cell Environ. 32:1272–1283.

Global Rice Science Partnership (GRiSP). 2013. Rice almanac, 4th ed. International Rice Research Institute, Los Baños, Philippines, 283 p.

Hutchinson, G. L., and G. P. Livingston. 1993. Use of chamber systems to measure trace gas fluxes. Pp. 63–78 in L. A. Harper, A. R. Mosier, J. M. Duxbury, and D. E. Rolston, eds. Agricultural ecosystem effects on trace gases and global climate change. ASA, CSSA, SSSA, Madison, WI.

Kukal, S. S., and G. C. Aggarwal. 2003. Puddling depth and intensity effects in rice-wheat system on a sandy loam soil, II Water use and crop performance. Soil Tillage Res. 74:37–45.

Lee, C. H., K. D. Park, K. Y. Jung, M. A. Ali, D. Lee, J. Gutierrez, et al. 2010. Effect of Chinese milk vetch (Astragalus sinicus L.) as a green manure on rice productivity and methane emission in paddy soil. Agric. Ecosyst. Environ. 138:343–347.

Li, X., J. Ma, Y. Yao, S. Liang, G. Zhang, H. Xu, et al. 2014. Methane and nitrous oxide emissions from irrigated lowland rice paddies after wheat straw application and midseason aeration. Nutr. Cycl. Agroecosys. 100:65–76.

Linquist, B. A., K. J. van Groenigen, M. A. Adviento-Borbe, C. Pittelkow, and C. van Kessel. 2012. An agronomic assessment of greenhouse gas emissions from major cereal crops. Global Change Biol. 18:194–209.

Linquist, B. A., M. M. Anders, M. A. A. Adviento-Borbe, R. L. Chaney, L. L. Nalley, E. F. F. Da Rosa, et al. 2015. Reducing greenhouse gas emissions, water use, and grain arsenic levels in rice systems. Global Change Biol. 21:407–417.

Liu, L., T. Chen, Z. Wang, H. Zhang, J. Yang, and J. Zhang. 2013. Combination of site-specific nitrogen management and alternate wetting and drying irrigation increases grain yield and nitrogen and water use efficiency in super rice. Field Crops Res. 154:226–235.

Luo, L., H. Mei, X. Yu, H. Liu, and F. Feng. 2011. Water-saving and drought-resistance rice and its development strategy. Chin. Sci. Bull. (Chinese Ver.) 56:804–811.

Lynch, J. P. 2013. Steep, cheap and deep: an ideotype to optimize water and N acquisition by maize root systems. Ann. Bot. 112:347–357.

Ma, J., E. Ma, H. Xu, K. Yagi, and Z. Cai. 2009. Wheat straw management affects CH_4 and N_2O emissions from rice fields. Soil Biol. Biochem. 41:1022–1028.

Ma, Y. C., X. W. Kong, B. Yang, X. L. Zhang, X. Y. Yan, J. C. Yang, et al. 2013. Net global warming potential and greenhouse gas intensity of annual rice-wheat rotations with integrated soil-crop system management. Agric. Ecosys. Environ. 164:209–219.

Maclean, J. L., D. C. Dawe, B. Hardy, and G. P. Hettel. 2002. Rice Almanac, 3rd edn. International Rice Research Institute, Los Banos, Philippines.

Millar, N., J. K. Ndufa, G. Cadisch, and E. M. Baggs. 2004. Nitrous oxide emissions following incorporation of improved-fallow residues in the humid tropics. Global Biogeochem. Cycl. 18:GB1032.

Minamikawa, K., and N. Sakai. 2006. The practical use of water management based on soil redox potential for decreasing methane emissions from a paddy field in Japan. Agric. Ecosyst. Environ. 116:181–188.

Mosier, A., A. Halvorson, C. Reule, and X. Liu. 2006. Net global warming potential and greenhouse gas intensity in irrigated cropping systems in northeastern Colorado. J. Environ. Qual. 35:1584–1598.

Osaki, M., T. Shinano, M. Matsumoto, T. Zheng, and T. Tadano. 1997. A root-shoot interaction hypothesis for high productivity of field crops. Soil Sci. Plant Nutr. 43:1079–1084.

Ponnamperuma, F. N. 1984. Straw as a source of nutrients for wetland rice. Pp. 117–136 in F. N. Ponnamperuma, ed. Organic matter and rice. Los Baños, Philippines, International Rice Research Institute.

Post, W. M., R. C. Izaurralde, L. K. Mann, and N. Bliss. 2001. Monitoring and verifying changes of organic carbon in soil. Clim. Change 51:73–99.

Raison, R. J. 1979. Modification of the soil environment by vegetation fires, with particular reference to nitrogen transformations: a review. Plant Soil 51:73–108.

Ramasamy, S., H. F. M. ten Berge, and S. Purushothaman. 1997. Yield formation in rice in response to drainage and nitrogen application. Field Crops Res. 51:65–82.

Sah, R. N., and S. D. S. Mikkelsen. 1983. Availability and utilization of fertilizer nitrogen by rice under alternate flooding; I. Kinetics of available nitrogen under rice culture. Plant Soil 75:221–226.

Shan, J., and X. Y. Yan. 2013. Effects of crop residue returning on nitrous oxide emissions in agricultural soils. Atmos. Environ. 71:170–175.

Stewart, C. E., K. Paustian, R. T. Conant, A. F. Plante, and J. Six. 2007. Soil carbon saturation: concept, evidence and evaluation. Biogeochem 86:19–31.

Wang, Y., C. Hu, B. Zhu, H. Xiang, and X. He. 2010. Effects of wheat straw application on methane and nitrous oxide emissions from purplish paddy fields. Plant Soil Environ. 56:16–22.

West, T. O., and W. M. Post. 2002. Soil organic carbon sequestration rates by tillage and crop rotation: a global data analysis. Soil Sci. Soc. America J. 66:1930–1946.

Xue, Y., H. Duan, L. Liu, Z. Wang, J. Yang, and J. Zhang. 2013. An improved crop management increases grain yield and nitrogen and water use efficiency in rice. Crop Sci. 53:271–284.

Yan, X., K. Yagi, H. Akiyama, and H. Akimoto. 2005. Statistical analysis of the major variables controlling methane emission from rice fields. Global Change Biol. 11:1131–1141.

Yang, J., J. Zhang, Z. Wang, Q. Zhu, and L. Liu. 2002. Abscisic acid and cytokinins in the root exudates and leaves and their relationship to senescence and remobilization of carbon reserves in rice subjected to water stress during grain filling. Planta 215:645–652.

Yang, J., J. Zhang, Z. Wang, Q. Zhu, and L. Liu. 2003. Involvement of abscisic acid and cytokinins in the senescence and remobilization of carbon reserves in wheat subjected to water stress during grain filling. Plant Cell Environ. 26:1621–1631.

Yang, C., L. Yang, Y. Yang, and Z. Ouyang. 2004. Rice root growth and nutrient uptake as influenced by organic manure in continuously and alternately flooded paddy soils. Agric. Water Manage. 70:67–81.

Yang, J., K. Liu, Z. Wang, Y. Du, and J. Zhang. 2007. Water-saving and high-yielding irrigation for lowland rice by controlling limiting values of soil water potential. J.

Integr. Plant Biol. 49:1445–1454.

Yang, L., Y. Wang, K. Kobayashi, J. Zhu, J. Huang, H. Yang, et al. 2008. Seasonal changes in the effects of free-air CO_2 enrichment (FACE) on growth, morphology and physiology of rice root at three levels of nitrogen fertilization. Global Change Biol. 14:1–10.

Yao, F., J. Huang, K. Cui, L. Nie, J. Xiang, X. Liu, et al. 2012. Agronomic performance of high-yielding rice variety grown under alternate wetting and drying irrigation. Field. Crop. Res. 126:16–22.

Yao, Z., X. Zheng, R. Wang, B. Xie, K. Butterbach-Bahl, and J. Zhu. 2013. Nitrous oxide and methane fluxes from a rice-wheat crop rotation under wheat residue incorporation and no-tillage practices. Atmos. Environ. 79:641–649.

Yoshida, S., D. Forno, J. Cock, and K. Gomez. 1976. Laboratory manual for physiological studies of rice. International Rice Research Institute, Los Baños, Philippines, 76 p.

Zhang, H., S. Zhang, J. Zhang, J. Yang, and Z. Wang. 2008. Postanthesis moderate wetting drying improves both quality and quantity of rice yield. Agron. J. 100:726–734.

Zhang, H., Y. Xue, Z. Wang, J. Yang, and J. Zhang. 2009. An Alternate wetting and moderate soil drying regime improves root and shoot growth in rice. Crop Sci. 49:2246–2260.

Zhang, H., T. Chen, Z. Wang, J. Yang, and J. Zhang. 2010. Involvement of cytokinins in the grain filling of rice under alternate wetting and drying irrigation. J. Exp. Bot. 61:3719–3733.

Zhang, H., H. Li, L. Yuan, Z. Wang, J. Yang, and J. Zhang. 2012. Post-anthesis alternate wetting and moderate soil drying enhances activities of key enzymes in sucrose-to-starch conversion in inferior spikelets of rice. J. Exp. Bot. 63:215–227.

Zou, J., Y. Huang, X. Zheng, and Y. Wang. 2007. Quantifying direct N_2O emissions in paddy fields during rice growing season in mainland China: dependence on water regime. Atmos. Environ. 41:8030–8042.

11

Sweet sorghum ideotypes: genetic improvement of stress tolerance

Sylvester Elikana Anami[1,2], Li-Min Zhang[1], Yan Xia[1], Yu-Miao Zhang[1], Zhi-Quan Liu[1] & Hai-Chun Jing[1]

[1]Key Laboratory of Plant Resources, Institute of Botany, Chinese Academy of Sciences, Beijing 100093, China
[2]Institute of Biotechnology Research, Jomo Kenyatta University of Agriculture and Technology, Nairobi, Kenya

Keywords
Abiotic stress, biotic stress, Herbicides, Sweet sorghum

Correspondence
Hai-Chun Jing, Key Laboratory of Plant Resources, Institute of Botany, Chinese Academy of Sciences, Beijing 100093, China.

E-mail: hcjing@ibcas.ac.cn

Funding Information
No funding information provided.

Abstract

Stress tolerance is a prerequisite for the success of biofuel production, which normally requires the use of marginal lands and nonfood biofuel feedstocks. Sorghum is known for its ability to withstand stress conditions, however, terminal stresses threaten its growth and development negatively impacting yield and sugar accumulation. It is crucial, therefore, that research aimed at developing sorghum resistance to stress factors should be pursued to expand the range of its growth to marginal and barren soils to meet the needs of a growing population, changing diets, and biofuel production. In this context, the leaf architectural trait of stay-green drought tolerance, in addition to salinity, cold, and aluminium toxicity and biotic stress tolerance and their genetic basis discussed in this review are expected to be available in future sweet sorghum ideotypes. Also highlighted is the key role of efficient management of farming systems, in particular the use of herbicides to control weeds, to ensure the sustainability of the sweet sorghum biomass productions.

Introduction

Abiotic and biotic stress limits plant growth, crop productivity, and biofuel production. Changes in precipitation patterns due to climate change and meteo-climatic variability have become a critical issue and a limiting factor for the crops under rain-fed systems and for the water resource when irrigation is applied. In arid and semiarid areas where water is limiting, the cultivation of irrigated energy crops could exacerbate the problem of competitiveness with food crops for the water use. Therefore, drought-tolerant energy crops should be the preferred choice in terms of both adaptation and environmental sustainability.

Sorghum originated from Africa and is currently the fifth most important cereal crop in the world and a staple crop for humans and other animals for food, feed, fodder, fiber, and fuel. Cultivation of sweet sorghum for the production of bioenergy is an attractive option to cope with the challenges of climate change to which adaptation is necessary in order to maintain good levels of productions (Berndes et al. 2003; Sims et al. 2006; Orlandini et al. 2007; Dalla Marta et al. 2014).

Sweet sorghum accumulate soluble sugars in the stalk at the expense of panicle production, these sugars could be mechanically extracted and directly fermented to obtain first generation bioethanol. Abiotic stress is a serious environmental obstacle for sugar production in sweet sorghum and strongly threatens biomass production and biofuel yields (Zegada-Lizarazu and Monti 2013). For instance, the structural carbohydrates (cellulose, hemicelluloses, and lignin) and biomass yields in sweet sorghum are significantly affected by drought stress (Zegada-Lizarazu and Monti 2013). In addition, during the 2010–2011 La Niña period in East Africa, drought led to sharp declines in the production of sorghum, a staple food in this semiarid region. For Somalia, the total sorghum production in 2011 was 25 kilotons, more than 80% below normal and the lowest for the last decade (Anyamba et al. 2014). However, sweet sorghum is considered water stress-resistant and suitable for arid and semiarid marginal areas (Staggenborg et al. 2008), due to morphophysiological characteristics which confer the drought tolerance (Zegada-Lizarazu and Monti 2013), and to the C_4 photosynthetic system which allows efficient CO_2 fixation and an outstanding dry matter accumulation (Mastrorilli et al. 1999).

The high sugar content, and therefore the sweetness in sweet sorghum attracts about 150 insect pests throughout its life cycle, which negatively impacts on biomass production (Guo et al. 2011). Common examples of pests that can severely damage a sorghum crop include the Miridae and Lygaeidae (Kruger et al. 2008), sorghum midge, Greenbug, fall armyworm, corn borers (Munson et al. 1993; Wu and Huang 2008; Damte et al. 2009), grasshoppers, sorghum shoot fly, corn rootworm, and sorghum aphid. Sorghum may also be affected by a number of diseases including anthracnose, down mildew, and Fusarium.

Climate change is associated with an increase in the frequency of heat stress, droughts, and floods (Kim et al. 2014) that negatively affect crop yields and biomass production and the ability of the climate smart sorghum to adapt and yield under such harsh environment. Resistance to abiotic and biotic stress is crucial in determining the sustainability of food and biofuel production in future (http://dialogues.cgiar.org/blog/millets-sorghum-climate-smart-grains-warmer-world/).

Here, we review the abiotic and biotic stress resistance, key traits expected to become available in new sweet sorghum ideotypes dedicated for biofuel production in the future and the control of these traits at the genetic level. We highlight the key role of a proper management of the farming systems, in particular, the use of herbicides to control weeds, to ensure the sustainability of the sweet sorghum biomass and biofuel productions.

QTLs related to abiotic and biotic stress tolerance in sorghum

Traditional breeding and QTL analysis have been applied for the identification of genes responsible for biotic and abiotic stress tolerance in crops plants (Collins et al. 2008; Takeda and Matsuoka 2008). Thus, 350 QTLs related to abiotic and biotic stress tolerance in sorghum (Table 1) were identified. These genetic loci have the potential to be utilized in developing superior sorghum ideotypes for various agroecological climates through marker-assisted breeding and genetic engineering once candidate genes have been fine mapped. In total, 51 and 182 loci were found to have known physical and genetic map positions, respectively. To have an idea of their chromosomal distribution in relation to biotic and abiotic stress traits they

control, an atlas map (Fig. 1) was generated. The outer rectangular marks indicate QTLs with known genetic position and the inner circular marks shows QTLs with known physical map positions on sorghum chromosomes. The next-generation sequencing and advanced metabolic profiling might impact the field of QTL analysis and facilitate the cloning of more genes responsible for tolerance to abiotic and biotic stresses. The ability to resequence a large number of F2 or recombinant inbred lines coupled with statistical linkage analysis could open the way for a very rapid and new type of marker-assisted mapping at the genome or metabolome level (Zheng et al. 2011).

Leaf stay-green drought resistance trait in sorghum

Sorghum is a drought-tolerant crop due to its ability to display morphological changes such as dense and deep root system, reduction in transpiration through leaf rolling and stomatal closure, and lowering of metabolic processes to near dormancy in response to terminal stress (Schittenhelm and Schroetter 2014). In fact, sorghum can survive prolonged dry periods, then 'resurrect' and resume growth once the soil moisture becomes available. However, depending on the severity, sorghum still suffers yield and biomass losses of up to 90% (House 1985). Recently, drought led to a sharp decline in agricultural production of sorghum up to 80% below the normal in Somalia during the 2010–2011 La Nina period (Anyamba et al. 2014). The impact of drought stress is greater during the grain-filling stage and causes premature leaf death, plant senescence, stalk lodging, and charcoal rot with poor yield of seed and stover. Cultivars tolerant to drought stress at postflowering stage are referred to as stay-green cultivars.

Stay-green is an integrated heritable drought adaptation trait characterized by a distinct green leaf phenotype during grain filling or postflowering under drought (Borrell et al. 2014a), Indeed, genetic studies show that QTLs for temperature and drought responses coincide with loci for leaf senescence, and in numerous examples of improvements in stress tolerance achieved by simultaneous selection for stay-green (Ougham et al. 2008; Vijayalakshmi et al. 2010; Jordan et al. 2012; Emebiri 2013). Stay-green trait is characterized either as cosmetic in which a lesion interferes with an early step in chlorophyll catabolism or functional in which the transition from the carbon capture period to the nitrogen mobilization (senescence) phase of canopy development is delayed, and/or the senescence syndrome proceeds slowly. Alteration in hormone metabolism and signaling, particularly affecting the networks involving cytokinins and ethylene, could contribute to the stay-green phenotype.

Table 1. The biofuel-associated traits of stay-green drought resistance trait, resistance to abiotic and biotic stresses, and their genetic determinants.

Traits	Trait category	No. of QTLs	QTL names	References
Abiotic stress tolerance	Stay-green/leaf senescence	1	Stg1 (green leaf area retention)	Subudhi et al. (2000), Xu et al. (2000), Kebede et al. (2001), Sanchez et al. (2002), Harris et al. (2007), Sabadin et al. (2012)
		1	Stg2 (green leaf area retention)	
		1	Stg3 (green leaf area retention)	
		1	Stg4 (green leaf area retention)	
		1	St2-1 (Stay green 2-1)	Sabadin et al. (2012)
		1	St2-2(Stay green 2-2)	
		1	St3 (Stay green 3)	
		1	St4 (Stay green 4)	
		1	St6 (Stay green 6)	
		1	St8 (Stay green 8)	
		1	St9 (Stay green 9)	
		1	St10 (Stay green 10)	
		1	Stg C.2 (Stay green C.2)	Kebede et al. (2001)
		1	Stg C.1 (Stay green C.1)	
		1	Stg B (Stay green B)	
		1	Stg E (Stay green E)	
		1	Stg D (Stay green D)	
		1	Stg A (Stay green A)	
		3	Ldg G (Lodging G), Ldg F (Lodging F), Ldg J (Lodging J)	
		4	Prf C(Preflowering drought tolerance C), Prf F(Preflowering drought tolerance F), Prf E(Preflowering drought tolerance E), Prf G(Preflowering drought tolerance G)	
		1	Stg F(Stay green F)	
		9	(% GL15) (green leaf area percentages)	Haussmann et al. (2002)
		14	(% GL30)green leaf area percentages)	
		13	(% GL45)green leaf area percentages)	
		6	t-E8/102, th19/50, tD9/103, t329/132, bB20/205, umc84	Tuinstra et al. (1997)
		1	SGA (Stay-green A)	Crasta et al. (1999)
		1	SGD (Stay-green D)	
		1	SGG (Stay-green G)	
		1	SGB (Stay-green B)	
		1	SG1.1 (Stay-green 1.1)	
		1	SG1.2 (Stay-green 1.2)	
		1	SGJ (Stay-green J)	
		1	MB6-84-TS136	Tao et al. (2000)
		1	TXS654-TXS943	
		1	ST1668-TXS558	
		1	CDO460-SSCIR165	
		2	QLsn.txs-B, QLsn.txs-Ea/Eb (Leaf senescence)	Feltus et al. (2006)
	SPAD at booting	3	QSpadb-dsr09-1, QSpadb-dsr06-1, QSpadb-dsr03	Reddy et al. (2014)
	SPAD values at maturity	1	QSpadm-dsr09-1	
	Green leaves at booting	4	QGlb-dsr01-1a, QGlb-dsr04-1, QGlb-dsr02-1, QGlb-dsr04-3	
	Green leaves at maturity	2	QGlm-dsr04-1, QGlm-dsr09-2	
	Per cent green leaves retained at maturity	2	QPglm-dsr04-2, QPglm-dsr09-2	
	Green leaf area at booting	2	QGlab-dsr10-1, QGlab-dsr02-1,	
	Green leaf area at maturity	1	QGlab-dsr02-2	

(Continued)

Table 1. Continued.

Traits	Trait category	No. of QTLs	QTL names	References
	Rate of leaf senescence	1	*QRls-dsr10-1*	
	Cold germinability/field emergency and early seedling vigor	16	*Germ30-1.1, Germ30-1.2, Germ30-2.1, Germ12-2.1, Germ12-9.2; Fearlygerm-1.2, Fearlygerm-7.1, Fearlygerm-9.2, Fearlygerm-9.3, Fearlygerm-1.1, Fearlygerm-4.1, Fearlygerm-9.1, Fearlygerm-9.3; Xtxp43, Xtxp51, Xtxp211*	Burow et al. (2011a,b),
	Aluminium tolerance	1	*Alt(SB)*	Magalhaes et al. (2007)
	Salinity stress (germination vigor, germination percentage, shoot height, root length, shoot fresh weight, root fresh weight, total fresh weight, shoot dry weight, root dry weight, total dry weight)	38	*qGV2-1, qGV2-2, qGV3; qGV1-1, qGV1-2, qGV4; qGP1, qGP2,qGP7-1, qGP7-2; qSH8, qSH1,qSH2, qSH4, qSH10;qRL1, qRL8, qRL3, qRL10-1, qRL10-2; qSFW8, qSFW9-1, qSFW4,qSFW9-2; qRFW6-1, qRFW2, qRFW6-2; qTFW6, qTFW9-1, qTFW1, qTFW4, qTFW9-2; qSDW4, qSDW9, qSDW6;qRDW3, qRDW6;qTDW6, qTDW8*	Wang et al. (2014a,b)
	Number of rhizomes/ subterranean rhizomes/ number of rhizome-derived shoots/ Overwintering	7	*pSB300a-pSBO88, pSB195-SH068, pSBJ02-pSB158; Overwintering2011A, Overwintering2011B; Ln2010RDS, Ln2010Dist*	Paterson et al. (1995), Washburn et al. (2013)
Subtotal		159		
Biotic stress tolerance	Midge disease egg count	2	*Flanking markers (ST698- RZ543, ST1017 -SG14)*	Tao et al. (2003)
	Midge disease pupal infestation	3	*Flanking markers (ST698- RZ543, ST1017 -SG14, TXS1931-SG37)*	
	Target leaf spot	1	*tls (Target leaf spot)*	Mohan et al. (2009)
	Zonate leaf spot	1	*Zls (Zonate leaf spot)*	
	Drechstera leaf plight	1	*Dls (Drechstera leaf plight)*	
	Rust resistance	8	*BNL5.09, TXS1625, RZ323, ISU102, ISU102, TXS2042, PSB47, TXS422*	Tao et al. (2003)
	Anthracnose resistance	14	*7 QTLs not named, Cg1, Locus 1-8, QAnt1, QAnt4, SC326-6, SCA 12, OPJ 0_{11437}*	Boora et al. (1998), Klein et al. (2001), Singh et al. (2006), Singh et al. (2006), Perumal et al. (2009), Mohan et al. (2010), Upadhyaya et al. (2013)
	Percentage ergot infection	9	*Not named*	Parh et al. (2008)
	Pollen quantity	5	*Not named*	
	Pollen viability	4	*Not named*	
	Greenbug resistance to biotype I and K, C, and K/ greenbug feeding	34	*B18-885, OPC01-880, Sb5-214, Sb1-10, SbAGB03, Sb6-84, SbAGA01, OPA08-1150, OPB12-795; Ssg1, Ssg2, Ssg3, Ssg4, Ssg5, Ssg6,Ssg7, Ssg8, Ssg9; (8 unnamed QTLs);QSsgr-09-01, QSsgr-09-02;Qstsgr-sbi09ii, Qstsgrsbi09iii, Qstsgr-sbi09i, Qstsgr-sbi09iv;Xtxp16–Starssbem162, Starssbem162–Starssbem265*	Agrama et al. (2002), Katsar et al. (2002), Nagaraj et al. (2005), Wu and Huang (2008), Punnuri et al. (2013)
	Head bug resistance/ damage	10	*SbRPG943-RZ630, RZ476-SbRPG872, SbRPG667-CDO580, BNL5.37-SbRPG749, BNL5.37-SbRPG749, BNL5.37-SbRPG749, CDO20-C223, RZ630-SbRPG826, RZ244b-SbRPG852, mAGB03-UMC139*	Deu et al. (2005)
	Leaf scotch	1	*QLsc.txs-B (Leaf scotch)*	Feltus et al. (2006)

(Continued)

Table 1. Continued.

Traits	Trait category	No. of QTLs	QTL names	References
	Stalk rot resistance(No of internode, length of infection, percent lodging)	8	*xtxp297, xtxp213, AC13, xtxp343, xtxp176, (3 unnamed QTL for lodging resitance*	Reddy et al. (2008), Felderhoff et al. (2012)
	Shoot fly leaf glossiness	8	*QGs.dsr-3, QGs.dsr-5, QGs.dsr-6, QGs.dsr-10; QGs.dsr-1, QGs.dsr-4.1, QGs.dsr-2, QGs.dsr-4.2*	Satish et al. (2009), Aruna et al. (2011)
	Shoot fly seedling vigor	8	*QSv.dsr-3, QSv.dsr-6.1, QSv.dsr-6.2, QSv.dsr-10; QSv.dsr-1.1, QSv.dsr-1.2, QSv.dsr-2, QSv.dsr-9*	Satish et al. (2009), Aruna et al. (2011)
	Shoot fly oviposition	7	*QEg21.dsr-1, QEg21.dsr-7, QEg21.dsr-9, QEg21.dsr-10, QEg28.dsr-5, QEg28.dsr-7, QEg28.dsr-10*	Satish et al. (2009)
	Shoot fly deadheart	13	*QDh.dsr-5, QDh.dsr-10.3, QDh.dsr-10.4; QDh.dsr-1.1, QDh.dsr-1.2, QDh.dsr-2, QDh.dsr-6.1, QDh.dsr-6.2, QDh.dsr-7.1, QDh.dsr-7.2, QDh.dsr-9, QDh.dsr-10.1, QDh.dsr-10.2*	Satish et al. (2009), Aruna et al. (2011)
	Shoot fly adaxial trichome density	4	*QTdu.dsr-10.1, QTdu.dsr-10.2; QTdu.dsr-7, QTdu.dsr-10*	Satish et al. (2009), Aruna et al. (2011)
	Shoot fly abaxial trichome density	7	*QTdl.dsr-1.1, QTdl.dsr-1.2, QTdl.dsr-4, QTdl.dsr-6, QTdl.dsr-10.1, QTdl.dsr-10.2; QTdl.dsr-3*	Satish et al. (2009), Aruna et al. (2011)
	Root and crown rot resistance (*Pc* locus)	1	*PC*	Nagy et al. (2007)
	Striga resistance	39	*38 QTLs Not named; lgs*	Haussmann et al. (2004), Satish et al. (2012)
	Resistance to down mildew	3	*bin 2.04/05, bin 3.04/05, bin 6.05*	Nair et al. (2005)
Subtotal		191		
Total QTLs		350		

Cytokinin production increases growth and yield by improving foliar stay-green indices under drought conditions and improving processes that impact grain filling and grain number (Wilkinson et al. 2012). This suggests that the stay-green phenotype could be achieved by the biotechnological expression of isopentenyltransferase (*IPT*) gene whose protein catalyzes the rate-limiting step in cytokinin biosynthesis. Indeed, transgenic tobacco plants overexpressing *IPT* gene produced more trans-zeatin, did not senesce, had water content of 86%, maintained photosynthetic activity, and resurrected upon rewatering (Rivero et al. 2010). In addition, members of the WRKY and NAC families, and an ever-expanding cast of additional senescence-associated transcription factors, are identified by mutations that result in stay-green phenotype (Thomas and Ougham 2014).

Retention of chlorophyll in leaves of stay-green genotypes is associated with enhanced capacity to continue normal grain fill and maintenance of the ability to undergo photosynthesis for longer periods under drought conditions, reduced lodging, high stem carbohydrate content and grain weight, and resistance to charcoal stem rot (McBee et al. 1983; Borrell et al. 2000b; Burgess et al.

2002; Jordan et al. 2012). Hybrids involving stay-green sources produced close to 47% more biomass between anthesis and maturity in comparison with senescent checks (Borrell et al. 2000a). Moreover, the stay-green cultivar Sorcoll-141/07 had high plant height and green leaf number that contribute to its high biomass (Yemata et al., 2014). Leaves of stay-green cultivars have higher nutritional quality (Jordan et al. 2012). The high leaf nitrogen content and the simultaneous prolonged photosynthesis associated with stay-green trait (Borrell et al. 2000b) are correlated with higher sugar production in sorghum (Serrão et al. 2012) suggesting that leaf nitrogen concentrations and increased photosynthetic capacity are indicators for predicting sugar production in sweet sorghum. Therefore, breeding for stay-green trait specifically associated with high stem carbohydrates and leaf nitrogen will undoubtedly boost biofuel production in sweet sorghum cultivars.

The sources of stay-green trait used in most of the genetic studies and associated breeding programmes are lines BTx642, formally (B35), SC 56, E36-1, and KS19 (Haussmann et al. 2002; Mahalakshmi and Bidinger 2002; Hash et al. 2003) and have been reported to have greater

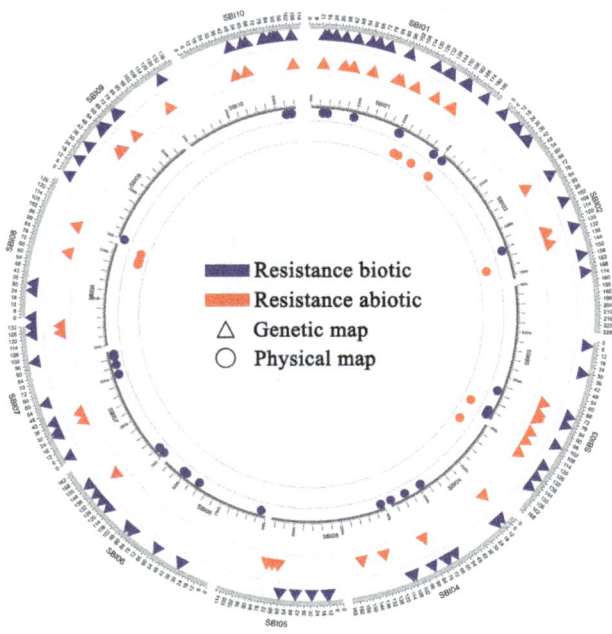

Figure 1. QTLs atlas map for biofuel-associated abiotic and biotic resistance traits in sorghum.

adaptation to drought stress through osmotic adjustment (Zhou et al. 2013). Indeed, the leaf relative water content of stay-green lines is much higher than those in nonstay-green lines, indicating that the stay-green lines keep the stalk transportation system functioning under severe drought conditions (Xu et al. 2000). Four loci *Stg1, Stg2, Stg3*, and *Stg4* (Table 1) controlling the stay-green trait in sorghum have consistently been identified across several environments in different mapping populations derived from crosses with BTx642, SC 56, E36-1, and KS19 lines, though each study reported varying phenotypic variation contributed by each QTL (Crasta et al. 1999; Subudhi et al. 2000; Tao et al. 2000; Xu et al. 2000; Kebede et al. 2001; Sanchez et al. 2002; Feltus et al. 2006; Harris et al. 2007; Sabadin et al. 2012). This suggests that although the ability of leaves to delay senescence has a genetic basis in sorghum, the expression of the character is strongly influenced by environmental cues (van Oosterom et al. 1996). QTLs controlling chlorophyll content colocated with the *Stg1, Stg2, Stg3*, and *Stg4* loci controlling stay-green (Subudhi et al. 2000) and (Xu et al. 2000). Therefore, chlorophyll content or loss of chlorophyll is a marker for the stay-green trait in sorghum during grain filling under drought.

The consistency of the four QTLs in various mapping populations across various environments and their combined phenotypic variation contribution of 54% to the stay-green drought trait suggest that they are stable and major QTLs for stay-green drought trait in sorghum

(Tanksley 1993; Xu et al. 2000; Sanchez et al. 2002). Fifteen (15) novel QTLs (Table 1) for various measures of stay-green trait have been identified using a genetic linkage map based on 245 F_9 Recombinant Inbred Lines (RILs) derived from a cross between M35-1 (more senescent) and B35 (less senescent) (Reddy et al. 2014). The phenotypic variation explained by each QTL ranged from 3.8 to 18.7%. Several other stay-green loci have also been reported (Table 1) but are generally unstable across environments (Crasta et al. 1999) with the potential to spontaneously reverts back to their parental phenotypes. The sorghum line E36-1, for instance, displays stay-green phenotype when grown under drought conditions in the field (van Oosterom et al. 1996), but not under well-watered conditions. (Tuinstra et al. 1997; Kebede et al. 2001; Thomas and Ougham 2014) identified six genetic loci controlling preflowering drought stress tolerance in sorghum from RILs derived from the crosses, SC 56 × Tx7000 and Tx7078 × B35, respectively with phenotypic variation contributed ranging between 15 to 40% under different environments, suggesting a strong genotype × environment (G × E) interaction at these loci.

Under water limited conditions, the stay-green alleles have been implicated to individually enhance grain and biomass yields in sorghum by modifying canopy development and water uptake patterns (Borrell et al. 2014b), indicating that stay-green phenotype and biomass yield could be achieved via the modification of root architecture (Mace et al. 2012), canopy development through reduction in tillering via increased size of lower leaves (Borrell et al. 2000a), or both. Thus far, breeders have transferred through marker-assisted backcrossing the stay-green trait into elite cultivars (Hash et al. 2003). The approach has been compromised because stress-related QTLs are dependent on the environmental conditions to which they were characterized (high G × E interaction) (Collins et al. 2008). In addition, different QTLs associated with stress-related traits can explain only a low percentage of the variation in the phenotype and that the effects of a favorable allele could not be transferred due to epistatic interactions (Peleg et al. 2009). Therefore, identifying QTLs of major effect that are independent of the particular genetic background and cloning the genes in the QTL could enhance breeding through biotechnology.

Functional analysis of the genes can be significantly aided through the application of reverse genetics approaches such as RNA interference (RNAi) and the type II CRISPR/Cas system (Jiang et al. 2013) in order to characterize the individual gene function(s). Emphasis should be given to forward genetics studies where the identified genes can be expressed in genotypes that have been already selected for their adaptation to stressful environmental conditions. The availability of the sorghum genome may aid in fine mapping of the candidate genes responsible for the stay-green trait through increasing marker density within the target chromosomal region in addition to increasing the number of segregating population for which phenotypic information linked to the QTL can be obtained.

Proline accumulates to high levels in many plant species in response to environmental stresses and its role has been extensively investigated under stress conditions (Verbruggen and Hermans 2008). Recently, the Arabidopsis proline dehydrogenase (*AtProDH2*) was found to be strongly expressed in senescent leaves and in roots (Funck et al. 2010), suggesting that proline could have a new role during plant developmental processes. Similar findings in oilseed rape (*Brassica napus*), demonstrated that the *BnaProDH2* genes are specifically expressed in vasculature in an age-dependent manner in roots and senescent leaves (Faës et al. 2014). Thus, indicating that such expression could be related to the provision of reducing power for cell degradation mechanisms when chloroplasts become dysfunctional, an early process in autophagy (Avila-Ospina et al. 2014). The catabolism of proline could also contribute to recycling of metabolites in senescent leaves and provides glutamate then glutamine, principal forms of nitrogen compounds transported in the phloem from senescing leaves to sink organs (Tilsner et al. 2005). Similar studies could be carried out in sorghum in order to link the role of proline in abiotic stress conditions and in the developmental process of leaf senescence and biomass production.

Molecular mechanisms involved in the response of sorghum to abiotic stress

MiRNA expression

MicroRNAs (miRNAs) are a recently discovered class of gene expression regulators that have also been linked to several plant stress responses (Sunkar et al. 2007; Rajwanshi et al. 2014; Zhai et al. 2014) and in the biosynthesis pathways of carbon, glucose, starch, fatty acid, and lignin and in xylem formation, which could aid in designing next-generation sweet sorghum for biomass and biofuel. Differentially expressed miRNAs involved in the regulation of transcription (*bZIPs*, *MYBs*, *HOXs*), signal transduction (phosphoesterases, kinases, phosphatases), carbon metabolism (*NADP-ME*), detoxification (*CYPs*, *GST*, *AKRs*), osmoprotection mechanisms (*P5CS*), and stability of protein membranes (*DHN1*, *LEA*, *HSPs*) were upregulated upon imposition of drought stress in a four-leaf-old sorghum genotype IS1945 (Pasini et al. 2014), indicating that these drought-related genes could be used to screen for potential drought tolerance in other sorghum genotypes including sweet sorghum. Indeed, rice miRNA 169 g,

upregulated during drought stress (Zhao et al. 2007), has five sorghum homologs (*sbi-MIR169c*, *sbi-MIR169d*, *sbi-MIR169.p2*, *sbi-MIR169.p6*, and *sbi-MIR169.p7*), suggesting that the miRNA may be involved in many different processes related to drought stress resistance. The computationally predicted targets of the sbi-MIR169 subfamily comprise members of the plant nuclear factor Y (*NF-Y*) B transcription factor family, linked to improved performance in Arabidopsis, and maize under drought stress (Nelson et al. 2007). *GmNFYA3* gene, a target of miR169, is a positive regulator of plant tolerance to drought stress and has potential applications in molecular breeding to enhance drought tolerance in crops (Ni et al. 2013).

Using a deep sequencing approach to generate a genome-wide transcriptome of foxtail millet after exposure to simulated drought stress, one long noncoding RNAs (lncRNAs) was found to share sequence conservation and colinearity with its counterpart in sorghum suggesting that the lncRNAs in sorghum have the potential to have an impact on drought-regulated gene expression (Qi et al. 2013). The analysis of *cis*-elements of miRNA targets including transcription factors, genes for chaperonins, and metabolic enzymes and other genes necessary for proper plant development provides molecular evidence for the possible involvement of miRNAs in the process of abiotic stress tolerance in sorghum, indicating that miRNAs could play an important role in water stress tolerance in future sorghum studies (Ram and Sharma 2013). Therefore, miRNA 169 could be an excellent target for the generation of drought-resistant sweet sorghum genotypes through genetic engineering.

Auxin-related genes

Auxin-related gene families in *Sorghum bicolor* have also been implicated in abiotic stress response. The Gretchen Hagen3 (GH3) *SbGH3* and lateral organ boundaries (LBD) (*SbLBD*) genes, expressed at low levels under natural condition, were highly induced by salt and drought stress consistent with their products being involved in both abiotic stresses. Three genes, *SbIAA1*, *SbGH3-13*, and *SbLBD32*, were highly induced under all the four treatments, Indole-3-Acetic Acid, brassinosteroids, salt, and drought. The analysis provided new evidence for role of auxin in stress response, implying there are cross talk between auxin, brassinosteroids, and abiotic stresses (Wang et al. 2010).

Moisture stress triggered the upregulation of more transcription factor genes of MADS-box, Auxin Responsive Factors, Heme Activator Protein 2, Multiprotein Bridging Factors, and Homeobox families in root tissues compared to shoot tissues in sorghum (Aglawe et al. 2012). Under ABA, salt and drought treatments, sorghum auxin transporters *SbPIN4*, *-5*, *-8*, *-9*, and *-11* were highly increased, whereas *SbPIN1-3*, *-6*, *-7*, and *-10* were almost inhibited by all three treatments (Shen et al. 2010). The expression levels of *SbLAX1*, *-2*, *-4*, and *-5* compared with *SbLAX3* in leaves were lower than those in roots when treated with ABA. However, the response of *SbLAX* genes to salt and drought stresses was irregular, with *SbLAX4* expression downregulated dramatically under the stresses. Interestingly, transcription of the *SbPGP* gene family was almost inhibited in roots under salt treatment. *SbPGP1*, *-2*, *-5*, *-13*, *-14*, and *-15* were induced in roots under ABA treatment, whereas *SbPGP2*, *-3*, *-4*, *-7*, *-12*, and *-23* were induced in leaves under salt or drought stress. Under salt and drought treatment, *SbPGP13*, *-15*, *-17*, *-18*, *-20*, *-21*, and *-24* were all downregulated in both leaves and roots. Exploiting RNA-Seq technology in combination with the sorghum genome sequence (Paterson et al. 2009) and the SorghumCyc metabolic pathways database, (Dugas et al. 2011) characterized the sorghum transcriptome and reexamined the differential expression of sorghum genes in response to exogenous ABA and osmotic stress (Buchanan et al. 2005). Fifty differentially expressed drought-responsive gene orthologs specific to sorghum were identified for which no function had been previously assigned either in maize, rice, or Arabidopsis and were enriched for ABREs and CGTCA-motifs, or motifs that are involved in responses to ABA.

Transcription factors

The ethylene response factor family, members of the APETALA2 (AP2)/ERF transcription factor superfamily, is known to play an important role in plant adaptation to biotic and abiotic stress (Lata et al. 2014) and 105 sorghum ERF (*SbERF*) genes, categorized into 12 groups (A-1 to A-6 and B-1 to B-6) based on their sequence similarity have been identified in sorghum (Yan et al. 2013) Glutathione reductases (GRs) are important components of the antioxidant machinery that plants use to respond against abiotic stresses. Phylogenetic analysis identified two chloroplast GRs in sorghum that could possibly have a role in the modulation of abiotic stress. Since chloroplasts GR are also targeted to mitochondria suggest a combined antioxidant mechanism in both chloroplasts and mitochondria (Wu et al. 2013). In addition, phylogenetic analysis of the rice heterotrimeric G-protein complexes Gα subunit revealed high homology with sorghum. The promoter sequence analysis of RGA1(I) confirms the presence of stress-related *cis*-regulatory elements viz. ABA, MeJAE, ARE, GT-1 boxes, and LTR suggesting its active and possible independent roles in abiotic stress signaling. Furthermore, transcript profiling of RGA1(I) showed upregulation following NaCl, cold and drought stress, but under an elevated temperature, its transcript was downregulated. Heavy

metal(loid)s stress showed rhythmic response in ABA stress and strong upregulation. These findings provide critical evidence for the active role of G-protein complexes in regulation of abiotic stresses in rice and possibly in sorghum and suggest that the Gα subunit of the heterotrimeric G-protein complexes could be exploited in the development of abiotic stress tolerance in sorghum. Genes coding for drought response element-binding (*DREB*) proteins regulate transcription of a large number of downstream genes involved in the plant response to abiotic stresses. An integration of abscisic acid, ethylene, auxin, and methyl jasmonate signaling was probably involved in regulating expression of the drought response through *DREB* transcription factors. The *SbEST8* gene was implicated to have a role in abiotic stress tolerance since imposition of drought resulted in rapid accumulation *SbEST8* mRNA in the germinating seeds of drought susceptible cultivar ICSV-272 (Dev Sharma et al. 2006).

Compatible solutes

The introduction of compatible solute synthesis pathway has emerged as a potential strategy for enhancing abiotic stress tolerance in crop plants (Rathinasabapathi 2000). The ability to synthesize and accumulate glycine betaine is widespread among angiosperms and is thought to contribute to salt and drought tolerance. Betaine aldehyde dehydrogenase *BADH1* and *BADH15* mRNA in sorghum were both induced by water deficit and their expression coincided with glycine betaine accumulation. The leaf water potential in stressed sorghum plants reached −2.3 MPa in the course of 17 days of water stress. Water deficit induced a 26-fold increase in glycine betaine levels and proline levels increased 108-fold (Wood et al. 1996). The upregulation of *Sorghum bicolor* glycine-rich RNA-binding protein designated as sbGR-RNP was induced by salinity and ABA and regulated by blue and red light, suggesting that there exists a cross talk between abiotic stress and light signaling in sorghum (Aneeta et al. 2002).

The mannitol biosynthetic pathway was engineered into *Sorghum bicolor* L. Moench cv. SPV462 with the *mtlD* gene encoding for mannitol-1-phosphate dehydrogenase from *E. coli*. The transgenic leaf segments were found to retain higher leaf water content when exposed to polyethylene glycol 8000 (−2.0 MPa) and maintained a 1.7- to 2.8-fold higher shoot and root growth, respectively, under NaCl stress (200 mmol/L) when compared to untransformed controls (Maheswari et al. 2010). These studies establish a role for a number of genes in modulating drought stress tolerance in sorghum. Therefore, functional characterization of these genes in sorghum including their overexpression or down regulation using genetic engineering could provide additional information as to their roles in broad abiotic stress tolerance.

Expression analysis of key stress inducible regulatory genes that play crucial roles in proline biosynthesis, *SbP5CS1* and *SbP5CS2*, revealed that the transcripts were upregulated after treatment of 10-day-old seedlings of sweet sorghum with drought, salt (250 mmol/L NaCl) and MeJA (10 μmol/L) indicating that the two genes could have the potential to be used in improving stress tolerance of sweet sorghum and other bioenergy feedstocks (Su et al. 2011).

Cold tolerance in sorghum

Soil temperatures below 15°C limit germination and seedling establishment for sorghum during early-season planting in temperate areas. Developing fast-growing sorghum seedlings is an important breeding goal for temperate climates since low spring time temperatures result in a prolonged juvenile development. In addition, this would allow expansion of sorghum to temperate region and for earlier planting in areas where it is being grown (Singh, 1985). In China, sorghum landrace, *kaoliang*, has poor agronomic characteristics though it exhibits higher seedling emergence and greater seedling vigor under cold conditions. The genetic basis of early-season cold tolerance in sorghum associated with germination, emergence, and vigor has been investigated and 15 QTLs have been identified (Table 1, (Burow et al. 2011a; Knoll and Ejeta 2008; Knoll et al. 2008). Using marker-assisted selection these desirable genomic regions can be introgressed into elite lines to improve early-season performance in sorghum. The quality of the messenger RNAs stored during embryo maturation on the mother plant, proteostasis, and DNA integrity play a major role in the germination phenotype. In addition, the sulfur amino acid metabolism pathway represents a key biochemical determinant of the commitment of the seed to initiate its development toward germination (Rajjou et al. 2012). Therefore, the characterization of molecular variables for germination and seed vigor under cold stress is expected to deliver new markers of seed quality that can be used in breeding programs and/or in biotechnological approaches to improve biomass yield in sweet sorghum. Further, a higher respiration rate is positively correlated with a higher germination rate and cultivars with higher respiration rate are likely to be resistant to early-season cold (Balota et al. 2010). Therefore, selection for a higher respiration rate can improve early-season vigor (germination, elongation, and growth rate in sorghum).

Rhizome formation trait is correlated and genetically linked to overwinter survival in sorghum (Washburn et al. 2013). The understanding of the genetic mechanisms

controlling overwintering has the potential to create perennial sorghums that can overwinter in climates where they previously could not. These overwintering sorghum types could be used for improvements in biofuel sorghum production by extending the period of biomass production and reducing production costs. In sorghum, rhizomatousness and overwintering are controlled by seven QTLs (Paterson et al. 1995; Washburn et al. 2013) Table 1). The QTLs were identified from a mapping population of a cross between BTx623 and *S. propinquum* and that regrowth after overwintering was associated with both rhizomatousness and tillering.

Salinity tolerance in sorghum

Salinity stress affects plant growth and productivity in many parts of the world and plants have developed adaptive responses to this external stress at the genetic level. For example, under sodium stress, *SbHKT1;4*, a member of the high-affinity potassium transporter gene family from *Sorghum bicolor*, functions to maintain optimal Na$^+$/K$^+$ balance (Wang et al. 2014b). Upon Na$^+$ stress *SbHKT1;4* expression was more strongly upregulated in salt-tolerant sorghum accessions, correlating with better balanced Na$^+$/K$^+$ ratio and enhanced plant growth. To gain insight into the genetic mechanism of salt tolerance at germination and seedling stage as a basis for improving salt tolerance in sorghum, (Wang et al. 2014a) identified 38 QTLs underlying salt tolerance (Table 1) from a 181 recombinant inbred lines derived from Shihong 137 and L-Tian. Six major QTLs with more than 10% phenotypic variation were detected at seedling stage under salt stress. These data indicate that the genetic mechanism for salt tolerance at germination and seedling stage in sorghum is different and that further research need to be done to identify genetic loci determining salt tolerance at different growth stages of sorghum during development.

Sorghum tolerance to Aluminium toxicity

Aluminum (Al) toxicity is an important limitation to food security in tropical and subtropical regions. In acidic soils, aluminum is solubilized into ionic forms (Al^{3+}), especially when the soil pH falls to lower than 5. This ionic form of Al is very toxic to plants, limiting the growth of roots either by inhibition of cell division, cell elongation, or by both. In this way, water and nutrient uptake by the roots is affected and as a consequence, plant growth and development is seriously hindered (Foy et al. 1993). Aluminum toxicity is, therefore, a major constraints for sorghum production in tropical and subtropical regions of the world (Doumbia et al. 1993, 1998). In addition to naturally occurring acid soils, agricultural practices may decrease soil pH, leading to yield losses due to Al toxicity. Elucidating the genetic and molecular mechanisms underlying sorghum Al tolerance is expected to accelerate the development of Al-tolerant cultivars. Using positional cloning, a gene encoding a member of the multidrug and toxic compound extrusion (MATE) family, an aluminum-activated citrate transporter, was identified as responsible for the major sorghum aluminum tolerance locus, Alt(Sb), on sorghum chromosome 3 (Magalhaes et al. 2007). These markers have been used by breeders to introgress rapidly the most favorable SbMATE alleles into sorghum germplasm, which is currently being field-tested in acid soils. Similar results have recently been demonstrated in maize where ZmMATE1 expression, controlled either by three copies of the target gene or by an unknown molecular mechanism, is responsible for Al tolerance mediated by QTL mapped on chromosome 6 (qALT6) (Guimaraes et al. 2014). Polymorphisms in regulatory regions of Alt(Sb) are likely to contribute to large allelic effects, acting to increase Alt(Sb) expression in the root apex of tolerant genotypes. Furthermore, aluminum-inducible Alt(Sb) expression is associated with induction of aluminum tolerance via enhanced root citrate exudation (Magalhaes et al. 2007). These information could allow scientist to identify superior Alt(Sb) haplotypes that can be incorporated via molecular breeding and biotechnology into acid soil breeding programs, thus helping to increase crop yields in developing countries where acidic soils predominate.

Resistance to biotic stresses

Insect pests

Sorghum biomass and sugar yield are severely affected by biotic stresses including about 150 insect pests with more than 100 of them occurring in Africa (Guo et al. 2011), and new parental lines having genes for various biotic stress tolerances have the potential to mitigate this negative effect. The most destructive pests are the lepidopteran stem borer (Chilo partellus) and the dipterans, midge (Stenodiplosis sorghicola) and shoot fly (Atherigona soccata). Given the wide host range of some of the insect pests, and low level of resistance in the cultivated germplasm against major sorghum pests such as stem borers, head bugs, and armyworms, it will be highly desirable to invoke molecular plant breeding approaches combining conventional plant resistance with novel genes from other sources such as Bacillus thuringiensis (Bt) toxic protein. Insecticidal crystal proteins (CRY) from Bacillus thuringiensis are very effective against the lepidopterans and dipterans. Bt and other genes including protease inhibitors, enzymes, secondary plant metabolites, and plant lectins

with insecticidal activities are being evaluated for eventual use in transforming cotton, maize, rice, sorghum, grain legumes, tobacco, potato, sugarcane, groundnuts and tomatoes and reducing losses due to these pests(Sharma et al. 2004; Visarada and Kishore 2007). A transgenic sorghum plant was generated carrying a synthetic gene, Bt cry1Ac, under the control of a wound-inducible promoter from a maize protease inhibitor gene (mpi) (Girijashankar et al. 2005). The transgenic sorghum had low levels of Bt protein of 1–8 ng/g of fresh leaf tissue. A moderate level of tolerance was reported, which in turn conferred partial protection against neonate larvae of the spotted stem borer (Chilo partellus). Transgenic sorghum plants expressing Bt cry1Ab gene displayed insect-resistance to pink rice borer (Sesamina inferens) (Zhang et al. 2009).

Sorghum midge is the most damaging pest of grain sorghum worldwide (Young and Teetes 1977). Though sweet sorghum accumulates sugar at the expense of grain, the damage caused on the grain of sweet sorghum by sorghum midge could impact biomass and sugar accumulations. At flowering, female midges oviposit into spikelets, and the larvae feed on the ovary during the following 2 weeks, resulting in the failure of kernel development. Using classical approach, over 40 sorghum cultivars resistant to midge have been identified (Sharma et al. 1999) and could be useful for use in resistance breeding programs and to mitigate against biomass and sugar lose in sweet sorghum. Two genetic mechanisms of midge resistance, antixenosis and antibiosis, have been resolved in a recombinant inbred population from the cross of sorghum lines ICSV745 × 90,562 (Tao et al. 2003). Two genetic regions (between loci ST698 and RZ543 of linkage group A and loci ST1017 and SG14 of linkage group G, respectively, were significantly associated with egg counts (antixenosis) and the degree of phenotypic variation explained by each region was 12% and 15%, respectively. Three genetic regions located on linkage group A, linkage group G, and linkage group J, respectively, were found to be associated with pupal infestation. The levels of phenotypic variations explained by each region are 8.8% and 15%, respectively. The other region associated with pupal counts is the interval between loci TXS1931 and SG37 on linkage group J. explained 33.9% of total variation in pupal counts (antibiosis). The identification of genes for different mechanisms of midge resistance will be particularly useful for exploring new sources of midge resistance and for gene pyramiding of different mechanisms for increased security in sorghum breeding through marker-assisted selection and for the development of agronomically superior sorghum hybrids (Tao et al. 2003). Indeed, a putative candidate gene (gm3) for the recessive gall midge resistance gene (gm3) in rice was identified using a mapping population consisting of 302 F_{10} recombinant inbred lines derived from the cross TN1 (susceptible)/RP2068-18-3-5 (Sama et al. 2014). Comparative genomics could, therefore, identify similar syntenic genomic regions in sorghum for incorporation into midge sorghum resistance breeding programmes.

Greenbug, *Schizaphis graminum* (Rondani) is one of the major insect pests of sorghum and can cause serious damage to sorghum plants, particularly in the US Great Plains. Identification of chromosomal regions responsible for greenbug resistance will facilitate both map-based cloning and marker-assisted breeding. A total of 36 QTLs have been identified affecting both resistance and tolerance to greenbug insect pest (Agrama et al. 2002; Katsar et al. 2002; Nagaraj et al. 2005; Wu and Huang 2008; Punnuri et al. 2013).

Mirid panicle-feeding bugs (head bugs), particularly *Eurystylus oldi* Poppius, are major pests of sorghum in sub-Saharan Africa (Ajayi et al. 2001) and could also affect biomass and sugar accumulation upon infecting sweet sorghum panicles. Three significant QTLs on linkage group C accounted for 13% of the phenotypic variation for reduction in thousands kernel weight trait. Nine additional genomic regions in sorghum were identified to have a role in controlling head bug resistance in sorghum (Deu et al. 2005) and one leaf scorch QTL, *QLsc.txs-B*, explained 8.5% of the genetic variance (Feltus et al. 2006).

The shoot fly is a pest of sorghum, especially in America and Australia, and the larvae of this insect cut the growing point of the growing apical shoot resulting in a deadheart symptom. Genetic variations in sorghum resistance to shoot fly have been detected and this polymorphism has been used to identify genetic loci-controlling resistance to shoot fly. Nine QTLs associated with the resistance to leaf glossiness with phenotypic variation explained by individual QTL ranging from 7.6 to 14.0% were identified (Satish et al. 2009; Aruna et al. 2011). Seven QTLs distributed on five chromosomes, two each on SBI-07 and SBI-10, one each on SBI-01, SBI-05, and SBI-09, controlling oviposition were identified and the phenotypic variation explained by individual QTL ranged from 5.0 to 19.0%. A major QTL for this trait was detected on chromosome SBI-10 near the marker Xnhsbm 1044, explaining 19.0 and 16.1% of the phenotypic variation for mean eggs on 21 days after seedling emergence and mean eggs on 28 days after seedling emergence. Six QTLs for deadheart trait, which is a direct measure of resistance, were distributed on three chromosomes with one each on SBI-05 and SBI-09, and four on SBI-10 were identified in (Satish et al. 2009) study. The phenotypic variation explained by the individual trait ranged from 5.5 to 15.0%. Two major QTLs, *QDh.dsr-10.2* (explaining 11.4% of the phenotypic variation) and *QDh.dsr-10.3*, explaining 15.0%

of the phenotypic variation, were located on chromosome SBI-10. However, (Aruna et al. 2011) identified 10 QTLs on six chromosomes, SBI-02, SBI-09, SBI-01, SBI-06, SBI-07, and SBI-10) controlling deadheart trait with individual QTL explaining 4.5 to 12.8% phenotypic variation. Two major QTLs, *QTdu.dsr-10.1* and *QTdu.dsr-10.2*, were detected for adaxial trichome density on chromosome SBI-10, explaining 15.7 and 33.0% of the phenotypic variation, while six QTLs (*QTdl.dsr-1.1*, *QTdl.dsr-1.2*, *QTdl.dsr-4*, *QTdl.dsr-6*, *QTdl.dsr-10.1*, *QTdl.dsr-10.2*) were detected for abaxial trichome density distributed on four chromosomes with two on SBI-01, one each on SBI-04 and SBI-06, and two on SBI-10. The phenotypic variation explained by individual QTL ranged from 5.2 to 22.7% (Satish et al. 2009). In (Aruna et al. 2011) study, two QTLs (*QTdu.dsr-7*, *QTdu.dsr-10*) distributed on two chromosomes (one each on SBI-07 and SBI-10) were identified for adaxial trichome density, explaining 4.3-44.1% of the phenotypic variation with a QTL on chromosome SBI-10 being a major QTL contributing for 44.1% of phenotypic variation. In addition, three QTLs controlling abaxial trichome density were identified on chromosomes SBI-03 and SBI-10 accounting for 5.0 to 24.1% of the phenotypic variation. Cloning of genetic these genomic loci underlying resistance to sorghum diseases and the understanding of the mechanisms of how pathogens circumvent the genetic resistance will contribute toward sustainable intensification of biomass production.

Foliar diseases

Sorghum is also negatively affected by foliar diseases, viz. anthracnose, target leaf spot, zonate leaf spot, Drechstera leaf blight, and rust. Sorghum anthracnose is caused by *Colletotrichum sublineolum* and is characterized by weakening the plant, severely reducing grain yield, and quality and biomass production. The disease is more prevalent and severe in warm and humid environments, where it causes substantial economic losses. The pathogen causes seedling blight, leaf blight, stalk rot, head blight, and grain molding, and thus limits both forage and grain production. Among these, foliar anthracnose is the most pronounced and devastating, especially on sweet sorghum cultivars directly impacting sugar production (Dalianis 1997). The *Cg1* anthracnose resistance dominant gene located at the distal region of linkage group SBI-05 has been mapped in sorghum cultivar SC748-5 using four AFLP markers (Perumal et al. 2009; Ramasamy et al. 2009). In planta and ex planta *C. sublineolum*, infection assays were carried out using 1-week-old seedlings and it was observed that transgenic line, KOSA-1, was found to be significantly more tolerant to anthracnose than the parent wild-type, KAT 412 (Akosambo-Ayoo et al. 2013).

Association analysis of a sorghum mini-core collection consisting of 242 diverse accessions identified eight loci (loci 1-8) linked to anthracnose resistance in sorghum (Upadhyaya et al. 2013) and found genes associated with anthracnose resistance. They include NB-ARC class of R genes (*Sb10 g021850, Sb10 g021860*) in locus 7 that share 20% homology to *Pib* (accession number BAA76281) which confers resistance to rice blast disease (Wang et al. 1999). Autophagy-related protein 3 (*Sb01 g029070*) in locus 6 coding for *SbATG3* gene is 77% identical and 85% similar to the tobacco homolog *ATG3* (*AAW80629*). Silencing *ATG3* in tobacco resulted in unrestricted TMV-induced hypersensitive cell death due to increased pathogen propagation (Liu et al. 2005). The sorghum loci *Sb08 g003690, Sb08 g003705, Sb08 g003710,* and *Sb08 g003720* on locus 4 code for harpin-induced *Hin 1* and is a well-known hypertensive response marker gene (Pontier et al. 1999). Overexpression of the Arabidopsis *Hin 1* homolog, *AtNHL3,* enhances resistance to infection by *Pseudomonas syringae* pv.tomato DC3000 in Arabidopsis (Varet et al. 2003). RAV transcription factor (*Sb01 g049150*) in locus 3 is also associated with anthracnose resistance. Silencing RAV homolog in tomato abolished the resistance to bacterial wilt caused by *Ralstonia solanacearum* (Li et al. 2011). In addition, overexpression in Arabidopsis enhanced resistance to infection by *Pseudomonas syringae* pv.tomato DC3000 and to osmotic stresses by high salinity and dehydration (Sohn et al. 2006). The oxysterol-binding protein *Sb01 g010720* in locus 5 was also found to have a role in the disease resistance pathway. It is homolog in tomato (*StOBP1*) was found to be induced rapidly by *Phytophthora infestans* (Avrova et al. 2004). In addition, four homologs of menthone:neomenthol reductase 1 (*MNR*) in locus 1 potentially were found and silencing the *MNR* in pepper (*Capsicum annuum*) significantly increased its susceptibility to *Xanthomonas campestris* pv *vesicatoria* and *Colletotrichum coccodes* infection (Choi et al. 2008). Overexpressing rice PR-5 enhances resistance to *Rhizoctonia solani*, the causal agent of sheath blight (Datta et al. 1999). In sorghum, protein expression level of one TLP, sormatin, correlates with resistance to grain mold (Bueso et al. 2000). Taken together it suggests that modulation of genes with a role in resistance pathway has the potential to provide simultaneous resistance to multiple biotic and abiotic stresses in sorghum. These genes potentially play a role in countering pathogen attack in sorghum through the hypersensitive response, the rapid death of plant cells at the site of pathogen infection. Therefore, these genes and markers may be developed into molecular tools for the genetic improvement of anthracnose resistance in sorghum. (Mohan et al. 2010) mapped four (*QAnt1 QAnt2, QAnt3, and QAnt4)* anthracnose resistance loci. *QAnt3* was also mapped by (Klein et al. 2001) and locus 8 from the (Upadhyaya et al. 2013) study was most likely *QAnt3* and locus 1 was close to *QAnt2* (Upadhyaya et al. 2013). A recessive anthracnose resistance gene in SC326-6 sorghum cultivar was mapped with a RAPD marker (Boora et al. 1998). Another recessive anthracnose resistance gene was mapped in G 73 sorghum cultivar with RAPD markers OPJ 0_{11437} at the same loci with SCAR marker SCJ 01 at 3.26 cM (Singh et al. 2006) and a RAPD-based SCAR marker SCA 12 at 6.03 cM (Singh et al. 2006). Anthracnose can be avoided by growing the sorghum in arid and semiarid environment. Additional genetic loci responsible for resistance to folia diseases in sorghum have been identified. Using 168 F7 recombinant inbred lines derived from a cross between 296 B (resistant) and IS18551 (susceptible) parents one major QTL with significant effects for each disease that colocated on SBI-06 was identified (Mohan et al. 2009). The variance explained by each QTL ranged from 12% to 50% with the QTL (*tls*) for target leaf spot explaining 50% of the total phenotypic variance. Similarly, one QTL each for zonate leaf spot (*zls*) and Drechstera leaf blight (*dls*) was identified as colocating with the QTL for target leaf spot disease. The QTL for Drechstera leaf blight explained 12%, while QTL for zonate leaf spot explained 16% of the phenotypic variance. The draft genome sequence of *Colletotrichum sublineola* has been presented and represents a new resource that will be useful for further research into the biology, ecology, and evolution of this key pathogen to find ways to mitigate its destructive ability on cultivated sorghum (Baroncelli et al. 2014).

Sorghum rust disease caused by *Puccinia purpurea* is important because its presence predisposes sorghum to other major disease problems like stalk rot and charcoal rot. Eight loci with significant effect on rust resistance have been identified (Tao et al. 1998) (Table 1). The percentage of the total phenotypic variation explained by each of these genomic regions varied from 6.8% to 42.6%.

Disease of the panicle

Sorghum ergot, caused predominantly by *Claviceps africana* is a significant threat to the sorghum industry worldwide and impacts juice and brix content in sweet sorghum (http://fenalce.org/archivos/SorFee.pdf). Ergot resistance in sorghum is controlled by many genes and that the pollen traits, pollen quantity, and pollen viability have moderate genetic correlation with percentage ergot infection. Nine genetic loci (Table 1) control percentage ergot infection in sorghum (Parh et al. 2008).

Stalk rot

Stalk rot caused by *Macrophomina phaseolina*, is also an economically important, soil-borne disease in major sorghum-growing areas across the world. It is associated with premature stem lodging and pith disintegration leading to inferior grain and fodder quality. Five QTLs were identified at Dharwad location and four QTLs at Bijapur locations for the component traits of stalk rot disease resistance (Reddy et al. 2008). Two QTLs associated with marker xtxp297, xtxp213 for number of internodes crossed on linkage group B, one QTL associated with marker AC13 for length of infection on linkage group D, and two QTLs associated with markers xtxp343, xtxp176 for per cent lodging on linkage group I accounted for 31.83, 10.76, and 18.90 per cent at Dharwad location and 14.87%, 10.47%, and 26.44% phenotypic variability at Bijapur location, respectively. The root and crown rot of sorghum known as milo disease is caused by the peritoxin produced by the saprophytic fungus *Periconia circinata* (Leukel 1948). The *PC* locus of sorghum (*Sorghum bicolor*) determines dominant sensitivity to a host-selective peritoxin. The *Pc* region was cloned by a map-based approach and found to contain three tandemly repeated genes with the structures of nucleotide-binding site-leucine-rich repeat (NBS-LRR) disease resistance genes (Nagy et al. 2007). The agronomically important gene *chi II*, encoding rice chitinase under the constitutive CaMV 35S promoter, has been transferred to sorghum for resistance to stalk rot (*Fusarium thapsinum*) (Krishnaveni et al. 2001). In addition, particle bombardment was used to genetically transform a sorghum genotype, KAT 412, with chitinase (harchit) and chitosanase (harcho) genes isolated from *Trichoderma harzianum*.

Resistance to *Striga* parasitism

Witchweed (*Striga* spp.) infestations are the greatest obstacle to sorghum [*Sorghum bicolor* (L.) Moench] grain and biomass production in many areas in Africa and Asia where they have a 20-100% yield reduction in any given season (Ejeta and Gressel 2007; Parker 2009). Sorghum coevolved with *Striga* in Africa and thus possesses intrinsic modicums of resistance that could be combined. Seeds of an acetolactate synthase (ALS) herbicide-tolerant sorghum hybrid mutant were treated with ALS-inhibiting herbicides before planting and the results showed that seeds treated with the highest herbicide rates had the fewest *Striga* attachments and the greatest delay in attachment (Tuinstra et al. 2009). Once the necessary sorghum genes are isolated and cloned, they could be transformed in a single, dominantly inherited construct containing a group of clustered genes, which would be a very effective strategy (Gressel 2010). Such resistance could easily be backcrossed into local varieties and land races preserving crop biodiversity, because it is inherited as a single dominant gene and not four separate recessive genes. Perhaps the resistance

genes from sorghum, once isolated, could be stacked with those responsible for Desmodium allelochemical production, along with resistance genes being found in cowpea and rice (Tuinstra et al. 2009), all into mini-chromosomes or into the genome at one locus. It would be very hard for the parasitic weeds to overcome such resistance and many crop species could be engineered with the same gene cluster. RNAi constructs encoding genes that suppress parasite-only metabolic pathways have been engineered into tomatoes (Aly et al. 2009) and the same strategy could be attempted in sorghum. In the field, drought stress and Striga infestation are rarely presented individually and sorghum plants are often subjected to a combination of stress types limiting its productivity. Identification of pathways and genetic loci directing specificity and crosstalk of sorghum responses combined with functional characterization of theses genetic signatures could lead to new targets for the enhancement of sorghum stress tolerance and identification of sorghum ideotypes specific to Africa.

Nevertheless, major QTLs for resistance of sorghum to the hemi-parasitic weed *Striga hermonthica* have been mapped in two recombinant inbred populations of $F_{3:5}$ lines developed from the crosses IS9830 × E36-1 and N13 × E36-1 (Haussmann et al. 2004) (Table 1). Sorghum cultivars resistant to Striga are known to produce low levels of strigolactone, a Striga germination stimulant. An in vitro assay for germination stimulant activity toward *Striga asiatica* in 354 recombinant inbred lines derived from SRN39 (low stimulant) × Shanqui Red (high stimulant), a single recessive gene *lgs* was precisely tagged and mapped (Satish et al. 2012) explaining about 40% of the phenotypic variance for area under the *Striga* number progress curve (Haussmann et al. 2004). So far, no QTL has been found to direct multiple disease resistance in sorghum. Given the selection pressure that many pathogens exert directly on natural plant populations and indirectly via variety improvement programs on crop plants, it is proposed that research should be focused on finding genetic loci responsible for multiple disease resistance as this has important implications for plant fitness. In maize, evidence of a locus conditioning resistance to multiple pathogens was found in bin 1.06 of the maize genome with the allele from inbred line 'Tx303' conditioning quantitative resistance to northern leaf blight (NLB) and qualitative resistance to Stewart's wilt and that *pan1* a gene conditioning susceptibility for NLB and Stewart's wilt was cloned (Jamann et al. 2014). Therefore, to reduce the risk of resistance breakdown and increase the levels of disease resistance in sorghum, new sources of disease resistance need to be explored to isolate and incorporate alternative mechanisms of resistance and to pyramid different resistance genes into commercial hybrids.

Weed control

Weeds cause a host of problems in agriculture, competing with crops for light, water, and nutrients, providing a reservoir for insects and diseases, and contaminating seedlots. Vegetative dispersal by rhizomes (underground stems) and seed dispersal by disarticulation of the mature inflorescence ("shattering") cause perennial monocots such as "johnsongrass" to rank among the world's most noxious weeds. Improvements in agricultural production have correlated well with the use of herbicides in controlling weeds. Transgenic glyphosate-resistant crops overexpressing 5-en olpyruvylshikimate-3-phosphate synthase (*cp4 epsps*) gene accelerated widespread use of glyphosate becoming the most widely used herbicide in world agriculture (Duke and Powles 2008) for its effectiveness in controlling recalcitrant weeds such as Johnsongrass. However, the increased utilization of these herbicides over a long period of time exerts selective pressure leading to widespread evolution of resistance in several weed species (Busi et al. 2013). For instance, metabolic resistance (enhanced metabolic capacity to detoxify herbicides) can be endowed by the increased activity of the endogenous cytochrome P450 mono-oxygenases, glucosyl transferases (GTs), glutathione S-transferases (GSTs), and/or other enzyme systems such as aryl acylamidase (Carey et al. 1995) that can metabolize herbicides (Yu and Powles 2014). Combined with lack of novel herbicides being brought to the market over the last 30 years and tougher registration and environmental regulations on herbicides have resulted in a loss of some herbicides, particularly in Europe, threatening crop production worldwide (Heap 1997).

Integrated weed management approach to control weed populations has been hailed as an effective tool in addition, reduce the environmental impact of individual weed management practices, increase cropping system sustainability, and reduce selection pressure for weed resistance to herbicides (Harker and O'Donovan 2013). Maize plants transformed with an aryloxyalkanoate dioxygenase (*AAD-1*) gene showed robust crop resistance to aryloxyphenoxypropionate herbicides over four generations and were also not injured by 2,4-dichlorophenoxyacetic acid (2,4-D) applications at any growth stage. Arabidopsis plants expressing *AAD-12* were resistant to 2,4-D as well as triclopyr and fluroxypyr, and transgenic soybean plants expressing *AAD-12* maintained field resistance to 2,4-D over five generations indicating that single *AAD* transgenes can provide simultaneous resistance to a broad repertoire of agronomically important classes of herbicides, including 2,4-D, with utility in both monocot and dicot crops which can help transgenes preserve the productivity and environmental benefits of herbicide-resistant crops (Wright et al. 2010). Recently, the use of multicopy transposons

bearing unfitness genes has been proposed in the management of weeds. Multicopy transposons rapidly disseminate through populations, appearing in ~100% of progeny unlike nuclear transgenes, which appear in a proportion of segregating populations (Gressel and Levy 2014). Here, weed populations could be generated that contain the unfitness gene under chemically or environmentally inducible promoters, activated after gene dissemination, or under constitutive promoters where the gene function is utilized only at special times (e.g., sensitivity to a herbicide), and thus are easily controllable. Efforts need to be accelerated to understand the genetic basis of weed resistance for employment of an RNAi approach to interfere with the expression of herbicide resistance genes in weeds (Sammons et al. 2012) and restore sensitivity of weeds to glyphosate (Green 2014). Enhancing crop competitiveness, for example, by genetic engineering with genes encoding phosphates, and the application of fertilizers with phosphites as the main source of phosphates is a key strategy for crops to outcompete weeds for essential nutrients (López-Arredondo and Herrera-Estrella 2012).

Perspectives: Sweet sorghum ideotypes with enhanced resistance to abiotic and biotic stresses

In the field, multiple abiotic stresses (drought, salinity, heat, cold, chilling, freezing, nutrient, high light intensity, ozone, and anaerobic stresses) and biotic stresses (insect pests and diseases) are presented. The performance of the plant, therefore, is affected by the degree of heterogeneity between stress levels, simultaneous occurrence of different stresses, the timing of the stress event with respect to the developmental stage of the plant and the intensity and duration of the stress (Mittler and Blumwald 2010). Plants respond differently to combined stresses as compared to their response to individual stress as it happens under laboratory conditions. The later response activates a specific program of gene expression relating to the exact environmental condition encountered. The responses are complex and could involve changes at transcriptomic, cellular, and physiological level. Genetic and genomic resources for sorghum breeding are available (Carpita and McCann 2008), and they offer an opportunity to employ multidisciplinary approaches involving traditional breeding and biotechnology to contribute to future improvements of sweet sorghum to adapt to both biotic and abiotic stresses. For instance, in tropical climate, drought stress is ranked most important followed by striga parasitism, and fungal and bacterial diseases in terms of limiting sorghum potential for growth and reproduction. Therefore, breeding for sorghum ideotypes tolerant to combined drought and Striga parasitism will be ideal for this region.

As discussed previously, resistance to abiotic and biotic stresses has been demonstrated through genetic engineering and classical breeding. Tolerance to both abiotic and biotic stresses has also been achieved. In maize, breeding programs have developed plants tolerant to drought and have additional resistance to the parasitic weed *Striga hermonthica* (Bänziger et al. 2006; Badu-Apraku and Yallou 2009). In Sudan, sorghum cultivars resistant to drought and striga infectivity have been developed through classical breeding (Nair Suliman personal communication), this suggests that hormone signaling pathways orchestrating the interaction between abiotic and biotic stresses are altered and in particular abscisic acid. This alteration could be interesting to breeders to further design sorghum ideotypes that can withstand a combination of stresses as presented in the field. In addition, research programs need to focus on developing tolerance to multiple stresses in order for the improved varieties to respond predictably under field condition.

In temperate environment, cool temperatures below 15°C during the early growing season limits optimal growth of sorghum, it is a key agronomic trait for warm season cereal crops such as sorghum. Breeding for sorghum ideotypes with improved early-season cold tolerance would be appropriate for temperate environments (Yu and Tuinstra 2001). So far, Chinese sorghum kaoliang, Shanqui Red (Knoll et al. 2008) and and F7 RIL population of RTx403xPI567946 (Burow et al. 2011b) have been found to exhibit higher emergence and greater seedling vigor under cool temperature than most breeding lines currently available. However, they lack desirable agronomic characteristics. Sorghum ideotypes' resistance to cold could be developed by introgression of desirable genes from Chinese landraces into elite lines through marker-assisted selection (Knoll et al. 2008; Burow et al. 2011a).

The use of herbicides is increasing in global crop production. Improved weed control with herbicides promotes fertilizer use and has the potential to improve crop yields in many developing countries in the near future (Gianessi 2013). Shattercane and *Sorghum halepense* (johnsongrass) are natural weeds for sorghum and resistant to most other herbicides used for their control (Heap 2014). In addition, they outcross with cultivated sorghum (Morrell et al. 2005; Muraya et al. 2011). This suggests that transgenes introduced to sorghum would readily introgress and be retained in these wild species, which often occur sympatrically with cultivated sorghum in Africa (Mutegi et al. 2010). Compared to other weed control strategies including manual hand weeding, herbicides are the key to sustainable crop production throughout the world, and, will remain the mainstay for weed control in the foreseeable future. Therefore, when developing sweet sorghum ideotypes for different ecological regions, the use of herbicides to control weeds should be considered. The rhizome formation trait is correlated and genetically linked to overwinter survival in sorghum. Genetic mechanisms controlling overwintering have the potential to minimize the risk of weediness and create perennial sorghums that can overwinter in climates where they previously could not (Paterson et al. 1995; Washburn et al. 2013). These perennial overwintering sorghum ideotypes could be used for improvements in biofuel production in sweet sorghum by extending the period of biomass production and reducing production costs.

Conflict of Interest

None declared.

References

Aglawe, S., B. Fakrudin, C. Patole, S. Bhairappanavar, R. Koti, and P. Krishnaraj. 2012. Quantitative RT-PCR analysis of 20 transcription factor genes of MADS, ARF, HAP2, MBF and HB families in moisture stressed shoot and root tissues of sorghum. Physiol. Mol. Biol. Plants 18:287–300.

Agrama, H., G. Widle, J. Reese, L. Campbell, and M. Tuinstra. 2002. Genetic mapping of QTLs associated with greenbug resistance and tolerance in *Sorghum bicolor*. Theor. Appl. Genet. 104:1373–1378.

Ajayi, O., H. Sharma, R. Tabo, A. Ratnadass, and Y. Doumbia. 2001. Incidence and distribution of the sorghum head bug, *Eurystylus oldi* Poppius (Heteroptera: Miridae) and other panicle pests of sorghum in West and Central Africa. Int. J. Trop. Insect Sci. 21:103–111.

Akosambo-Ayoo, L., M. Bader, H. Loerz, and D. Becker. 2013. Transgenic sorghum (*Sorghum bicolor* L. Moench) developed by transformation with chitinase and chitosanase genes from *Trichoderma harzianum* expresses tolerance to anthracnose. Afr. J. Biotechnol. 10:3659–3670.

Aly, R., H. Cholakh, D. M. Joel, D. Leibman, B. Steinitz, A. Zelcer, et al. 2009. Gene silencing of mannose 6-phosphate reductase in the parasitic weed *Orobanche aegyptiaca* through the production of homologous dsRNA sequences in the host plant. Plant Biotechnol. J. 7:487–498.

Aneeta, S. N. N. Tuteja, and S. Kumar Sopory. 2002. Salinity- and ABA-induced up-regulation and light-mediated modulation of mRNA encoding glycine-rich RNA-binding protein from *Sorghum bicolor*. Biochem. Biophys. Res. Commun., 296:1063–1068.

Anyamba, A., J. L. Small, S. C. Britch, C. J. Tucker, E. W. Pak, C. A. Reynolds, et al. 2014. Recent weather extremes and impacts on agricultural production and vector-borne disease outbreak patterns. PLoS ONE 9:e92538.

Aruna, C., V. Bhagwat, R. Madhusudhana, V. Sharma, T. Hussain, R. Ghorade, et al. 2011. Identification and validation of genomic regions that affect shoot fly resistance in sorghum [Sorghum bicolor (L.) Moench]. Theor. Appl. Genet. 122:1617–1630.

Avila-Ospina, L., M. Moison, K. Yoshimoto, and C. Masclaux-Daubresse. 2014. Autophagy, plant senescence, and nutrient recycling. J. Exp. Bot. doi:10.1093/jxb/eru039.

Avrova, A. O., N. Taleb, V. M. Rokka, J. Heilbronn, E. Campbell, I. Hein, et al. 2004. Potato oxysterol binding protein and cathepsin B are rapidly up-regulated in independent defence pathways that distinguish R gene-mediated and field resistances to Phytophthora infestans. Mol. Plant Pathol. 5:45–56.

Badu-Apraku, B., and C. Yallou. 2009. Registration of-resistant and drought-tolerant tropical early maize populations TZE-W Pop DT STR C and TZE-Y Pop DT STR C. J. Plant Regist. 3:86–90.

Balota, M., W. Payne, S. Veeragoni, B. Stewart, and D. Rosenow. 2010. Respiration and its relationship to germination, emergence, and early growth under cool temperatures in sorghum. Crop Sci. 50:1414–1422.

Bänziger, M., P. S. Setimela, D. Hodson, and B. Vivek. 2006. Breeding for improved abiotic stress tolerance in maize adapted to southern Africa. Agric. Water Manag. 80:212–224.

Baroncelli, R., J. M. Sanz-Martín, G. E. Rech, S. A. Sukno, and M. R. Thon. 2014. Draft genome sequence of Colletotrichum sublineola, a destructive pathogen of cultivated sorghum. Genome Announc. 2:e00540-14.

Berndes, G., M. Hoogwijk, and R. van den Broek. 2003. The contribution of biomass in the future global energy supply: a review of 17 studies. Biomass Bioenergy 25:1–28.

Boora, K. S., R. Frederiksen, and C. Magill. 1998. DNA-based markers for a recessive gene conferring anthracnose resistance in sorghum. Crop Sci. 38:1708–1709.

Borrell, A. K., G. L. Hammer, and A. C. Douglas. 2000a. Does maintaining green leaf area in sorghum improve yield under drought? I. Leaf growth and senescence. Crop Sci. 40:1026–1037.

Borrell, A. K., G. L. Hammer, and R. G. Henzell. 2000b. Does maintaining green leaf area in sorghum improve yield under drought? II. Dry matter production and yield. Crop Sci. 40:1037–1048.

Borrell, A. K., J. E. Mullet, B. George-Jaeggli, E. J. Van Oosterom, G. L. Hammer, P. E. Klein, et al. 2014a. Drought adaptation of stay-green sorghum is associated with canopy development, leaf anatomy, root growth, and water uptake. J. Exp. Bot., 65:6137–6139

Borrell, A. K., E. J. Oosterom, J. E. Mullet, B. George-Jaeggli, D. R. Jordan, P. E. Klein, et al. 2014b. Stay-green alleles individually enhance grain yield in sorghum under drought by modifying canopy development and water uptake patterns. New Phytol. 203:817–30.

Buchanan, C. D., S. Lim, R. A. Salzman, I. Kagiampakis, D. T. Morishige, B. D. Weers, et al. 2005. Sorghum bicolor's transcriptome response to dehydration, high salinity and ABA. Plant Mol. Biol. 58:699–720.

Bueso, F. J., R. D. Waniska, W. L. Rooney, and F. P. Bejosano. 2000. Activity of antifungal proteins against mold in sorghum caryopses in the field. J. Agric. Food Chem. 48:810–816.

Burgess, M. G., C. Rush, G. Piccinni, and G. Schuster. 2002. Relationship between charcoal rot, the stay-green trait, and irrigation in grain sorghum. Phytopathology 92:S10.

Burow, G., J. J. Burke, Z. Xin, and C. D. Franks. 2011a. Genetic dissection of early-season cold tolerance in sorghum (Sorghum bicolor (L.) Moench). Mol. Breeding 28:391–402.

Burow, G., Z. Xin, C. Franks, J. Burke, and P. Hi. 2011b. Genetic enhancement of cold tolerance to overcome a major limitation in sorghum. American Seed Trade Association Conference Proceedings.

Busi, R., M. M. Vila-Aiub, H. J. Beckie, T. A. Gaines, D. E. Goggin, S. S. Kaundun, et al. 2013. Herbicide-resistant weeds: from research and knowledge to future needs. Evol. Appl. 6:1218–1221.

Carey, V., S. O. Duke, R. E. Hoagland, and R. E. Talbert. 1995. Resistance Mechanism of Propanil-Resistant Barnyardgrass 1. Absorption, Translocation, and Site of Action Studies. Pestic. Biochem. Physiol. 52:182–189.

Carpita, N. C., and M. C. McCann. 2008. Maize and sorghum: genetic resources for bioenergy grasses. Trends Plant Sci. 13:415–420.

Choi, H. W., B. G. Lee, N. H. Kim, Y. Park, C. W. Lim, H. K. Song, et al. 2008. A role for a menthone reductase in resistance against microbial pathogens in plants. Plant Physiol. 148:383–401.

Collins, N. C., F. Tardieu, and R. Tuberosa. 2008. Quantitative trait loci and crop performance under abiotic stress: where do we stand? Plant Physiol. 147:469–486.

Crasta, O., W. Xu, D. Rosenow, J. Mullet, and H. Nguyen. 1999. Mapping of post-flowering drought resistance traits in grain sorghum: association between QTLs influencing premature senescence and maturity. Mol. Gen. Genet. 262:579–588.

Dalianis, C.1997. Productivity, sugar yields, ethanol potential and bottlenecks of sweet sorghum in European Union. Pp. 65–79 in D. Li, ed. Proceedings of the 1st International sweet sorghum conference. Ed: Dajue Li, 1997. 65–79

Dalla Marta, A., M. Mancini, F. Orlando, F. Natali, L. Capecchi, and S. Orlandini. 2014. Sweet sorghum for bioethanol production: crop responses to different water stress levels. Biomass Bioenergy, 64:211–219.

Damte, T., B. B. Pendleton, and L. K. Almas. 2009. Cost-benefit analysis of sorghum midge, Stenodiplosis sorghicola 1 (Coquillett)-resistant sorghum hybrid research and development in Texas. Southwest. Entomol. 34:395–405.

Datta, K., R. Velazhahan, N. Oliva, I. Ona, T. Mew, G. Khush, et al. 1999. Over-expression of the cloned rice thaumatin-like protein (PR-5) gene in transgenic rice plants enhances environmental friendly resistance to Rhizoctonia solani causing sheath blight disease. Theor. Appl. Genet. 98:1138–1145.

Deu, M., A. Ratnadass, M. Hamada, J. Noyer, M. Diabatedagger, and J. Chantereau. 2005. Quantitative trait loci for head-bug resistance in sorghum. Afr. J. Biotechnol. 4:247–250.

Dev Sharma, A., S. Kumar, and P. Singh. 2006. Expression analysis of a stress-modulated transcript in drought tolerant and susceptible cultivars of sorghum (Sorghum bicolor). J. Plant Physiol., 163:570–576.

Doumbia, M., L. Hossner, and A. Onken. 1993. Variable sorghum growth in acid soils of subhumid West Africa. Arid Land Res. Manag. 7:335–346.

Doumbia, M., L. Hossner, and A. Onken. 1998. Sorghum growth in acid soils of West Africa: variations in soil chemical properties. Arid Land Res. Manag. 12:179–190.

Dugas, D. V., M. K. Monaco, A. Olson, R. R. Klein, S. Kumari, D. Ware, et al. 2011. Functional annotation of the transcriptome of Sorghum bicolor in response to osmotic stress and abscisic acid. BMC Genom. 12:514.

Duke, S. O., and S. B. Powles. 2008. Glyphosate: a once-in-a-century herbicide. Pest Manag. Sci. 64:319–325.

Ejeta, G., and J. Gressel. 2007. Integrating new technologies for Striga control: towards ending the witch-hunt. World Scientific, Publishing Co. Pte Ltd, 5 Tol Tuck Link, Singapore, pp. 3–16.

Emebiri, L. C. 2013. QTL dissection of the loss of green colour during post-anthesis grain maturation in two-rowed barley. Theor. Appl. Genet. 126:1873–1884.

Faës, P., C. Deleu, A. Aïnouche, F. Le Cahérec, E. Montes, V. Clouet, et al. 2014. Molecular evolution and transcriptional regulation of the oilseed rape proline dehydrogenase genes suggest distinct roles of proline catabolism during development. Planta 241: 403-419

Feltus, F., G. Hart, K. Schertz, A. Casa, S. Kresovich, S. Abraham, et al. 2006. Alignment of genetic maps and QTLs between inter-and intra-specific sorghum populations. Theor. Appl. Genet. 112:1295–1305.

Felderhoff, T. J., et al. "QTLs for Energy-related Traits in a Sweet× Grain Sorghum [(L.) Moench] Mapping Population." Crop Science 52.5 (2012):2040–2049.

Foy, C., T. Jr Carter, J. Duke, and T. Devine. 1993. Correlation of shoot and root growth and its role in selecting for aluminum tolerance in soybean. J. Plant Nutr., 16:305–325.

Funck, D., S. Eckard, and G. Müller. 2010. Non-redundant functions of two proline dehydrogenase isoforms in Arabidopsis. BMC Plant Biol. 10:70.

Gianessi, L. P. 2013. The increasing importance of herbicides in worldwide crop production. Pest Manag. Sci. 69:1099–1105.

Girijashankar, V., H. Sharma, K. K. Sharma, V. Swathisree, L. S. Prasad, B. Bhat, et al. 2005. Development of transgenic sorghum for insect resistance against the spotted stem borer (Chilo partellus). Plant Cell Rep., 24:513–522.

Green, J. M.. 2014. Current state of herbicides in herbicide-resistant crops. Pest Manag. Sci. 70:1351–1357.

Gressel, J. 2010. Needs for and environmental risks from transgenic crops in the developing world. New Biotechnol. 27:522–527.

Gressel, J., and A. A. Levy. 2014. Use of multi-copy transposons bearing unfitness genes in weed control: four example scenarios. Plant Physiol. 166:1221–31.

Guimaraes, C. T., C. C. Simoes, M. M. Pastina, L. G. Maron, J. V. Magalhaes, R. C. Vasconcellos, et al. 2014. Genetic dissection of Al tolerance QTLs in the maize genome by high density SNP scan. BMC Genom. 15:153.

Guo, C., W. Cui, X. Feng, J. Zhao, and G. Lu. 2011. Sorghum insect problems and Managementf. J. Integr. Plant Biol. 53:178–192.

Harker, K. N., and J. T. O'Donovan. 2013. Recent weed control, weed management, and integrated weed management. Weed Technol. 27:1–11.

Harris, K., P. Subudhi, A. Borrell, D. Jordan, D. Rosenow, H. Nguyen, et al. 2007. Sorghum stay-green QTL individually reduce post-flowering drought-induced leaf senescence. J. Exp. Bot. 58:327–338.

Hash, C., A. Bhasker Raj, S. Lindup, A. Sharma, C. Beniwal, R. Folkertsma, et al. 2003. Opportunities for marker-assisted selection (MAS) to improve the feed quality of crop residues in pearl millet and sorghum. Field. Crop. Res., 84:79–88.

Haussmann, B., V. Mahalakshmi, B. Reddy, N. Seetharama, C. Hash, and H. Geiger. 2002. QTL mapping of stay-green in two sorghum recombinant inbred populations. Theor. Appl. Genet. 106:133–142.

Haussmann, B., D. Hess, G. Omanya, R. Folkertsma, B. Reddy, M. Kayentao, et al. 2004. Genomic regions influencing resistance to the parasitic weed Striga hermonthica in two recombinant inbred populations of sorghum. Theor. Appl. Genet. 109:1005–1016.

Heap, I. 1997. International survey of herbicide-resistant weeds. Western Society of Weed Science (USA), 1997.

Heap, I. 2014. Global perspective of herbicide-resistant weeds. Pest Manag. Sci. 70:1306–1315.

House, L. R. 1985. A guide to sorghum breeding. International Crops Research Institute for the Semi-Arid Tropics Patancheru, India.

Jamann, T. M., J. A. Poland, J. M. Kolkman, L. G. Smith, and R. J. Nelson. 2014. Unraveling genomic complexity at a quantitative disease resistance locus in Maize. Genetics 198:333–344.

Jiang, W., H. Zhou, H. Bi, M. Fromm, B. Yang, and D. P. Weeks. 2013. Demonstration of CRISPR/Cas9/sgRNA-mediated targeted gene modification in Arabidopsis, tobacco, sorghum and rice. Nucleic Acids Res., 41, e188

Jordan, D., C. Hunt, A. Cruickshank, A. Borrell, and R. Henzell. 2012. The relationship between the stay-green trait and grain yield in elite sorghum hybrids grown in a range of environments. Crop Sci. 52:1153–1161.

Katsar, C. S., A. H. Paterson, G. L. Teetes, and G. C. Peterson. 2002. Molecular analysis of sorghum resistance to the greenbug (Homoptera: Aphididae). J. Econ. Entomol. 95:448–457.

Kebede, H., P. Subudhi, D. Rosenow, and H. Nguyen. 2001. Quantitative trait loci influencing drought tolerance in grain sorghum (Sorghum bicolor L. Moench). Theor. Appl. Genet. 103:266–276.

Kim, K.-H., E. Kabir, and S. Ara Jahan. 2014. A review of the consequences of global climate change on human health. J. Environ. Sci. Health Part C 32:299–318.

Klein, R., R. Rodriguez-Herrera, J. Schlueter, P. Klein, Z. Yu, and W. Rooney. 2001. Identification of genomic regions that affect grain-mould incidence and other traits of agronomic importance in sorghum. Theor. Appl. Genet. 102:307–319.

Knoll, J., and G. Ejeta. 2008. Marker-assisted selection for early-season cold tolerance in sorghum: QTL validation across populations and environments. Theor. Appl. Genet. 116:541–553.

Knoll, J., N. Gunaratna, and G. Ejeta. 2008. QTL analysis of early-season cold tolerance in sorghum. Theor. Appl. Genet. 116:577–587.

Krishnaveni, S., J. Joeung, S. Muthukrishnan, and G. Liang. 2001. Transgenic sorghum plants constitutively expressing a rice chitinase gene show improved resistance to stalk rot. J. Genet. Breed. 55:151–158.

Kruger, M., J. van den Berg, and H. du Plessis. 2008. Diversity and seasonal abundance of sorghum panicle-feeding Hemiptera in South Africa. Crop Prot. 27:444–451.

Lata, C., A. K. Mishra, M. Muthamilarasan, V. S. Bonthala, Y. Khan, and M. Prasad. 2014. Genome-wide investigation and expression profiling of AP2/ERF transcription factor superfamily in Foxtail Millet (Setaria italica L.). PLoS ONE 9:e113092.

Leukel, R. 1948. Periconia circinata and its relation to Milo disease. J. Agric. Res. 77:201–222.

Li, J.-G., J. Cao, F.-F. Sun, D.-D. Niu, F. Yan, H.-X. Liu, et al. 2011. Control of tobacco mosaic virus by PopW as a result of induced resistance in tobacco under greenhouse and field conditions. Phytopathology 101:1202–1208.

Liu, Y., M. Schiff, K. Czymmek, Z. Tallóczy, B. Levine, and S. Dinesh-Kumar. 2005. Autophagy regulates programmed cell death during the plant innate immune response. Cell 121:567–577.

López-Arredondo, D. L., and L. Herrera-Estrella. 2012. Engineering phosphorus metabolism in plants to produce a dual fertilization and weed control system. Nat. Biotechnol. 30:889–893.

Mace, E., V. Singh, E. van Oosterom, G. Hammer, C. Hunt, and D. Jordan. 2012. QTL for nodal root angle in sorghum (Sorghum bicolor L. Moench) co-locate with QTL for traits associated with drought adaptation. Theor. Appl. Genet. 124:97–109.

Magalhaes, J. V., J. Liu, C. T. Guimaraes, U. G. Lana, V. M. Alves, Y.-H. Wang, et al. 2007. A gene in the multidrug and toxic compound extrusion (MATE) family confers aluminum tolerance in sorghum. Nat. Genet. 39:1156–1161.

Mahalakshmi, V., and F. R. Bidinger. 2002. Evaluation of stay-green sorghum germplasm lines at ICRISAT. Crop Sci. 42:965–974.

Maheswari, M., Y. Varalaxmi, A. Vijayalakshmi, S. Yadav, P. Sharmila, B. Venkateswarlu, et al. 2010. Metabolic engineering using mtlD gene enhances tolerance to water deficit and salinity in sorghum. Biol. Plant. 54:647–652.

Mastrorilli, M., N. Katerji, and G. Rana. 1999. Productivity and water use efficiency of sweet sorghum as affected by soil water deficit occurring at different vegetative growth stages. Eur. J. Agron. 11:207–215.

McBee, G., R. Waskom, and R. Creelman. 1983. Effect of senescene on carbohydrates in sorghum during late Kernel Maturity states. Crop Sci. 23:372–376.

Mittler, R., and E. Blumwald. 2010. Genetic engineering for modern agriculture: challenges and perspectives. Annu. Rev. Plant Biol. 61:443–462.

Mohan, S., R. Madhusudhana, K. Mathur, C. Howarth, G. Srinivas, K. Satish, et al. 2009. Co-localization of quantitative trait loci for foliar disease resistance in sorghum. Plant Breeding 128:532–535.

Mohan, S. M., R. Madhusudhana, K. Mathur, D. Chakravarthi, S. Rathore, R. N. Reddy, et al. 2010. Identification of quantitative trait loci associated with resistance to foliar diseases in sorghum [Sorghum bicolor (L.) Moench]. Euphytica 176:199–211.

Morrell, P., T. Williams-Coplin, A. Lattu, J. Bowers, J. Chandler, and A. Paterson. 2005. Crop-to-weed introgression has impacted allelic composition of johnsongrass populations with and without recent exposure to cultivated sorghum. Mol. Ecol. 14:2143–2154.

Munson, R. E., J. A. Schaffer, and E. W. Palm. 1993. Sorghum aphid pest management. University of Missouri-Extension.

Muraya, M. M., E. Mutegi, H. H. Geiger, S. M. de Villiers, F. Sagnard, B. M. Kanyenji, et al. 2011. Wild sorghum from different eco-geographic regions of Kenya display a mixed mating system. Theor. Appl. Genet. 122:1631–1639.

Mutegi, E., F. Sagnard, M. Muraya, B. Kanyenji, B. Rono, C. Mwongera, et al. 2010. Ecogeographical distribution of wild, weedy and cultivated Sorghum bicolor (L.) Moench in Kenya: implications for conservation and crop-to-wild gene flow. Genet. Resour. Crop Evol. 57:243–253.

Nagaraj, N., J. C. Reese, M. R. Tuinstra, C. M. Smith, P. St. Amand, M. Kirkham, et al. 2005. Molecular mapping of sorghum genes expressing tolerance to damage by greenbug (Homoptera: Aphididae). J. Econ. Entomol., 98:595–602.

Nagy, E. D., T.-C. Lee, W. Ramakrishna, Z. Xu, P. E. Klein, P. Sanmiguel, et al. 2007. Fine mapping of the Pc locus of Sorghum bicolor, a gene controlling the reaction to a fungal pathogen and its host-selective toxin. Theor. Appl. Genet. 114:961–970.

Nair, S. K., et al. "Identification and validation of QTLs conferring resistance to sorghum downy mildew (Peronosclerospora sorghi) and Rajasthan downy mildew (P. heteropogoni) in maize." Theoretical and applied genetics 110.8 (2005):1384–1392.

Nelson, D. E., P. P. Repetti, T. R. Adams, R. A. Creelman, J. Wu, D. C. Warner, et al. 2007. Plant nuclear factor Y (NF-Y) B subunits confer drought tolerance and lead to improved corn yields on water-limited acres. Proc. Natl Acad. Sci. 104:16450–16455.

Ni, Z., Z. Hu, Q. Jiang, and H. Zhang. 2013. GmNFYA3, a target gene of miR169, is a positive regulator of plant tolerance to drought stress. Plant Mol. Biol. 82:113–129.

van Oosterom, E., R. Jayachandran, and F. Bidinger. 1996. Diallel analysis of the stay-green trait and its components in sorghum. Crop Sci. 36:549–555.

Orlandini, S., M. Mancini, and A. Dalla Marta. 2007. Sistema per la realizzazione di una filiera corta per la produzione di energia da biomasse agricole. Proceedings of the XXXVII Convegno Nazionale della Società Italiana di Agronomia, Catania. 13-14.

Ougham, H., I. Armstead, C. Howarth, I. Galyuon, I. Donnison, and H. Thomas. 2008. The genetic control of senescence revealed by mapping quantitative trait loci. Ann. Plant Rev. Senes. Proc. Plants 26:171.

Parh, D., D. Jordan, E. Aitken, E. Mace, P. Jun-Ai, C. McIntyre, et al. 2008. QTL analysis of ergot resistance in sorghum. Theor. Appl. Genet. 117:369–382.

Parker, C. 2009. Observations on the current status of Orobanche and Striga problems worldwide. Pest Manag. Sci. 65:453–459.

Pasini, L., M. Bergonti, A. Fracasso, A. Marocco, and S. Amaducci. 2014. Microarray analysis of differentially expressed mRNAs and miRNAs in young leaves of sorghum under dry-down conditions. J. Plant Physiol. 171:537–548.

Paterson, A. H., K. F. Schertz, Y.-R. Lin, S.-C. Liu, and Y.-L. Chang. 1995. The weediness of wild plants: molecular analysis of genes influencing dispersal and persistence of johnsongrass, Sorghum halepense (L.) Pers. Proc. Natl Acad. Sci. 92:6127–6131.

Paterson, A. H., J. E. Bowers, R. Bruggmann, I. Dubchak, J. Grimwood, H. Gundlach, et al. 2009. The Sorghum bicolor genome and the diversification of grasses. Nature 457:551–556.

Peleg, Z., T. Fahima, T. Krugman, S. Abbo, D. Yakir, A. B. Korol, et al. 2009. Genomic dissection of drought resistance in durum wheat× wild emmer wheat recombinant inbreed line population. Plant, Cell Environ. 32:758–779.

Perumal, R., M. A. Menz, P. J. Mehta, S. Katile, L. A. Gutierrez-Rojas, R. R. Klein, et al. 2009. Molecular mapping of Cg1, a gene for resistance to anthracnose (Colletotrichum sublineolum) in sorghum. Euphytica 165:597–606.

Pontier, D., S. Gan, R. M. Amasino, D. Roby, and E. Lam. 1999. Markers for hypersensitive response and senescence show distinct patterns of expression. Plant Mol. Biol. 39:1243–1255.

Punnuri, S., Y. Huang, J. Steets, and Y. Wu. 2013. Developing new markers and QTL mapping for greenbug resistance in sorghum [Sorghum bicolor (L.) Moench]. Euphytica 191:191–203.

Qi, X., S. Xie, Y. Liu, F. Yi, and J. Yu. 2013. Genome-wide annotation of genes and noncoding RNAs of foxtail millet in response to simulated drought stress by deep sequencing. Plant Mol. Biol. 83:459–473.

Rajjou, L., M. Duval, K. Gallardo, J. Catusse, J. Bally, C. Job, et al. 2012. Seed germination and vigor. Annu. Rev. Plant Biol. 63:507–533.

Rajwanshi, R., S. Chakraborty, K. Jayanandi, B. Deb, and D. A. Lightfoot. 2014. Orthologous plant microRNAs: microregulators with great potential for improving stress tolerance in plants. Theor. Appl. Genet. 127:2525–2543.

Ram, G., and A. D. Sharma. 2013. In silico analysis of putative miRNAs and their target genes in sorghum (Sorghum bicolor). Int. J. Bioinform. Res. Appl. 9:349–364.

Ramasamy, P., M. Menz, P. Mehta, S. Katilé, L. Gutierrez-Rojas, R. Klein, et al. 2009. Molecular mapping of Cg1, a gene for resistance to anthracnose (Colletotrichum sublineolum) in sorghum. Euphytica 165:597–606.

Rathinasabapathi, B. 2000. Metabolic engineering for stress tolerance: installing osmoprotectant synthesis pathways. Ann. Bot. 86:709–716.

Reddy, P. S., B. Fakrudin, S. Punnuri, S. Arun, M. Kuruvinashetti, I. Das, et al. 2008. Molecular mapping of genomic regions harboring QTLs for stalk rot resistance in sorghum. Euphytica 159:191–198.

Reddy, N. R. R., M. Ragimasalawada, M. M. Sabbavarapu, S. Nadoor, and J. V. Patil. 2014. Detection and validation of stay-green QTL in post-rainy sorghum involving widely adapted cultivar, M35-1 and a popular stay-green genotype B35. BMC Genom. 15:909.

Rivero, R. M., J. Gimeno, A. van Deynze, H. Walia, and E. Blumwald. 2010. Enhanced cytokinin synthesis in tobacco plants expressing PSARK: IPT prevents the degradation of photosynthetic protein complexes during drought. Plant Cell Physiol. 51:1929–1941.

Sabadin, P., M. Malosetti, M. Boer, F. Tardin, F. Santos, C. Guimaraes, et al. 2012. Studying the genetic basis of drought tolerance in sorghum by managed stress trials and adjustments for phenological and plant height differences. Theor. Appl. Genet. 124:1389–1402.

Sama, V., N. Rawat, R. Sundaram, K. Himabindu, B. S. Naik, B. Viraktamath, et al. 2014. A putative candidate for the recessive gall midge resistance gene gm3 in rice identified and validated. Theor. Appl. Genet. 127:113–124.

Sammons, D. R., D. Wang, P. Morris, B. Duncan, G. Griffith, and D. Findley. 2012 Strategies for countering herbicide resistance. Abstracts of papers of the American Chemical Society. Amer Chemical Soc 1155 16TH ST, NW, Washington, DC 20036 USA.

Sanchez, A., P. Subudhi, D. Rosenow, and H. Nguyen. 2002. Mapping QTLs associated with drought resistance in sorghum (Sorghum bicolor L. Moench). Plant Mol. Biol. 48:713–726.

Satish, K., G. Srinivas, R. Madhusudhana, P. Padmaja, R. N. Reddy, S. M. Mohan, et al. 2009. Identification of quantitative trait loci for resistance to shoot fly in sorghum [Sorghum bicolor (L.) Moench]. Theor. Appl. Genet. 119:1425–1439.

Satish, K., Z. Gutema, C. Grenier, P. J. Rich, and G. Ejeta. 2012. Molecular tagging and validation of microsatellite markers linked to the low germination stimulant gene (lgs) for Striga resistance in sorghum [Sorghum bicolor (L.) Moench]. Theor. Appl. Genet. 124:989–1003.

Schittenhelm, S., and S. Schroetter. 2014. Comparison of drought tolerance of maize, sweet sorghum and sorghum-sudangrass hybrids. J. Agron. Crop Sci. 200:46–53.

Serrão, M., M. Menino, J. Martins, N. Castanheira, M. Lourenço, I. Januário, et al. 2012. Mineral leaf composition of sweet sorghum in relation to biomass and sugar yields under different nitrogen and salinity conditions. Commun. Soil Sci. Plant Anal. 43:2376–2388.

Sharma, H., S. Mukuru, K. Hari Prasad, E. Manyasa, and S. Pande. 1999. Identification of stable sources of resistance in sorghum to midge and their reaction to leaf diseases. Crop Prot., 18:29–37.

Sharma, H. C., K. K. Sharma, and J. H. Crouch. 2004. Genetic transformation of crops for insect resistance: potential and limitations. Crit. Rev. Plant Sci. 23:47–72.

Shen, C., Y. Bai, S. Wang, S. Zhang, Y. Wu, M. Chen, et al. 2010. Expression profile of PIN, AUX/LAX and PGP auxin transporter gene families in Sorghum bicolor under phytohormone and abiotic stress. FEBS J. 277:2954–2969.

Sims, R. E., A. Hastings, B. Schlamadinger, G. Taylor, and P. Smith. 2006. Energy crops: current status and future prospects. Glob. Change Biol. 12:2054–2076.

Singh SP. 1985. Sources of cold tolerance in grain sorghum. Can J Plant Sci. 65:251–257

Singh, M., K. Chaudhary, H. Singal, C. Magill, and K. Boora. 2006. Identification and characterization of RAPD and SCAR markers linked to anthracnose resistance gene in sorghum [Sorghum bicolor (L.) Moench]. Euphytica 149:179–187.

Sohn, K. H., S. C. Lee, H. W. Jung, J. K. Hong, and B. K. Hwang. 2006. Expression and functional roles of the pepper pathogen-induced transcription factor RAV1 in bacterial disease resistance, and drought and salt stress tolerance. Plant Mol. Biol. 61:897–915.

Staggenborg, S. A., K. C. Dhuyvetter, and W. Gordon. 2008. Grain sorghum and corn comparisons: yield, economic, and environmental responses. Agron. J. 100:1600–1604.

Su, M., X.-F. Li, X.-Y. Ma, X.-J. Peng, A.-G. Zhao, L.-Q. Cheng, et al. 2011. Cloning two P5CS genes from bioenergy sorghum and their expression profiles under abiotic stresses and MeJA treatment. Plant Sci. 181:652–659.

Subudhi, P., D. Rosenow, and H. Nguyen. 2000. Quantitative trait loci for the stay green trait in sorghum (Sorghum bicolor L. Moench): consistency across genetic backgrounds and environments. Theor. Appl. Genet. 101:733–741.

Sunkar, R., V. Chinnusamy, J. Zhu, and J.-K. Zhu. 2007. Small RNAs as big players in plant abiotic stress responses and nutrient deprivation. Trends Plant Sci. 12:301–309.

Takeda, S., and M. Matsuoka. 2008. Genetic approaches to crop improvement: responding to environmental and population changes. Nat. Rev. Genet. 9:444–457.

Tanksley, S. D. 1993. Mapping polygenes. Annu. Rev. Genet. 27:205–233.

Tao, Y., D. Jordan, R. Henzell, and C. McIntyre. 1998. Identification of genomic regions for rust resistance in sorghum. Euphytica 103:287–292.

Tao, Y., R. Henzell, D. Jordan, D. Butler, A. Kelly, and C. McIntyre. 2000. Identification of genomic regions associated with stay green in sorghum by testing RILs in multiple environments. Theor. Appl. Genet. 100:1225–1232.

Tao, Y., A. Hardy, J. Drenth, R. Henzell, B. Franzmann, D. Jordan, et al. 2003. Identifications of two different mechanisms for sorghum midge resistance through QTL mapping. Theor. Appl. Genet. 107:116–122.

Thomas, H., and H. Ougham. 2014. The stay-green trait. J. Exp. Bot., 65:3889–3900.

Tilsner, J., N. Kassner, C. Struck, and G. Lohaus. 2005. Amino acid contents and transport in oilseed rape (*Brassica napus* L.) under different nitrogen conditions. Planta 221:328–338.

Tuinstra, M. R., E. M. Grote, P. B. Goldsbrough, and G. Ejeta. 1997. Genetic analysis of post-flowering drought tolerance and components of grain development in *Sorghum bicolor* (L.) Moench. Mol. Breeding 3:439–448.

Tuinstra, M. R., S. Soumana, K. Al-Khatib, I. Kapran, A. Toure, A. van Ast, et al. 2009. Efficacy of herbicide seed treatments for controlling infestation of sorghum. Crop Sci. 49:923–929.

Upadhyaya, H. D., Y.-H. Wang, R. Sharma, and S. Sharma. 2013. Identification of genetic markers linked to anthracnose resistance in sorghum using association analysis. Theor. Appl. Genet. 126:1649–1657.

Varet, A., B. Hause, G. Hause, D. Scheel, and J. Lee. 2003. The Arabidopsis NHL3 gene encodes a plasma membrane protein and its overexpression correlates with increased resistance to *Pseudomonas syringae* pv. tomato DC3000. Plant Physiol. 132:2023–2033.

Verbruggen, N., and C. Hermans. 2008. Proline accumulation in plants: a review. Amino Acids 35:753–759.

Vijayalakshmi, K., A. K. Fritz, G. M. Paulsen, G. Bai, S. Pandravada, and B. S. Gill. 2010. Modeling and mapping QTL for senescence-related traits in winter wheat under high temperature. Mol. Breeding 26:163–175.

Visarada, K., and N. Kishore 2007. Improvement of Sorghum through transgenic technology. *Information System for Biotechnology News Report (Virginia tech, US)* pp. 1-3.

Wang, Z. X., M. Yano, U. Yamanouchi, M. Iwamoto, L. Monna, H. Hayasaka, et al. 1999. The Pib gene for rice blast resistance belongs to the nucleotide binding and leucine-rich repeat class of plant disease resistance genes. Plant J. 19:55–64.

Wang, S., Y. Bai, C. Shen, Y. Wu, S. Zhang, D. Jiang, et al. 2010. Auxin-related gene families in abiotic stress response in *Sorghum bicolor*. Funct. Integr. Genomics 10:533–546.

Wang, H., G. Chen, H. Zhang, B. Liu, Y. Yang, L. Qin, et al. 2014a. Identification of QTLs for salt tolerance at germination and seedling stage of *Sorghum bicolor* L Moench. Euphytica 196:117–127.

Wang, T. T., Z. J. Ren, Z. Q. Liu, X. Feng, R. Q. Guo, B. G. Li, et al. 2014b. SbHKT1; 4, a member of the high-affinity potassium transporter gene family from Sorghum bicolor, functions to maintain optimal Na⁺/K⁺ balance under Na⁺ stress. J. Integr. Plant Biol. 56:315–332.

Washburn, J. D., S. C. Murray, B. L. Burson, R. R. Klein, and R. W. Jessup. 2013. Targeted mapping of quantitative trait locus regions for rhizomatousness in chromosome SBI-01 and analysis of overwintering in a *Sorghum bicolor*× *S. propinquum* population. Mol. Breeding 31:153–162.

Wilkinson, S., G. R. Kudoyarova, D. S. Veselov, T. N. Arkhipova, and W. J. Davies. 2012. Plant hormone interactions: innovative targets for crop breeding and management. J. Exp. Bot. 63:3499–3509.

Wood, A. J., H. Saneoka, D. Rhodes, R. J. Joly, and P. B. Goldsbrough. 1996. Betaine aldehyde dehydrogenase in sorghum (molecular cloning and expression of two related genes). Plant Physiol. 110:1301–1308.

Wright, T. R., G. Shan, T. A. Walsh, J. M. Lira, C. Cui, P. Song, et al. 2010. Robust crop resistance to broadleaf and grass herbicides provided by aryloxyalkanoate dioxygenase transgenes. Proc. Natl Acad. Sci. 107:20240–20245.

Wu, Y., and Y. Huang. 2008. Molecular mapping of QTLs for resistance to the greenbug *Schizaphis graminum* (Rondani) in *Sorghum bicolor* (Moench). Theor. Appl. Genet. 117:117–124.

Wu, T.-M., W.-R. Lin, Y.-T. Kao, Y.-T. Hsu, C.-H. Yeh, C.-Y. Hong, et al. 2013. Identification and characterization of a novel chloroplast/mitochondria co-localized glutathione reductase 3 involved in salt stress response in rice. Plant Mol. Biol. 83:379–390.

Xu, W., P. K. Subudhi, O. R. Crasta, D. T. Rosenow, J. E. Mullet, and H. T. Nguyen. 2000. Molecular mapping of QTLs conferring stay-green in grain sorghum (*Sorghum bicolor* L. Moench). Genome 43:461–469.

Yan, H., L. Hong, Y. Zhou, H. Jiang, S. Zhu, J. Fan, et al. 2013. A genome-wide analysis of the ERF gene family in sorghum. Genet. Mol. Res. 12:2038–2055.

Yemata, G., Fetenel, M., Assefa, A and Tesfaye, K (2014) Evaluation of the agronomic performance of stay green and farmer preferred sorghum (Sorghum bicolor (L) Moench) varieties at Kobo North Wello zone, Ethiopia. Sky Journal of Agricultural Research Vol. 3 240–248

Young, W., and G. Teetes. 1977. Sorghum entomology. Annu. Rev. Entomol. 22:193–218.

Yu, Q., and S. B. Powles. 2014. Metabolism-based herbicide resistance and cross-resistance in crop weeds: a threat to herbicide sustainability and global crop production. Plant Physiol. 166:1106–1118.

Yu, J., and M. R. Tuinstra. 2001. Genetic analysis of seedling growth under cold temperature stress in grain sorghum. Crop Sci. 41:1438–1443.

Zegada-Lizarazu, W., and A. Monti. 2013. Photosynthetic response of sweet sorghum to drought and re-watering at different growth stages. Physiol. Plant. 149:56–66.

Zhai, J., Y. Dong, Y. Sun, Q. Wang, N. Wang, F. Wang, et al. 2014. Discovery and analysis of microRNAs in Leymus chinensis under saline-alkali and drought stress using high-throughput sequencing. PLoS ONE 9:e105417.

Zhang, M., Q. Tang, Z. Chen, J. Liu, H. Cui, Q. Shu, et al. 2009. [Genetic transformation of Bt gene into sorghum

(*Sorghum bicolor* L.) mediated by *Agrobacterium tumefaciens*]. Chin. J. Biotechnol. 25:418–423.

Zhao, B., R. Liang, L. Ge, W. Li, H. Xiao, H. Lin, et al. 2007. Identification of drought-induced microRNAs in rice. Biochem. Biophys. Res. Commun. 354:585–590.

Zheng, L.-Y., X.-S. Guo, B. He, L.-J. Sun, Y. Peng, S.-S. Dong, et al. 2011. Genome-wide patterns of genetic variation in sweet and grain sorghum (*Sorghum bicolor*). Genome Biol. 12:R114.

Zhou, Y., D. Wang, Z. Lu, N. Wang, Y. Wang, F. Li, et al. 2013. [Impacts of drought stress on leaf osmotic adjustment and chloroplast ultrastructure of stay-green sorghum]. Ying Yong Sheng Tai Xue Bao 24:2545–2550.

12

Harnessing diversity from ecosystems to crops to genes

Vicky Buchanan-Wollaston[1], Zoe Wilson[2], François Tardieu[3], Jim Beynon[1] & Katherine Denby[4]

[1]School of Life Sciences, University of Warwick, Gibbet Hill, Coventry CV4 7AL, U.K.
[2]School of Biosciences, University of Nottingham, Sutton Boningon Campus, Sutton Bonington, Leicestershire LE12 5RD, U.K.
[3]INRA Laboratoire d'Ecophysiologie des Plantes sous Stress Environnementaux (LEPSE), Montpellier, France
[4]Department of Biology, University of York, Heslington, York YO10 5DD, U.K.

Keywords
Food Security, Crop Stress Resilience, climate change, sharing Plant Resources, Data sharing

Correspondence
Katherine Denby, Department of Biology, University of York, Wentworth Way, York YO10 5DD UK
E-mail: katherine.denby@york.ac.uk

Abstract

To feed humanity, while maintaining a stable and diverse biosphere, crop science needs to adapt to an open research environment where genetic resources and the data demonstrating the environments in which they are effective are freely shared. The challenge faced is to expand crop production on a reduced land area, due to environmental degradation caused by human encroachment and climate change, while maintaining biodiversity. Individual researchers are discovering alleles and genetic combinations that are effective in certain environments but not in others. These data and alleles are useful globally to speed progress in breeding for similar environments while not wasting time on ineffective genotypes. However, currently, there are significant barriers to the sharing of genetic resources and their underpinning data, which must be overcome if we are to sustain the planet for future generations.

Introduction/State of Play

Current problems/challenges

More than a third of the world's land is currently used for agriculture, and future expansion of this industry threatens the stability and survival of the wider ecosystem. To sustain the world's expanding population, which is estimated to increase from 7 to 11 billion in the next 50 years, it is essential to generate more food from each hectare of arable land; however, this needs to be done in an environmentally sustainable manner.

Climate change poses a significant additional challenge, resulting in hotter and drier conditions in many parts of the world, as well as the increased likelihood of unpredictable weather extremes. Crop plants and their growing conditions will need to be more plastic; capable of coping with a variety of different environmental conditions. Maintaining and exploiting biodiversity (plants, insects, microorganisms, etc.) using well-organized management practices is essential to sustainably maintain and expand productivity.

Global solutions to stress resilient cropping systems are not exempt from the 'data storm' that is overwhelming all of biology. The development of new technologies, such as high-throughput sequencing and phenotyping, has led to an increasing number of large-scale and complex data sets. Integration and effective mining of these data sets is difficult. In particular, field and greenhouse experiments dedicated to plant responses to environmental conditions are by definition not reproducible, because the combination of environmental conditions in one experiment will never be experienced again by a particular set of genotypes. Data sets are generated on different technical platforms and the number of environmental variables even between growth chamber experiments is significant. Reuse of these data sets (for example, reanalysis based on different biological hypotheses and/or new methods) provides countless opportunities for new biological insights and achievements, but this requires data to be easily discoverable, as well as clearly annotated and explained, and in a format to easily enable use and integration with other data sets or analysis methods.

The ability to share data globally, and integrate and utilize such data in a variety of analysis pipelines and approaches, overcoming nonbiological variation and varying amounts of noise, will greatly influence the efficiency and impact of biological research around the world, and play a key role in attempts to feed the world in the face of changing environments by linking crop genotype with the agro-ecosystem to predict phenotype.

Agro-ecosystems

Understanding the wide diversity of ecosystems available for crop production is a key challenge. Manipulation and exploitation of agro-ecosystems should help to expand crop productivity while minimizing the resources used; for example, to maximize the use of available water it should be applied when the plant will make maximum use of it (Tardieu 2012), which requires an understanding of water deficit scenarios for different crops. Alternate wetting and drying, which results in a mild water deficit, has been shown to save water and maintain yields in rice (Yao et al. 2012; Price et al. 2013), and other crops such as maize and cotton (Kang et al. 1998; Tang et al. 2010). In the Shiyang river project in China, an intensive effort to limit nitrogen and reduce water use using methods such as drip irrigation has resulted in sustained or improved yields with reduced environmental impact (Du et al. 2015).

A genotype that has been selected for favorable performance in one environment does not necessarily perform equally well in a different environment, and the introduced alleles may potentially have a negative effect. It is important to understand the mechanisms by which plants respond to environmental stress; for example, different varieties may be susceptible to water deficit at different times of development (Tardieu and Tuberosa 2010; Tardieu 2012). A better understanding of ecosystem scenarios mapped with suitable genotypes may allow more appropriate varieties to be selected. The timing of planting (e.g., growing autumn instead of spring wheat) and soil manipulation even when not growing crops, can have a significant effect on water use and availability. Removal of summer weed cover in Australia resulted in retained water levels in the soil and significant subsequent increase in crop yield (https://grdc.com.au/Media-Centre/Ground-Cover/Ground-Cover-Issue-106-Sept-Oct-2013/Productivity-gains-there-for-the-taking?). Other agronomic approaches, such as intercropping, are also advantageous in certain conditions (Stoltz and Nadeau 2014), while problems with disease caused by crop monoculture could be addressed by mixing several genotypes of the crop in the same field (Zhu et al. 2000).

Crop resources

The genetic diversity of traditional varieties, modern cultivars, and wild relatives is crucial for crop improvement and food production, and also to act as a buffer for adaptation and resilience in the face of climate change. In recent years, however, there has been a strong tendency for farmers worldwide to abandon their multiple local varieties and landraces for genetically uniform, high-yielding varieties (http://www.fao.org/nr/cgrfa/cthemes/plants/en/; van den Wouw et al. 2010). This means that, currently, approximately 75% of the genetic diversity of crops may have been lost. Breeding for yield under good conditions means that stress resilience genes are not necessarily selected. An example of this problem was clearly shown in Bengal, where modern high-yielding varieties of rice were no match for the traditional varieties following the instant salinization of soil caused by a hurricane in 2009 (https://www.independentsciencenews.org/un-sustainable-farming/valuing-folk-crop-varieties/). For sustainable production under variable conditions, breeding for resilience of yield rather than maximal yield under optimal conditions is needed, which may require the reintroduction of such lost alleles.

Humans use only around 150–300 of the approximately 80,000 known edible plant species. Three of these – rice, maize, and wheat – contribute nearly 60% of our intake of plant-based calories and protein. The exploitation of additional crop species that are more resilient to certain growth environments could be key to expanding the productivity of the most challenging agronomic areas, both in terms of yield but also nutrient quality. Research into orphan crops such as teff, millet, cassava, sweet potato, and bambara groundnut are crucial for strengthening regional agriculture and improving nutrition (Crops for the Future http://www.cropsforthefuture.org).

Genetic improvement

Many research projects around the world have generated genetic markers and mapping populations for phenotyping and quantitative trait loci (QTL) identification in the major crops such as rice, maize and wheat. For example, CIMMYT's extensive international wheat improvement program is developing genomics for precision breeding in wheat using techniques such as high-throughput phenotyping, the collation of genetic resources (half a million wheat lines are available worldwide), and interspecific hybridization. QTL have been identified for heat and drought tolerance in several crops, and commercial drought-tolerant maize varieties have been developed, using marker-assisted breeding (e.g. Artesian hybrid maize from Syngenta and AquaMax from Pioneer (Tollefson 2010)).

There is an increasing gap between science and breeding. Research cannot deliver directly to the farmer without including breeding companies; therefore, large, collaborative projects are essential. For example, the DROPS project (http://www6.inra.fr/dropsproject, http://cordis. europa.eu/project/rcn/95052_en.html) is a multi-scale, multi-environment project on drought tolerance that exploits natural diversity, phenotyping platforms, and field analysis. It involves collaborators from 11 countries, including 11 public organizations and five large seed companies, which are involved in development of the method, as well as the results. It is strongly interdisciplinary, involving modelers and statisticians, as well as plant breeders, molecular geneticists, and biochemists. It aims to combine precise crop modeling and genome prediction with environmental influences (Tardieu 2012; Tardieu and Tuberosa 2010).

Gene identification

Many research projects across the globe aim to genetically improve stress resilience in plants, using both the model plant Arabidopsis and several crop species. Many of these projects are at the fundamental research level. Knowledge of individual genes that can be manipulated to confer tolerance to a single stress is relatively advanced; for example, multiple genes conferring drought tolerance (reviewed for rice in Todaka et al. (2015)) and salt tolerance (e.g. *Nax2* in wheat (Munns et al. 2012) and *SALT3* in soybean (Guan et al. 2014)) have been identified, and the *SUB1* gene was found to control submergence tolerance in rice (Xu et al. 2006)). Also, the *ALT1* gene, which confers aluminum tolerance in sorghum, could be highly beneficial to use in Africa (Ryan et al. 2011). So far, few individual genes have been used to develop transgenic, stress-tolerant commercial lines, though DroughtGuard maize from Monsanto (which contains the cold shock protein *cspB* from the bacterium *Bacillus subtilis*) is one example.

Sharing data sets

High quality data relating to environmental stress responses in plants is extremely expensive to generate, and is obtained from specific species under specific conditions. The complexity and cost of generating good data tend to make organizations and scientists protective of those data, limiting their impact and value by restricting their availability. There is an abundance of less complete data sets, many of which would make valuable contributions to building up more robust collections and conclusions. No-one has the resources to test all genotypes or agronomic techniques in multiple situations, and reusing data provides a means to extend what is possible. Globally, we want to know: in what environments does a particular genotype succeed?

Cultural changes in the availability of data, and the ways in which they are queried, are required if effective outcomes are to be delivered. It is important that there is a shift toward quantitative data, and that data can be easily shared in a reliable and informed manner. The continual growth in data requires the linking of multiple data stores around the world, as well as the development of appropriate best-practice guidelines to adequately describe data so that computational methods can be used to query and/or discover data in different databases/structures. Several initiatives are working toward enhancing description of data and hence its accessibility including NCBI (Barrett et al. 2012), with the Bio Project database and submission portal, and the development of recommendations for metadata in plant phenotyping experiments (Krajewski et al. 2015).

Utilizing negative results

Furthermore, most current data release is via publication in scientific journals; this leads to an emphasis on positive results. It is likely that such positive results are less common than negative results, but negative results are never made available. However, we consider as results per se the lack of response of a given set of genotypes in a given environmental scenario and/or with the chosen physiological traits with the chosen protocol. Although some publishers are attempting to establish journals for the publication of negative results, it is debatable whether researchers will spend the time writing up a negative results publication; despite the buzz surrounding Elsevier's *New Negatives in Plant Science*, the journal was discontinued in September 2016, just 1 year after its first issue. Simple and fast deposition in a database may be more likely to succeed. Open data stores are increasingly popular and could be used in this way, along with data-only journals in which high quality, curated data sets can be described without the requirement of scientific interpretation (Leonelli et al. 2013). The availability of negative results could be a game changer in international plant breeding, as it would stop continuous retesting of genotypes that do not work, and save time and money on a global scale, so speeding the development of stress tolerant local varieties.

Training the next generation of data scientists

To fully exploit new technologies for generating genome and phenotypic data, the plant research/breeding community needs to attract (and/or train) a new generation of researchers who are skilled at extracting relevant knowledge from diverse large-scale datasets, and can combining diverse data sets to provide new insights. This is not just people with bioinformatics skills – most bioinformatics-trained scientists have used a limited number of tools, approaches and data sets. Bioinformaticians usually have skills in looking at genome data, creating novel ways of displaying data, and using computational tools, but not in developing new innovative methods for merging and analyzing data. We need people or teams with an amalgamation of computer science, statistical and biological knowledge to be able to explore and mine these datasets effectively to solve global food production challenges caused by population growth and environmental induced stress (see e.g., DROPS Project. http://www6.inra.fr/dropsproject, http://cordis.europa.eu/project/rcn/95052_en.html). Above all, we need a generation of researchers interested in quantitative analyses based on practical solutions that define *when and where* specific genotypes will thrive and not only interested in nice stories to satisfy reporting requirements for funding agencies. There is a real need to bring innovative quantitative scientists into the sustainable crop production arena; often too much money is spent on technology, and not enough on creating the innovative scientists who will use these data for the benefit of mankind.

Effective and innovative use of plant data will require dialog between different groups of expertise, and a focusing of effort on key problems and challenges. Such dialog initiatives exist within Big Data (for example, the Alan Turing Institute in the UK or, in the plant science community, projects such as EU DROPS (Millet et al. 2016), CyVerse in the USA (http://www.cyverse.org) and the Agrimetrics Big Data Centre of Excellence in the United Kingdom (http://www.agrimetrics.co.uk/), which brings together capability in data science and smart analytics with agrifood research expertise to drive exploitation of agrifood data though not necessarily focused on crop science. Additional crop-focused initiatives include the French programmes Amaizing and Breedwheat (http://www.amaizing.fr; http://www.breedwheat.fr).

The ultimate goal of this field would be the ability to design plant varieties/genotypes that would thrive in certain environmental and agronomic conditions. A plant breeder should be able to define the environment in which they wish a plant to grow, and be able to mine the genetic variants that will generate such a crop plant with the optimum characteristics to produce a stable yield (Hammer et al. 2006). The time is ripe to move away from old ways of working, to combine accurately described and shared genetic resources with new data structures together with a generation of scientists to interact with them, which will enable plant breeding in the 21st century.

Actions

Enhance information gathering, access and reuse

There are multiple individual projects worldwide that clearly contribute to our knowledge of mechanisms to improve crop stress resilience at all scales, from agroecosystems to individual genes. Working with other relevant organizations, the Global Plant Council (http://globalplant-council.org/) which is described in an Editorial article in this volume, could help to make the outputs of these projects more easily accessible to a global audience. This short-term goal is not straight-forward, especially on a global scale, but initiatives do exist. For example, UK's Collaborative Open Plant Omics (COPO) project is finding ways to make it easier for researchers to annotate and deposit data (http://copo-project.org), POPCorn in the US is an online resource providing access to distributed and diverse maize data (Cannon et al. 2011, http://dx.doi.org/10.1155/2011/923035) and the Genomes to Field (G2F) initiative that has released its data through CyVerse (http://www.cyverse.org/news/genomes-environment-dataset-now-publicly-accessible; http://www.genomes2fields.org).

Actions that GPC might take include:
1. Current status: Landscaping the different initiatives in this area and facilitating communication between them to develop bridges, enhance outputs and thus enable more effective global data reuse, while respecting and valuing different approaches and their comparative advantages, and showcasing successes to inspire and inform future activities.
2. Environmental conditions: Identifying minimum requirements and standards for the annotation of environments associated with a given experiment. These are not just "metadata" and are intrinsically part of phenotypic datasets. "Minimum datasets" have been defined in projects such as EU DROPS (http://www6.inra.fr/dropsproject, http://cordis.europa.eu/project/rcn/95052_en.html) or EPPN (Tardieu 2013)
3. Collating methods and protocols to be included in information systems, including the methods that have been used for image analyses or sensor calibration
4. Genotypes: what is being used, and what is publically available for testing and breeding?
5. Information on modeling: surveying the methods that are being used, and how these can be made widely accessible.
6. Data on orphan crops and their associated growth characteristics: gather available data to guide recommendations on which crops should be funding priorities.

7. Germplasm resources: increase knowledge of what has been developed and help to enable access. A directory of germplasm resources could be created, listing the resources that each facility has (including contact details), and this could be promoted and distributed via GPC channels and members.

8. Case studies from different communities: facilitating discussion between communities, with the aim of developing a simple, easy-to-use set of criteria that will be more likely to be adopted by experimentalists, while at the same time enabling data scientists to analyze the likelihood of specific outcomes and combine the maximum amount of available data.

9. Data integration and reuse projects: helping to define what is likely to have significant impact, including publicizing success stories in this area and disseminating resources about game-changing projects to prevent duplication, and trigger new ideas, and initiatives in gap areas.

10. Facilitate a cultural change in the way that plant stress resilience research is structured; switching from essentially qualitative to quantitative approaches, thus moving away from the traditional emphasis on telling a nice qualitative "story" that can be published in a high impact factor journal.

11. Training courses: landscaping existing courses, mirroring e-courses, and facilitating local workshops in different languages to help train and build capacity in scientific experts applying quantitative approaches to global datasets.

Building on available information

As an organization with links to plant scientists and professional societies from six continents, the GPC is well placed to undertake landscaping surveys and information-gathering activities that help to build on and link up existing silos of knowledge. For example, the GPC might coordinate efforts to:

1. Identify and prioritize the most devastating stresses, and combinations of stress factors, to assess the biggest challenges. Taking drought as an example, what are the most common dynamics of drought (e.g., the time at which plants experience drought, extent of drought)?

2. Collate knowledge on the diverse agro-ecosystems and map their similarities worldwide (Tardieu 2013). What is the limiting factor at each location? How conserved are these factors between different available data sets?

3. Identify suitable cropping systems for the different environments identified. Match cropping systems to environments using local knowledge, identify geographically distinct but agro-ecologically similar areas and propose

cropping systems for these additional areas globally. The identification of common environments and stresses around the world will provide opportunities for sharing germplasm.

4. Enable the sharing of germplasm. It would be a very valuable achievement if GPC could improve this major bottleneck, facilitating the testing of germplasm bred for a particular environment in one country in similar environments across the globe.

5. Utilize the modeling performed by the Intergovernmental Panel on Climate Change to predict the most likely stress scenarios for different regions in the future, and where will be most impacted in terms of temperature, water, etc.

6. Integrate international research to generate a set of optimal practices in cropping systems.

Provision of characterized genetic resources

A key activity the GPC could undertake is to enable well coordinated comparisons of selected genotypes, tested over a long time scale, and under different combinations of stresses in geographically diverse natural environments. For example, the GPC might coordinate efforts as follows:

1. Pull together a small panel of genetically diverse lines for each target crop and make these available to researchers worldwide to allow phenotyping of known genetic materials under different agronomic environments. Data comparisons will allow germplasm information to be linked to stress resilience (individual or combinations of stress) and other traits (See Borrell and Reynolds 2017, this volume).

2. Model the outcomes and use these data to predict yield potential under variable or known adverse conditions, as well as determine the cost of plasticity and resilience.

3. Define links between genotypes/phenotypes and agricultural practices; help to promote the cultural change that is required to facilitate greater exchange and sharing of expertise, resources and data.

Improve data availability

Data are only of any value if they are available and contain sufficient annotation to ensure their utility. The current transition from a traditional, secretive approach to data to working on globally available resources can be influenced by research funders and scientific journals. The community must approach them to demand adherence to minimum data standards, to widely distributed data information systems, and to the availability of negative, as well as positive data. Together with European and International initiatives (www.fao.org/nr/gaez/en; http://www.plant-phenotyping.org; http://www.plant-phenotyping-network.eu/eppn/home; http://emphasis.plant-phenotyping.eu), ways in which the GPC could help in this regard are highlighting on a global scale the critical issues of:

1. Building, using and interacting with IT infrastructures;
2. Forming appropriate biological questions using this infrastructure;
3. The ability to extract meaningful results and predictions from data; by collecting international experience in what has and has not worked.
4. Identifying where innovation is needed, whether it is already available in other fields (e.g. data-search, finance), and what resources are needed to transfer the technologies, approaches and mindset.

Securing long-term global funding

These issues of scientific culture, data availability, and the ability to ask the relevant questions are a major challenge that, although very difficult, need to be addressed urgently. The scale of the issue is such that, although small steps can be made immediately, to make the phase shift that would impact global food security may require a globally funded initiative. GPC could structure such a program aimed at crop stress resilience and propose it to major independent charitable funders.

Acknowledgements

This paper is based on outcomes from a Stress Resilience Symposium held in Brazil in October 2015 organized by the Global Plant Council and Society for Experimental Biology. The authors would like to thank the Society for Experimental Biology for funding support for this symposium.

Conflict of Interest

None declared.

References

Barrett, T., K. Clark, R. Gevorgyan, V. Gorelenkov, E. Gribov, I. Karsch-Mizrachi, et al. 2012. BioProject and BioSample databases at NCBI: facilitating capture and organization of metadata. Nucleic Acids Res. 40:D57–D63.

Cannon, E. K., S. M. Birkett, B. L. Braun, S. Kodavali, D. M. Jennewein, A. Yilmaz, et al. 2011. POPcorn: an online resource providing access to distributed and diverse maize project data. Int. J. Plant Genomics http://dx.doi.org/10.1155/2011/923035

Du, T., S. Kang, J. Zhang, and W. J. Davies. 2015. Deficit irrigation and sustainable water-resource strategies in agriculture for China's food security. J. Exp. Biol. 66:2253–2269.

Guan, R., Y. Qu, Y. Guo, L. Yu, Y. Liu, J. Jiang, et al. 2014. Salinity tolerance in soybean is modulated by natural variation in GmSALT3. Plant J. 80:937–950.

Hammer, G., M. Cooper, F. Tardieu, S. Welch, B. Walsh, F. van Eeuwijk, et al. 2006. Models for navigating biological complexity in breeding improved crop plants. Trends Plant Sci. 11:587–593.

Kang, S., Z. Liang, W. Hu, and J. Zhang. 1998. Water use efficiency of controlled alternate irrigation on root-divided maize plants. Agric. Water Manag. 38:69–76.

Krajewski, P., D. Chen, H. Ćwiek, A. van Dijk, F. Fiorani, P. Kersey, et al. 2015. Towards recommendation for metadata and data handling in plant phenotyping. J. Exp. Bot. 66:5417–5427.

Leonelli, S., N. Smirnoff, J. Moore, C. Cook, and R. Bastow. 2013. Making open data work for plant scientists. J. Exp. Bot. 4109–4117.

Millet, E., C. Welcker, W. Kruijer, S. Negro, S. Nicolas, S. Praud, et al. 2016. Genome-wide analysis of yield in Europe: allelic effects as functions of drought and heat scenarios. Plant Physiol. 172:749–764.

Munns, R., R. A. James, B. Xu, A. Athman, S. J. Conn, C. Jordans, et al. 2012. Wheat grain yield on saline soils is improved by an ancestral Na(+) transporter gene. Nat. Biotechnol. 30:360–364.

Price, A. H., G. J. Norton, D. E. Salt, O. Ebenhoeh, A. Meharg, C. Meharg, et al. 2013. Alternate wetting and drying irrigation for rice in Bangladesh: is it sustainable and has plant breeding something to offer? Food Energy Secur. 2:120–129.

Ryan, P. R., S. D. Tyerman, T. Sasaki, T. Furuichi, Y. Yamamoto, W. H. Zhang, et al. 2011. The identification of aluminium-resistance genes provides opportunities for enhancing crop production on acid soils. J. Exp. Bot. 62:9–20.

Stoltz, E., and E. Nadeau. 2014. Effects of intercropping on yield, weed incidence, forage quality and soil residual N in organically grown forage maize (Zea mays L.) and faba bean (Vicia faba L.). Field. Crop. Res. 169:21–29.

Tang, L., Y. Li, and J. Zhang. 2010. Partial rootzone irrigation increases water use efficiency, maintains yield and enhances economic profit of cotton in arid area. Agric. Water Manag. 97:1527–1533.

Tardieu, F. 2012. Any trait or trait-related allele can confer drought tolerance: just design the right drought scenario. J. Exp. Bot. 63:25–31.

Tardieu, F. 2013. Plant response to environmental conditions: assessing potential production, water demand, and negative effects of water deficit. Front. Physiol. 4:1–11.

Tardieu, F., and R. Tuberosa. 2010. Dissection and modelling of abiotic stress tolerance in plants. Curr. Opin. Plant Biol. 13:206–212.

Todaka, D., K. Shinozaki, and K. Yamaguchi-Shinozaki. 2015. Recent advances in the dissection of drought-stress regulatory networks and strategies for development of drought-tolerant transgenic rice plants. Front. Plant Sci. 6:84.

Tollefson, J. 2010. Drought-tolerant maize gets US debut. Nature 469:144.

van den Wouw, M., C. Kik, T. van Hintum, R. van Reuren, and B. Visser. 2010. Genetic erosion in crops: concept, research result and challenges. Plant Genet. Resour. 8:1–15.

Xu, K., X. Xu, T. Fukao, P. Canlas, R. Maghirang-Rodriguez, S. Heuer, et al. 2006. Sub1A is an ethylene-response-factor-like gene that confers submergence tolerance to rice. Nature 442:705–708.

Yao, F., J. Huang, K. Cui, L. Nie, J. Xiang, X. Liu, et al. 2012. Agronomic performance of high-yielding rice variety grown under alternate wetting and drying irrigation. Field. Crop. Res. 126:16–22.

Zhu, Y., H. Chen, J. Fan, Y. Wang, Y. Li, J. Chen, et al. 2000. Genetic diversity and disease control in rice. Nature 406:718–722.

13

Reducing the acrylamide-forming potential of wheat

Tanya Y. Curtis & Nigel G. Halford

Plant Biology and Crop Science Department, Rothamsted Research, Harpenden, Hertfordshire AL5 2JQ, UK

Keywords

Acrylamide, asparagine, bread, food safety, free amino acids, processing contaminants, reducing sugars, sulfur fertilization, wheat.

Correspondence

Tanya Y. Curtis, Plant Biology and Crop Science Department, Rothamsted Research, Harpenden, Hertfordshire AL5 2JQ, United Kingdom.
E-mail: tanya.curtis@rothamsted.ac.uk

Funding Information

No funding information provided.

Abstract

Acrylamide is a Class 2a carcinogen that was discovered in a variety of popular foods, including baked cereal products, in 2002. The predominant route for its formation is from free asparagine and reducing sugars in the Maillard reaction, with free asparagine concentration being the main determinant of acrylamide-forming potential in cereal products. The European Commission set "indicative" levels for acrylamide in food in 2011 and 2013, and is currently reviewing its options for further measures. Agronomic and genetic approaches to reducing the acrylamide-forming potential of wheat include the evaluation of existing varieties for low asparagine accumulation in the grain, ensuring adequate sulfur fertilization in relation to nitrogen supply, developing an understanding of the genetic control of asparagine metabolism, and identifying quantitative trait loci or molecular markers for low asparagine accumulation in the grain. Asparagine concentration in grain is affected by environmental factors (E), genetic factors (G), and interactions between the two (G × E). This paper reviews the continuing efforts being made to reduce the acrylamide-forming potential of wheat, and to increase awareness of the issue among wheat breeders, farmers, and the food industry.

Introduction

Acrylamide is a neurotoxin and probable (Class 2a) carcinogen in humans (Schulze and Siegers 2004; Taeymans et al. 2005). According to the opinion on threshold of toxicological concern (Dybing and Sanner 2003), the no-observed-effect level (NOEL) is 0.2 mg/kg body weight (bw). The lower limit of this range is called the benchmark dose lower confidence limit (BMDL) (Parzefall 2008); a BMDL of 0.17 mg/kg bw/day has been set for tumor-inducing effects, and 0.43 mg/kg bw/day for neurological effects (http://www.efsa.europa.eu/sites/default/files/corporate_publications/files/acrylamide150604.pdf). The European Food Safety Authority (EFSA) Expert Panel on Contaminants in the Food Chain (EFSA 2015b) stated that the margins of exposure for acrylamide indicate a concern for neoplastic effects based on animal evidence (EFSA Panel on Contaminants in the Food Chain [EFSA 2015b]). The European Commission had already issued "indicative" levels for the presence of acrylamide in food in 2011, and revised them downward for many products in 2013 (European Commission 2013). Furthermore, in response to the CONTAM report, the Commission is currently reviewing additional options for risk management measures.

Acrylamide Formation

Acrylamide forms during cooking and processing at temperatures above 120°C, usually during the processes of frying, roasting, and baking, and levels higher than 2000 ppb have been reported in some foods produced from high-sugar and free asparagine-containing raw materials. It is formed during the Maillard reaction, a series of nonenzymatic reactions between free (nonprotein) amino acids and reducing sugars when all desirable flavors are formed. The Maillard reaction occurs in three stages: during the first stage, free amino acids and other amino groups react with reducing sugars to form carbonyl compounds (Mottram et al. 2002; Stadler et al. 2002; Stadler 2005). The reaction is initiated by the condensation of the carbonyl (C=O) group of a reducing sugar with the amino group, producing a Schiff base. If the sugar is an aldose, the Schiff base cyclises to give an N-substituted aldosylamine, such as glucosylamine from glucose.

Acid-catalyzed rearrangement of the aldosylamine gives a 1,2-enaminol, which is in equilibrium with its ketotautomer, an *N*-substituted 1-amino-2-deoxyketose: these are known as Amadori rearrangement products. Ketoses, such as fructose, give related Heyns rearrangement products by similar pathways.

In the second stage of the reaction, the Amadori and Heyns rearrangement products undergo enolization, deamination, dehydration, and fragmentation, giving rise to sugar dehydration and fragmentation products containing one or more carbonyl groups, including heterocyclic furfurals, furanones, and pyranones. These carbonyl compounds can undergo condensation reactions with amino groups and other components present at this stage of the Maillard reaction, resulting in the formation of many different flavor compounds (International Agency For Research On and International Agency For Research On 1994). An important reaction of carbonyl compounds is Strecker degradation: the deamination and decarboxylation of an amino acid to give an aldehyde in which the α-carbon of the amino acid is converted to an aldehyde group, the reaction also yielding an α-aminoketone. Acrylamide is formed in a Strecker-type reaction involving sugar-derived carbonyl compounds and asparagine (Zyzak et al. 2003; Mottram 2007; Koehler et al. 2008).

During the third and final stage of the Maillard reaction, a thermally induced reaction between the carbonyl compounds, amino acids, and their degradation derivatives produces the desired flavor compounds and melanoidin pigments (Mottram et al. 2002).

Acrylamide in Food

The following food categories are the main contributors to acrylamide exposure in the diet (European Food Safety Authority 2011, EFSA 2015a,b): potato crisps, French fries, bread, breakfast cereals, biscuits, other cereal-based snacks, battered fried foods, popcorn, coffee, ginger bread, and chocolate. Recently, it was found that there is also acrylamide in processed olives (Casado and Montano 2008; Javier Casado et al. 2010, 2013, 2014). Acrylamide formation depends on the levels of its precursors (free asparagine and reducing sugars) in the raw material and the processing methods that are used, and the concentrations of free asparagine and reducing sugars are affected by specific variety, location, and management of the crop (Curtis 2010; Curtis et al. 2010a,b, Curtis et al. 2009).

Risk Assessment by International Authorities

The latest report from EFSA's Expert Panel on Contaminants in the Food Chain (EFSA) described the risk characterization for neoplastic effects using as a reference point the BMDL of 0.17 mg/kg bw/day of acrylamide in food, although it considered the neurological, reproductive, and developmental effects of acrylamide not to be a concern at the current levels of dietary exposure (EFSA 2015a). "Since MOEs calculated are substantially lower than the value of 10,000, the CONTAM Panel concluded that, although the available human studies have not demonstrated acrylamide to be a human carcinogen, the MOEs across survey and age groups indicate a concern with respect to neoplastic effects" (Food Standard Agency 2012). The Food and Agriculture Organisation of the United Nations and the World Health Organisation (FAO/WHO) Joint Expert Committee of Food Additives (JECFA) has also concluded that the presence of acrylamide in the human diet is a concern (World Health Organization 2006).

EFSA monitors data supplied by EU Member States on acrylamide in food, and the figures for cereal-based products from 2007 to 2012 are given in Table 1. In 2011 and 2013, the European Commission set "indicative" levels for acrylamide in different food categories (European Commission 2013), based on the results of this screening exercise, and these are also shown in Table 1. Indicative levels are not maximum levels or an indication of safety or lack of it, rather they are the levels that the Commission believes the food industry should be able to achieve, based on the screening data.

Coverage of Acrylamide in the Media

The coverage of the acrylamide issue in the press has not reflected the efforts of the food industry to mitigate the acrylamide issue and to decrease the levels of acrylamide found in food.

Baby food is of greatest interest because a baby's body weight is lower than that of adults. The European Commission therefore set a relatively low indicative level for cereal-based baby foods of 100 ppb in 2011 and reduced it to 50 ppb in 2013. Table 1 shows that the maximum level of acrylamide found in cereal-based baby foods has been well above the 50 ppb mark throughout the screening period. It is important to note that NOELs for babies will be considerably lower than for an adult because of the difference in body weight. For a body weight of, for example, 12 months old baby of 7.1 kg (growth chart), the NOEL would be 1.42 mg, whereas for a 15-g portion of breakfast cereal (even based on the maximum level of 2072 μg/kg in Table 1), the acrylamide content would be 31.08 μg. Even if no other acrylamide-containing food were eaten in that day, the MOE would be approximately 40, not the 10,000 favored by EFSA.

Table 1. Acrylamide concentrations measured in cereal-based foods in Europe from 2007 to 2012 (EFSA 2015a,b) and indicative levels set by the European Commission (European Food Safety Authority 2011).

Acrylamide levels (µg/kg)

Food type	Mean					Max					Indicative levels (µg/kg)	
	2007	2008	2009	2010/2011	2012	2007	2008	2009	2010/2011	2012	2011	2013
Biscuits												
Crackers	291–292	203–206	195–208	275	333	1526	1042	1320	473	1062	**500**	**500**
Infant	197–204	98–110	88–108	110	86	2300	1200	430	598	470	**250**	**200**
Unspecified	299–303	213–223	128–140	625	289	4200	1940	2640	1574	5849		**500**
Wafers	206–210	251–254	244–246	154	389	1378	2353	725	154	1300		**500**
Bread												
Bread crisp	221–226	229–231	219–223	197	249	2430	1538	860	326	1863	**150**	**450**
Bread soft	54–68	31–46	27–37	15	30	910	528	364	37	425		**80**
Unspecified	172–190	45–231	54–76	14		2565	86	1460	51			**150**
Breakfast cereals	130–150	140–156	132–142	149	138	1600	2072	1435	325	1290	**400**	**200–400**
Baby food	48–69	35–51	55–70	18	51	353	660	710	68	578	**100**	**50**

Bold values are the indicative levels recognized by the EU Commission.

Industry and Research Outcomes

Significant efforts have been made by the food industry in recent years toward reducing levels of acrylamide in its products. Approaches include selecting varieties which contain low levels of acrylamide precursors (free asparagine and reducing sugars); removing precursors before processing (Friedman 2003, 2005; Howie et al. 2006, 2007; Friedman and Levin 2008); using the enzyme asparaginase to hydrolyze asparagine to aspartic acid and ammonia prior to cooking or processing; controlling processing conditions such as pH, temperature, time, processing, and storage atmosphere to minimize acrylamide formation; and adding food ingredients that have been reported to inhibit acrylamide formation, such as amino acids, antioxidants, nonreducing carbohydrates, garlic compounds, protein, and metal salts. For the reduction of acrylamide formation in biscuits on an industrial scale the following measures were published by Graf et al.: replacement of ammonium hydrogen carbonate by sodium hydrogen carbonate which reduced the acrylamide content by about 70%; use of a sucrose solution instead of inverted sugar syrup (glucose and fructose) had a similar effect (sucrose will participate in the Maillard reaction, but only if it is first hydrolyzed through enzymatic, thermal, or acid-catalyzed reaction [Vleeschouwer et al. 2009]), while the addition of extra tartaric acid reduced the acrylamide content by approximately 30%. These results showed that mitigation on an industrial scale, based on the optimization of baking agent, reducing sugars, and organic acid, is feasible (Graf et al. 2006). The overall reduction presented in this paper was from 3200 ppb (µg/kg) acrylamide to 960 ppb (µg/kg) (this is 70%).

Methods for reducing acrylamide formation have been compiled and reviewed in the "Acrylamide Toolbox" produced by Food Drink Europe (2014). The Toolbox consists of four pillars: agronomy, recipe, processing, and final preparation. The agronomic advice concerns the amount of reducing sugars and free asparagine in the raw material; recipe refers to basic formulae, ingredients, and product form; processing deals with thermal input and moisture, the addition of asparaginase, pretreatments, finished product color, texture, and flavor; while the term final preparation includes instruction and consumer guidance.

The main objective for the food industry is to continue working to reduce acrylamide levels to as low as reasonably achievable (the ALARA principle). Table 1 shows that the mean level of acrylamide in foods such as wafers, breakfast cereals, and baby foods has decreased substantially between 2007 and 2012, whereas some other food categories only show modest results. This is because many of the acrylamide mitigation tools are food system specific and show large variations in effectiveness across food categories. An example would be pretreatment with the enzyme asparaginase. This works well in foods that have an "aqueous" preparation step, but does not work well in foods that are produced with limited moisture content. Therefore, the food industry has a continuing need for the development of acrylamide mitigation tools that are more universal in nature and can be applied across food categories and in the home. Research on producing wheat with a low concentration of free asparagine in the grain targets this need. It is critical to continue with research in this direction in order to have tools and concepts in place for the future. The challenges facing the food industry require more detailed research and understanding of acrylamide precursors in different elite wheat varieties in order to identify agronomic, environmental, or genetic factors likely to influence acrylamide formation in the end product (Curtis 2010).

Acrylamide Level Uptake Due to Cereal Products

Table 2 shows the acrylamide intake due to cereal-based products in various European countries (Curtis and Halford 2014, EFSA 2015a,b). In France, Germany, and Sweden, a major contributor to dietary acrylamide intake is bread, while in the United Kingdom, the contribution of bread and cereal products overall is lower. This reflects differences in dietary preferences rather than the acrylamide levels in the products between the four countries, with more fried potato products being consumed in the United Kingdom. Similarly, muesli and crisp bread are high contributors in Sweden because of the popularity of those foods in that country.

Bread contains relatively low levels of acrylamide, but due to its large consumption it is a main contributor to total dietary intake. It is important to note that some bread is consumed as toast and acrylamide levels in the product before toasting are a lot lower in comparison to the levels in the toasted bread. For example, Granby et al. reported that levels of <5 μg/kg in a soft bread slice rose to 11–161 μg/kg in a toasted slice of bread depending on the coloration (Granby et al. 2008). This highlights the problem of how foods are cooked in the home, and the need to educate consumers on ways to reduce acrylamide formation, something to which consumers in the United Kingdom at least have so far been unreceptive.

In the United Kingdom, the contribution of acrylamide from cereal products is highest in bread, followed by biscuits and breakfast cereals, but these products are not consumed every single day and consumer preferences are changing anyway as alternative foods become more readily available and popular. Detailed annual statistics on family food and drink purchases, for example, demonstrate a trend of declining white bread consumption. Standard unsliced white bread sales, for example, have

decreased by 15% since 2011 and 9% since 2013, while white bread, soft grain, sliced and unsliced, sales have decreased by 156% since 2011, although they have levelled out from 2013. On the other hand, the consumption of total other products such as takeaway bread in pre-prepared sandwiches from takeaway outlets has increased by 38% since 2011 and 7% since 2013. Additionally, consumption of takeaway breads has increased 15% since 2011 and a little decline of 5% since 2013 (https://www.gov.uk/government/statistical-data-sets/family-food-datasets).

Asparagine as a Major Precursor of Acrylamide Formation in Food: Factors Affecting Asparagine Accumulation

Free asparagine is the main amino compound precursor of acrylamide formation in food, as confirmed by studies using isotopically labeled asparagine which showed that the three C atoms and single N atom of acrylamide were all derived from asparagine (Zyzak et al. 2003). In addition, asparagine concentration has been shown to be the main limiting factor for acrylamide formation in wheat (Muttucumaru et al. 2008) and rye (Curtis et al. 2010a; Postles et al. 2013).

Asparagine is one of the main amino acids involved in nitrogen accumulation and transport in plants, together with glutamine (Lea et al. 2007). It is the major transport compound in the xylem from the root to the leaves and in the phloem from the leaves to the developing seeds in a range of plants (Lea et al. 2007), although this is not the case for potato (Muttucumaru et al. 2014) and has not been established for wheat. Asparagine is relatively inert and therefore particularly suited to the role of a nitrogen transport and storage compound. Asparagine is also one of the building blocks of wheat and rye seed proteins. It is present in γ-gliadin and secalin (1.3–1.47%), α-gliadin and secalin (2.54–2.68%), ω-gliadin and secalin (0.75–0.77%), and low-molecular-weight glutenin subunits (LMW subunits) (0.7–1.08%), but not in high-molecular-weight subunits (HMW subunits) (Khan 2007). It is one of the nonessential amino acids in the diet.

Table 2. Contribution of cereal products (%) to dietary acrylamide intake for adults in selected European countries (Curtis and Halford 2014).

| Country | Food group | | | | | |
	Biscuits	Crisp bread	Bread	Breakfast cereal	Muesli	Total
France	7.6	5.3	25.7	1.3	1.0	40.9
Germany	6.1	4.0	32.0	1.2	2.1	45.4
Sweden	5.0	9.7	11.9	1.5	13.1	41.2
United Kingdom	6.3	2.0	15.0	5.0	3.6	31.9

Efforts to Reduce Acrylamide-Forming Potential

Efforts to reduce free asparagine accumulation in wheat grain have involved the following strategies:

1. Identification of existing varieties with low grain asparagine concentrations.
2. Identification of genotypes with low grain asparagine concentration that are not current commercial varieties but could be incorporated into breeding programmes.
3. Development of a comprehensive understanding of asparagine metabolism (including the use of mathematical modeling).
4. Elucidation of genetic (G) and environmental (E) factors (including crop management) that affect asparagine accumulation in the grain.
5. Understanding of the relationship between asparagine concentration, total grain sulfur and nitrogen, and acrylamide formation.
6. Identification of quantitative trait loci (QTL), genes, and markers for use by plant breeders to produce very low acrylamide varieties.

Genetic Differences Between Wheat Varieties

Comparison of four doubled haploid (DH) lines and Spark and Rialto parental lines

To establish the differences between doubled haploid (DH) lines from a Spark × Rialto mapping population provided by the John Innes Centre Wheat Genetics Group, Spark and Rialto parental lines plus SR3 (a low asparagine DH line), SR41 (a high free asparagine DH line), and SR7 together with SR107 (both intermediary asparagine DH lines) were grown under controlled conditions in a glasshouse (Curtis et al. 2009). Free amino acids were extracted, derivatized, and analyzed by gas chromatography mass spectrometry (GC-MS). Statistical analysis showed the lines differed significantly in concentration of free asparagine, aspartic acid, glycine, and valine. The main contributors to the total free amino acid pool were asparagine, aspartic acid, and glutamic acid, and the difference between SR3 as a low free asparagine genotype and SR41, the most abundant free asparagine genotype, was the ratio between aspartic acid and asparagine, and the concentration of the total free amino acid pool. In SR3, the concentration of the total free amino acid pool was 8.5 mmol/kg, whereas in SR41 it was 12.5 mmol/kg, while the concentration of free asparagine in SR3 was 1.68 mmol/kg and aspartic acid was 3.26 mmol/kg. In contrast, in SR41, the concentration of free asparagine was 3.23 mmol/kg and aspartic

acid was 3.95 mmol/kg. This implied that the contrast between the DH lines could possibly be explained by differences in asparagine synthetase activity, resulting from changes in gene expression, protein turnover, or enzyme activity (Curtis et al. 2009).

Varietal differences were also observed in additional studies by Corol et al. (2016), who analyzed 150 bread wheat varieties grown at a single site in 2005. The varieties were separated by asparagine content into low asparagine 0.32–0.43 mg/g dry matter (dm) (correlates to 2.42–3.25 mmol/kg) (cvs Chinese Spring, Palesio, Blasco, Mv-Emese, Bilancia, Granbel, Soissons, Nomade, Valoris, and Alba) and high asparagine 1.50–1.56 mg/g dm (correlates to 11.3534–11.8076 mmol/kg) (cvs Fleischmann, Spark, Kirkpinar 79). Varieties with asparagine contents of 1.28–1.40 mg/g dm were Mexique 50, Renan, Bankuti 1201, Alanasskaja, while Kirac 66, Qualital and Blue/A had 1.10–1.25 mg/g dm. In our analyses (Curtis et al. 2009), cv Spark had a concentration of free asparagine in our tested samples between 2.54 mmol/kg (under a sulfur-sufficient regime) and 62.02 mmol/kg (under a sulfur-deficient regime) compared to that used by Corol et al., where cv Spark had a free asparagine content of 11.35 mmol/kg and was classified as a high asparagine variety. That level of free asparagine suggests that the sample was obtained from wheat that was grown with insufficient sulfur, or there were other stresses preventing it from achieving its usual level of asparagine (2.54 mmol/kg).

Structure and Expression of the Asparagine Synthase Gene Family of Wheat

Identification of the sites of synthesis and accumulation of free asparagine in the grain under sulfur-deficient and sulfur-sufficient conditions is an important objective. There is evidence that free asparagine is predominantly accumulated in the embryo and aleurone layer under sulfur sufficiency, but accumulates at high levels in the endosperm under sulfur-deficient conditions (Shewry et al. in press). This is reflected to some extent in the expression of asparagine synthetase genes (Gao et al. 2016). There are four asparagine synthetase genes in wheat, *TaASN1-4* (Gao et al. 2016), although *TaASN4* has only been identified in genome data and has not been analyzed in detail. Of the other three, the expression of *TaASN2* in the embryo during mid-development dwarfs the expression of any of the genes in any other tissue, although it is also expressed at relatively high levels in the endosperm, even when the wheat is well supplied with sulfur. Indeed, *TaASN1* appears to be the most responsive to both sulfur deficiency and nitrogen

Table 3. Asparagine synthetase genes in different plant species.

Gene name	Species	Comment	References
AS (cDNA clone)	Asparagus officinalis	The study showed that AS plays different roles to asparagine synthetase genes studied in other plants and is induced in harvested asparagus spears in response to carbohydrate stress	Davies and King (1993)
ASN1	Arabidopsis thaliana	The study provided experimental confirmation that phytochrome plays a role in the transmission of light signals to repress accumulation of Arabidopsis ASN1 mRNA Light and metabolic control of amide amino acid biosynthesis was demonstrated Light was shown to repress the synthesis of asparagine, which therefore accumulates only in the tissues of dark-adapted plants	Lam et al. (1994)
ASN1, ASN2 and ASN3	Arabidopsis thaliana	ASN1 gene expression was shown to be mainly in the stem, leaves, and flowers. ASN1 and ASN2 showed reciprocal regulation in response to light: levels of ASN2 mRNA (extremely low in dark) were rapidly increased in a light treatment, whereas ASN1 expression was repressed by light. The levels of ASN1 and ASN2 mRNA were also affected by organic nitrogen in the form of glutamate, glutamine, and asparagine	Lam et al. (1998a)
AS	Maize (Zea mays)	An exogenous supply of metabolizable sugars downregulated gene expression, while nonmetabolizable sugars induced gene expression. Effects of nitrogen metabolite supply and stress conditions indicated that gene expression might be under metabolic control in maize root tips	Chevalier et al. (1996)
AS	Rice (Oryza sativa)	Immunoblotting revealed a high content of asparagine synthetase protein in the leaf sheath at the second position from the fully expanded top leaf and in grains at the middle stage of ripening. Accumulation of mRNA for AS was also observed in these organs. During the ripening of the spikelets, the AS protein contents increased during the first 21 days after flowering, then declined rapidly	Nakano et al. (2000)
ZmAsnS1, ZmAsnS, ZmAsnS3, and ZmAsnS4	Maize (Zea mays)	Four asparagine synthetase genes, TaASN1–4, were identified and shown to be differentially expressed The asparagine synthetase enzymes were shown to be kinetically distinct	Todd et al. (2008) Duff et al. (2011)
TaASN1 and TaASN2	Wheat (Triticum aestivum)	Salinity and osmotic stress caused rapid accumulation of TaASN1 transcript in both shoots and roots. The expression levels of TaASN2 were different from TaASN1 as the levels were only increased by addition of abscisic acid (ABA) at 24 h exposure and there was no response under osmotic or salinity stress	Wang et al. (2005)
DIN6 (dark-inducible-6 = asparagine synthetase (ASN1)	Arabidopsis thaliana	The expression of asparagine synthetase gene is induced 256 ± 1.2-fold by sucrose nonfermenting-1-related protein kinase (SnRK) 1.1 DIN6 is glutamine-dependent asparagine synthetase (ASN1)	Baena-Gonzalez et al. (2007)
TaASN1-4	Wheat (Triticum aestivum)	The expression of three genes, TaASN1–3, was studied in different tissues and in response to nitrogen and sulfur supply. The expression of TaASN2 in the embryo and endosperm during mid to late grain development was the highest of any of the genes in any tissue, but TaASN1 showed more response to nitrogen feeding and sulfur deficiency. TaASN4 was identified from recent genome data but was not studied in detail	Gao et al. (2016)
HvASN1-5	Hordeum vulgare	Five ASN genes were sequenced and characterized in this paper and were shown to be differentially expressed. The paper only discusses HvASN1, HvASN3, HvASN4, and HvASN5. The HvASN1, HvASN3, and HvASN5 were repressed by aging and low-nitrate conditions. All four genes were induced by leaf senescence	Avila-Ospina et al. (2015)

supply (Byrne et al. 2012; Gao et al. 2016). There is evidence that the sulfur response of TaASN1 involves the protein kinase, general control nonderepressible-2 (GCN2) (Byrne et al. 2012), and the role of GCN2 and a putative regulatory element (Gao et al. 2016) that is identical to the N-motif or GCN4-like regulatory motif of storage protein genes but this requires further investigation. There are also four asparagine synthetase genes in maize (*Zea mays*) and barley (*Hordeum vulgare*) and, similarly wheat genes, these too are differentially expressed (Todd et al. 2008). The kinetic parameters of the enzymes encoded by three of the maize genes also show significant differences (Duff et al. 2011). To conclude, asparagine synthetase genes and enzymes are main factors defining asparagine accumulation and acrylamide formation in wheat grain.

The fact that wheat, barley, and maize all have a complement of four asparagine synthetase genes suggests that this may be typical of the cereals. Other species have also been shown to have multiple asparagine synthetase genes. The first to be characterized at the molecular level were *AS1* and *AS2* from pea (*Pisum sativum*) (Goruzzi 1990; Tsai and Coruzzi 1990, 1991). They were shown to encode proteins of 66.3 and 65.6 kDa, respectively, with 50–55% amino acid sequence identity with human asparagine synthetase and highly conserved glutamine binding sites (Met-Cys-Gly-Ile) (Goruzzi 1990). Tsai et al. (1990) showed that both *AS1* and *AS2* exist as single copies in peas and northern analyses revealed that both are dark induced, particularly in mature plants. Subsequently, three genes, *ASN1*, *ASN2*, and *ASN3*, were identified in Arabidopsis (Lam et al. 1994) and shown to be differentially expressed tissue specifically and in response to stress stimuli, light, and sucrose (Lam et al. 1998b). Light, for example, represses expression of *ASN1* in a phytochrome-dependent manner, whereas expression of *ASN2* is extremely low in the dark but rapidly induced by light treatment. The expression of both *ASN1* and *ASN2* is also affected by the supply of organic nitrogen in the form of glutamate, glutamine, or asparagine (43). Details of these and other asparagine synthetase studies are given in Table 3.

Sulfur Content in the Grain, Asparagine Accumulation, and Acrylamide Formation

Reaction to sulfur deficiency of DH lines

Analyses showed that the lowest free asparagine line was SR3, but it still demonstrated a large and significant increase of free asparagine content in response to sulfur deficiency: 26 mmol/kg under sulfur-deficient conditions compared with 1.6 mmol/kg under sulfur-sufficient conditions. Spark on the other hand had the most dramatic increase in free asparagine, from 2.8 mmol/kg to over 61 mmol/kg (Fig. 1). After canonical variate analysis (CVA), the main contributors in canonical variate 1 (CV1) were again asparagine, alanine, and aspartic acid, and in CV2 were alanine, glycine, and phenylalanine (data not shown) (Curtis et al. 2009).

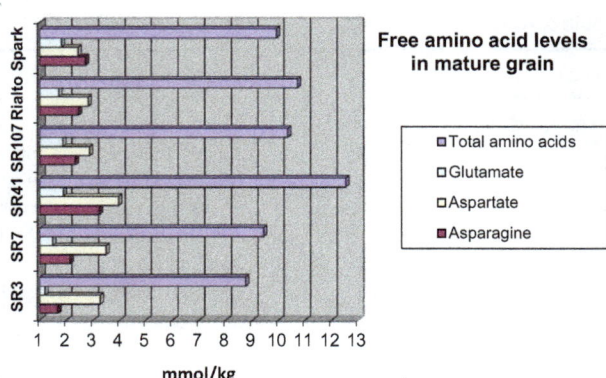

Figure 1. Free asparagine (red), glutamate (light blue), and aspartic acid (yellow), and total amino acid concentration levels (blue) in mature grain of wheat varieties Spark and Rialto and four DH lines from a Spark × Rialto mapping population. Concentrations are in mmol/kg (Curtis et al. 2009).

Relationship between asparagine concentration and acrylamide formation in whole grain flour samples from wheat

To establish the relationship between asparagine concentration and acrylamide formation in wholegrain wheat matrixes, whole wheat grain was milled and 0.5 g analyzed for total amino acid. Fractions of the samples were then baked dry and analyzed for acrylamide, the correlation calculated from these data showed a range of possible asparagine concentrations, and quadratic correlation between asparagine and acrylamide levels gave $R^2 = 0.9945$. Most of the samples analyzed that were grown under sulfur-sufficient conditions were in the range between 1 and 3.5 mmol/kg (Fig. 2). The highest levels of asparagine and therefore increased acrylamide-forming potential were the samples grown under acute sulfur-deficient conditions (Fig. 2). Asparagine concentration was closely linked with

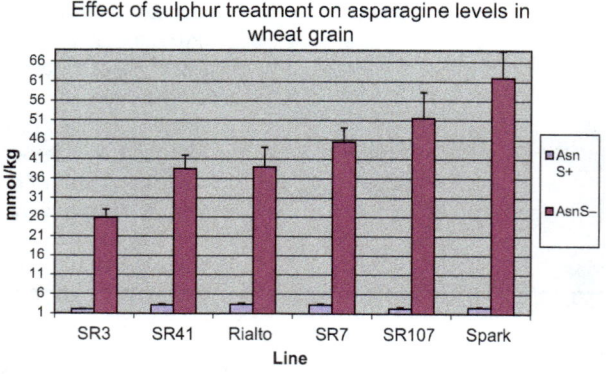

Figure 2. Asparagine concentration (mmol/kg) under sulfur-sufficient (blue AsnS+) and sulfur-deficient (red AsnS−) treatments of Spark and Rialto parental varieties and four DH lines: SR3, SR41, SR7, and SR107.

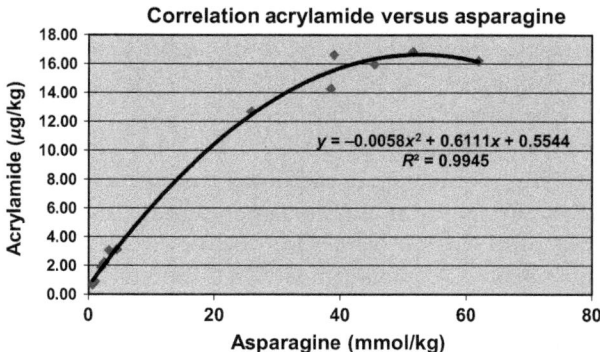

Figure 3. Free asparagine concentration (mmol/kg) in wheat (*Triticum aestivum*) grain plotted against acrylamide formed in heated flour (Curtis et al. 2009).

acrylamide formation (Fig. 3). For example, if the wheat grain contains 1.9 mmol/kg asparagine, the acrylamide-forming potential will be 1.69 μg/kg (0.042 μmol/kg), which is well below the indicative levels set by the European Commission (Curtis et al. 2009).

The formula predicting the acrylamide concentration based on the asparagine content is as follows: $y = -0.0058x^2 + 0.6111x + 0.5544$.

Relationship between acrylamide and sulfur in wheat grain

To establish a correlation between acrylamide content and sulfur content in the grain, the same samples with already measured asparagine and acrylamide contents were analyzed for total sulfur and nitrogen content. Acrylamide levels measured in samples grown under sulfur-sufficient, sulfur-deficient, and control conditions were negatively correlated with sulfur content in the grain (Fig. 4). The samples with the highest concentration of acrylamide were the samples with the least sulfur in the grain. For example,

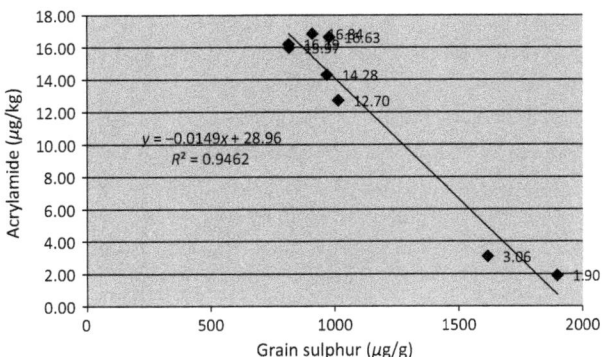

Figure 4. Total grain sulfur (mg/g [ppm]) plotted against acrylamide formed in heated flour (μg/kg [ppb]) (Curtis et al. 2009).

samples with 16.84 μg/kg acrylamide had 800 μg/g sulfur. This is a very important finding as it would suggest that the higher the sulfur in the grain, the lower the acrylamide formed in the grain will be (Fig. 4). However, there will be a level when the addition of more sulfur does not result in a change in the level of asparagine. As asparagine is needed for germination the level is unlikely to ever fall to zero (Curtis et al. 2009).

Previously, the high levels of acrylamide 2600 and 5200 μg/kg were measured in wheat flour of sulfur-deprived treatment in comparison to 600–900 μg/kg acrylamide content of normal levels of sulfate-fertilized wheat (Muttucumaru et al. 2006). Independent research of wheat cultivar "Star" in Germany was tested to determine the impact of sulfur fertilization on asparagine accumulation and acrylamide formation levels in the flour. In sufficient sulfur-fertilized wheat, the free asparagine was determined at 0.03–0.4 g/kg and significantly higher 3.9–5.7 g/kg in sulfur-deficient wheat sample. The high levels of acrylamide and 3-aminoproionamide (40–76 mg/kg) were also observed in the low-sulfur samples (1.7–3.1 mg/kg) (Granvogl et al. 2007). Additional evidences for elevated levels of acrylamide in response to higher asparagine were presented in 2008 in paper by Elmore et al., where acrylamide was up to six times higher in sulfur-deprived wheat flour in comparison to the wheat flour obtained from sulfur-sufficient wheat grain (Elmore et al. 2008).

Conclusions From This Study

It was concluded that the most important factor controlling asparagine concentration was sulfur deficiency, although there were also significant effects of G, E, and G × E (Curtis et al. 2009). Therefore, a key aspect of acrylamide mitigation would be the avoidance of sulfur deficiency during wheat cultivation. Even small proportions of sulfur-deficient grain would lead to a large increase in acrylamide formation during baking. Screening existing varieties for low asparagine accumulation and further improvement by plant breeding could also be part of the solution.

Grain Storage Protein and Sulfur Content in Wheat

Prolamins, known as grain storage proteins, play a major role in the distribution and accumulation of sulfur and nitrogen in the grain. A high level of available nitrogen results in an increased proportion of prolamin storage proteins. If no additional sulfur is provided, then the proportion of sulfur-poor prolamins and HMW prolamins increases and the proportion of sulfur-rich prolamins decreases (Shewry et al. 2001). The wheat grain consists of two predominant types of grain storage proteins: monomeric gliadins and polymeric glutenins, comprising 60–80% of the total protein content of the mature grain (Shewry et al. 2001). Gliadins are classified into ω 1, 2-, ω5-, α/β-, and γ-gliadins, and the glutenins into high-molecular-weight glutenin subunits (HMW-GS) and low-molecular weight glutenin subunits (LMW-GS). Low sulfur-containing proteins are also called sulfur-poor proteins: HMW-GS, some LMW-GS, and ω 1, 2-, ω5-gliadins are called sulfur-poor because they differ in their content of the sulfur-containing amino acids cysteine and methionine in their poly amino acids chains (Shewry et al. 1997). To understand the mechanisms of sulfur regulation of storage protein gene expression Dai et al. investigated different treatments of sulfur and nitrogen availability (Dai et al. 2015). In their results they reported that the sulfur-deficient grain contained 28% less total sulfur than the control grain. The kinetic analyses also showed that sulfur-rich and sulfur-poor proteins accumulated at different rates depending on the nitrogen and sulfur supply, the most important results being that under sulfur-deficient conditions the levels of S-poor grain storage proteins were greater per grain at maturity (Dai et al. 2015). Another interesting observation was that by adding additional sulfur midway through grain filling the mass per grain increased, ultimately relieving the sulfur deficiency. After the nutritional shift the mass of sulfur-rich grain storage proteins increased gradually (Dai et al. 2015).

One additional finding by Dai et al. was that the expression of sulfur-poor genes was tightly regulated at the transcriptional level. They also confirmed that nitrogen and sulfur availability affected the free amino acid pools. When sulfur was added at the mid grain-filling stage, the effects of sulfur deficiency were reversed and levels of asparagine, aspartate, lysine, arginine, tyrosine, leucine, and valine quickly decreased, whereas the levels of glutathione (GSH) and glutathione disulfide (GSSG) increased to the levels observed under normal nitrogen and sulfur conditions (Dai et al. 2015).

The timing of regulatory events in response to N and S supply is also very important. Dai et al. also found that the increase in most amino acids in response to sulfur supply started midway through grain filling (~600°C dpa) thus indicating the critical point in grain response to sulfur deficiency leading to more efficient application of sulfur fertilizer in the field. It was also shown that S supply at 490°C days after a thesis can also efficiently mitigate S-deficiency (Dai et al. 2015).

Effects of sulfur fertilization on acrylamide-forming potential of wheat

The effect of sulfur fertilization on the acrylamide-forming potential in wheat was measured in a group of samples provided by Prof Steve McGrath (Rothamsted Research) and grown at Woburn (UK), Rosemaund (Hereford), and Frostender (Suffolk) (UK) in 2010 and 2011 under five sulfur treatments at 0, 12.5, 25, 50, and 75 kg/ha SO_3. Analysis of the levels of free amino acids and acrylamide gave a recommended application rate of 50 kg per hectare SO_3. This is at the upper end of the recommended application rate given in the Fertilizer Manual (RB209) for wheat used for the biscuit and breakfast cereal market. These data were published in HGCA information sheet on sulfur for cereals and oilseed rape (http://cereals.ahdb. org.uk/media/357116/is28-sulphur-for-cereals-and-oilseed-rape.pdf).

Advice to Farmers, Breeders, and Food Processors

The data suggest that there are significant differences between wheat varieties with respect to their acrylamide-forming potential, the limiting factor being the level of free asparagine concentration in the grain. This is the parameter on which varietal selection should be based and a trait that should be incorporated into breeding programmes.

The identification of the QTL for free asparagine concentration has so far been unsuccessful but may be possible with different mapping populations. One potential target for breeders could be the *TaANS2* gene.

E factors have a significant effect on free asparagine accumulation in wheat, on their own and in combination with varietal differences (G × E). It is therefore important that varieties are tested over a range of environmental conditions.

Management of wheat production is an important additional factor. Sulfur deficiency in particular causes a massive accumulation of free asparagine in wheat grain which should be avoided: an application rate of 50 kg/ha SO_3 is recommended, more if the soil is already sulfur deficient.

The use of nitrogen fertilizer increases the free asparagine and total free amino acid concentrations in wheat causing a concomitant increase in the acrylamide-forming potential (Sandli et al. 1993; Kingston-Smith et al. 2006; Hoegy et al. 2013). This, however, is only the case when nitrogen fertilization is in much greater proportion than sulfur. Nitrogen fertilizer is required to maintain the yield and quality of the crop, but excessive application should be avoided. Ensuring that other minerals are available to the crop may mitigate the effect of excessive nitrogen. Some or all of these points may apply to other cereals.

Conclusions

The acrylamide issue is a difficult problem facing the food industry in Europe and worldwide.

Avoidance of sulfur deficiency is essential: even small amounts of sulfur-deficient grain entering the food chain should be avoided. The application of 50 kg/ha of SO_3 equivalent (20 kg/ha of sulfur) is currently advisable depending on the soil type.

Plant breeders must take on board this issue or risk losing market share to those who do. However, many of the compounds which make a food product appealing to the consumer are formed by similar pathways to acrylamide, so changes in free amino acid concentrations may affect desirable tastes and aromas.

There are significant health benefits associated with eating cereal (particularly wholegrain) products and these must be retained while acrylamide levels are decreased.

Acrylamide is a good news story! It is new knowledge, not a new risk, and its discovery should enable food to be made safer. However, managing how this issue is covered by the media and therefore perceived by the public is important: facts should not be misinterpreted, misunderstood, or overexaggerated.

Conflict of Interest

Tanya Curtis is currently supported by a consortium of companies from the wheat supply chain.

References

Avila-Ospina, L., A. Marmagne, J. Talbotec, K. Krupinska, and C. Masclaux-Daubresse. 2015. The identification of new cytosolic glutamine synthetase and asparagine synthetase genes in barley (*Hordeum vulgare* L.), and their expression during leaf senescence. J. Exp. Bot. 66:2013–2026.

Anjum, F. M., M. R. Khan, A. Din, M. Saeed, I. Pasha, and M. U. Arshad. 2007. Wheat gluten: High molecular weight glutenin subunits – structure, genetics, and relation to dough elasticity. J. Food Sci. 72:R56–R63.

Baena-Gonzalez, E., F. Rolland, J. M. Thevelein, and J. Sheen. 2007. A central integrator of transcription networks in plant stress and energy signalling. Nature 448:938–10.

Byrne, E. H., I. Prosser, N. Muttucumaru, T. Y. Curtis, A. Wingler, S. Powers, et al. 2012. Overexpression of GCN2-type protein kinase in wheat has profound effects on free amino acid concentration and gene expression. Plant Biotechnol. J. 10:328–340.

Casado, F. J., and A. Montano. 2008. Influence of processing conditions on acrylamide content in black ripe olives. J. Agric. Food Chem. 56:2021–2027.

Chevalier, C., E. Bourgeois, D. Just, and P. Raymond. 1996. Metabolic regulation of asparagine synthetase gene expression in maize (*Zea mays* L.) root tips. Plant J. 9:1–11.

Corol, D. I., C. Ravel, M. Rakszegi, G. Charmet, Z. Bedo, M. H. Beale, P. R. Shewry, and J. L. Ward. 2016. (1) H-NMR screening for the high-throughput determination of genotype and environmental effects on the content of asparagine in wheat grain. Plant Biotechnol. J. 14:128–139.

Curtis, T. Y. 2010. Genetic and environmental factors controlling acrylamide formation in wheat and rye products. *Thesis* 198.

Curtis, T., and N. G. Halford. 2014. Food security: the challenge of increasing wheat yield and the importance of not compromising food safety. Ann. Appl. Biol. 164:354–372.

Curtis, T. Y., N. Muttucumaru, P. R. Shewry, M. A. J. Parry, S. J. Powers, J. S. Elmore, et al. 2009. Effects of genotype and environment on free amino acid levels in wheat grain: implications for acrylamide formation during processing. J. Agric. Food Chem. 57:1013–1021.

Curtis, T. Y., S. J. Powers, D. Balagiannis, J. S. Elmore, D. S. Mottram, M. A. J. Parry, et al. 2010a. Free amino acids and sugars in rye grain: implications for acrylamide formation. J. Agric. Food Chem. 58:1959–1969.

Dai, Z. W., A. Plessis, J. Vincent, N. Duchateau, A. Besson, M. Dardevet, et al. 2015. Transcriptional and metabolic alternations rebalance wheat grain storage protein accumulation under variable nitrogen and sulfur supply. Plant J. 83:326–343.

Davies, K. M., and G. A. King. 1993. Isolation and characterization of a cDNA clone for a harvest-induced asparagine synthetase from *Asparagus-Officincalis* L. Plant Physiol. 102:1337–1340.

Duff, S. M. G., Q. G. Qi, T. Reich, X. Y. Wu, T. Brown, J. H. Crowley, et al. 2011. A kinetic comparison of asparagine synthetase isozymes from higher plants. Plant Physiol. Biochem. 49:251–256.

Dybing, E., and T. Sanner. 2003. Risk assessment of acrylamide in foods. Toxicol. Sci. 75:7–15.

EFSA. 2015a. Acrylamide in food is a public health concern. European Food Safety Authority. Available at: https://www.efsa.europa.eu/en/press/news/150604.

EFSA. 2015b. Scientific opinion on acrylamide in food. EFSA J. 13:4104–4104.

Elmore, J. S., J. K. Parker, N. G. Halford, N. Muttucumaru, and D. S. Mottram. 2008. Effects of plant sulfur nutrition on acrylamide and aroma compounds in cooked wheat. J. Agric. Food Chem. 56:6173–6179.

European Commission 2013. Commission recommendation of 8 November 2013 on investigations into the levels of acrylamide in food. European Commission. Available at: http://eur-lex.europa.eu/LexUriServ/LexUriServ.do?uri=OJ:L:2013:301:0015:0017:EN:PDF.

European Food Safety Authority. 2011. EFSA report on data collection: future directions. EFSA J. 2010(8):1533–1568.

Food Drink Europe. 2014. Acrylamide toolbox 2014. Food Drink Europe, Brussels, Belgium.

Food Standards Agency, Food Survey Information Sheet. 2012. A rolling programme of surveys on process contaminants in the UK retail foods, Acrylamide & Furan: Survey 4, 1–47.

Friedman, M. 2003. Chemistry, biochemistry, and safety of acrylamide. A review. J. Agric. Food Chem. 51:4504–4526.

Friedman, M. 2005. Biological effects of Maillard browning products that may affect acrylamide safety in food. Chem. Saf. Acrylamide Food 561:35–156.

Friedman, M., and C. E. Levin. 2008. Review of methods for the reduction of dietary content and toxicity of acrylamide. J. Agric. Food Chem. 56:6113–6140.

Gao, R., T. Y. Curtis, S. J. Powers, H. Xu, J. Huang, and N. G. Halford. 2016. Food safety: structure and expression of the asparagine synthetase gene family of wheat. J. Cereal Sci. 68:122–131.

Goruzzi, F.-Y. T. A. G. M. 1990Dark-induced and organ-specific expression of two asparagine synthetase genes in *Pisum sativum*. ENBO J. 9:323–332.

Graf, M., T. M. Amrein, S. Graf, R. Szalay, F. Escher, and R. Amado. 2006. Reducing the acrylamide content of a semi-finished biscuit on industrial scale. LWT Food Sci. Technol. 39:724–728.

Granby, K., N. J. Nielsen, R. V. Hedegaard, T. Christensen, M. Kann, and L. H. Skibsted. 2008. Acrylamide–asparagine relationship in baked/toasted wheat and rye breads. Food Addit. Contam. Part A Chem. Anal. Control Expo. Risk Assess. 25:921–929.

Granvogl, M., H. Wieser, P. Koehler, S. von Tucher, and P. Schieberle. 2007. Influence of sulfur fertilization on the amounts of free amino acids in wheat. Correlation with baking properties as well as with 3-aminopropionamide and acrylamide generation during baking. J. Agric. Food Chem. 55:4271–4277.

Hoegy, P., M. Brunnbauer, P. Koehler, K. Schwadorf, J. Breuer, J. Franzaring, et al. 2013. Grain quality characteristics of spring wheat (*Triticum aestivum*) as affected by free-air CO2 enrichment. Environ. Exp. Bot. 88:11–18.

Howie, J. K., P. Y. T. Lin, and D. V. Zyzak. 2006. *Method for reducing acrylamide in foods comprising reducing the level of reducing sugars, foods having reduced levels of acrylamide, and article of commerce.*.

Howie, J. K., P. Y. T. Lin, and D. V. Zyzak. 2007. *Method for reduction of acrylamide in cocoa products, cocoa products having reduced levels of acrylamide, and article of commerce.*

International Agency for Research on, C. & International Agency for Research on, C.. 1994. IARC Monographs on the Evaluation of Carcinogenic Risks to Humans, Vol. 60. Some industrial chemicals. *IARC Monographs on the Evaluation of Carcinogenic Risks to Humans; Some industrial chemicals.*

Javier Casado, F., A. Higinio Sanchez, and A. Montano. 2010. Reduction of acrylamide content of ripe olives by selected additives. Food Chem. 119:161–166.

Javier Casado, F., A. Montano, D. Spitzner, and R. Carle. 2013. Investigations into acrylamide precursors in sterilized table olives: evidence of a peptic fraction being responsible for acrylamide formation. Food Chem. 141:1158–1165.

Javier Casado, F., A. Montano, and R. Carle. 2014. Contribution of peptides and polyphenols from olive water to acrylamide formation in sterilized table olives. LWT Food Sci. Technol. 59:376–382.

Kingston-Smith, A. H., A. L. Bollard, and F. R. Minchin. 2006. The effect of nitrogen status on the regulation of plant-mediated proteolysis in ingested forage; an assessment using non-nodulating white clover. Ann. Appl. Biol. 149:35–42.

Koehler, P., M. Granvogl, H. Wieser, and P. Schieberle. 2008. Asparagine concentration and acrylamide formation potential in wheat flour as affected by sulfur fertilization. American Association of Cereal Chemists, Inc (AACC), St Paul, USA; 133–136.

Lam, H. M., S. S. Y. Peng, and G. M. Coruzzi. 1994. Metabolic regulation of the gene encoding glutamine-dependent asparagine synthetase in *Arabidopsis thaliana*. Plant Physiol. 106:1347–1357.

Lam, H.-M., M.-H. Hsieh, and G. Coruzzi. 1998a. Reciprocal regulation of distinct asparagine synthetase genes by light and metabolites in *Arabidopsis thaliana*. Plant J. 16:345–353.

Lam, H. M., M. H. Hsieh, and G. M. Coruzzi. 1998b. Reciprocal regulation of distinct asparagine synthetase genes by light and metabolites in *Arabidopsis thaliana*. Plant J. 16:345–353.

Lea, P. J., L. Sodek, M. A. J. Parry, R. Shewry, and N. G. Halford. 2007. Asparagine in plants. Ann. Appl. Biol. 150:1–26.

Mottram, D. S. 2007. The Maillard reaction: source of flavour in thermally processed foods. *in* Flavours and Fragrances. Ralf Günter Berger, editor, Springer Berlin Heidelberg, 269–283.

Mottram, D. S., B. L. Wedzicha, and A. T. Dodson. 2002. Acrylamide is formed in the Maillard reaction. Nature 419:448–449.

Muttucumaru, N., N. G. Halford, J. S. Elmore, A. T. Dodson, M. Parry, P. R. Shewry, et al. 2006. Formation of high levels of acrylamide during the processing of flour derived from sulfate-deprived wheat. J. Agric. Food Chem. 54:8951–8955.

Muttucumaru, N., J. S. Elmore, T. Curtis, D. S. Mottram, M. A. J. Parry, and N. G. Halford. 2008. Reducing acrylamide precursors in raw materials derived from wheat and potato. J. Agric. Food Chem. 56:6167–6172.

Muttucumaru, N., S. J. Powers, J. S. Elmore, A. Briddon, D. S. Mottram, and N. G. Halford. 2014. Evidence for the complex relationship between free amino acid and sugar concentrations and acrylamide-forming potential in potato. Ann. Appl. Biol. 164:286–300.

Nakano, K., T. Suzuki, T. Hayakawa, and T. Yamaya. 2000. Organ and cellular localization of asparagine synthetase in rice plants. Plant Cell Physiol. 41:874–880.

Parzefall, W. 2008. Minireview on the toxicity of dietary acrylamide. Food Chem. Toxicol. 46:1360–1364.

Postles, J., S. J. Powers, J. S. Elmore, D. S. Mottram, and N. G. Halford. 2013. Effects of variety and nutrient availability on the acrylamide-forming potential of rye grain. J. Cereal Sci. 57:463–470.

Sandli, N., M. M. Svenning, K. Rosnes, and O. Junttila. 1993. Effects of nitrogen supply on frost-resistance, nitrogen-metabolism and carbohydrate content in white clover (Trifolium repens). Physiol. Plant. 88:661–667.

Schulze, J., and C. P. Siegers. 2004. Combination toxicity of acrylamide. Toxicol. Appl. Pharmacol. 197:267–267.

Shewry, P. R., F.-J. Zhao, G. B. Gowa, N. D. Hawkins, J. L. Ward, M. H. Beale, N. G. Halford, M. A. Parry, and J. Abécassis. 2009. Sulphur nutrition differentially affects the distribution of asparagine in wheat grain. J. Cereal Sci. 50:407–409.

Shewry, P. R., A. S. Tatham, and P. Lazzeri. 1997. Biotechnology of wheat quality. J. Sci. Food Agric. 73:397–406.

Shewry, P. R., A. S. Tatham, and N. G. Halford. 2001. Nutritional control of storage protein synthesis in developing grain of wheat and barley. Plant Growth Regul. 34:105–111.

Stadler, R. H. 2005. Acrylamide formation in different foods and potential strategies for reduction. Chem. Saf. Acrylamide Food 561:157–169.

Stadler, R. H., I. Blank, N. Varga, F. Robert, J. Hau, P. A. Guy, et al. 2002. Acrylamide from Maillard reaction products. Nature 419:449–450.

Taeymans, D., A. Andersson, P. Ashby, I. Blank, P. Gonde, P. van Eijck, et al. 2005. Acrylamide: update on selected research activities conducted by the European food and drink industry. J. AOAC Int. 88:234–241.

Todd, J., S. Screen, J. Crowley, J. Peng, S. Andersen, T. Brown, et al. 2008. Identification and characterization of four distinct asparagine synthetase (AsnS) genes in maize (Zea mays L.). Plant Sci. 175:799–808.

Tsai, F. Y., and G. M. Coruzzi. 1990. Dark-induced and organ-specfic expression of 2 asparagine synthetase genes in Pisum-sativum. EMBO J. 9:323–332.

Vleeschouwer, K. D., I. V. D. Plancken, A. V. Loey, and M. E. Hendrickx. 2009. Role of precursors on the kinetics of acrylamide formation and elimination under low moisture conditions using a multiresponse approach – Part I: effect of the type of sugar. Food Chem. 114:116–126.

Wang, H. B., D. C. Liu, J. Z. Sun, and A. M. Zhang. 2005. Asparagine synthetase gene TaASN1 from wheat is up-regulated by salt stress, osmotic stress and ABA. J. Plant Physiol. 162:81–89.

World Health Organization. 2006. Safety evaluation of certain contaminants in food. Prepared by the Sixty-fourth meeting of the Joint FAO/WHO Expert Committee on Food Additives (JECFA). FAO Food Nutr. Pap. 82:1–778.

Zyzak, D. V., R. A. Sanders, M. Stojanovic, D. H. Tallmadge, B. L. Eberhart, D. K. Ewald, et al. 2003. Acrylamide formation mechanism in heated foods. J. Agric. Food Chem. 51:4782–4787.

14

Stress resilience in crop plants: strategic thinking to address local food production problems

William J. Davies[1] & Jean-Marcel Ribaut[2]

[1]The Lancaster Environment Centre, Lancaster University, Bailrigg, Lancaster LA1 4YQ, UK
[2]Generation Challenge Programme (GCP) c/o CIMMYT, Carretera Mexico-Veracruz, El Batan, Texcoco, Estado de Mexico, Mexico

Keywords
crop plants, local food production problems, strategic thinking, Stress resilience.

Correspondence
William J. Davies, The Lancaster Environment Centre, Lancaster University, Bailrigg, Lancaster LA1 4YQ, UK.
E-mail: w.davies@lancaster.ac.uk

Funding Information
The authors are grateful to the Society for Experimental Biology for financial support.

Abstract

There are many ways to assess or define the stress resilience of crop production, but ultimately the resilience of systems (and communities), i.e., an ability to survive and prosper, is driven by profitability. Here we review challenges for those who seek to bring about beneficial change in practice or policy as we translate novel crop science research findings into impacts on the food supply chain. While advances in plant and crop science are relevant to this challenge, the context of application is crucial here and this will mean that many other considerations, discussed below, will potentially moderate the impact on crop growth and yield of what could be the introduction of very significant breakthroughs in genetic gain. This paper considers opportunities for plant scientists seeking to address the world's growing food security challenge by exploiting new understanding of the basis of crop stress resilience. Ultimately the local challenge is to increase the resilience of cropping systems and rural communities.

Introduction: The Challenge and a Local Response

There are many ways to assess or define the stress resilience of crop production, but ultimately the resilience of systems (and communities), i.e., an ability to survive and prosper, is driven by profitability. Here, we review challenges for those who seek to bring about beneficial change in practice or policy as we translate novel crop science research findings into impacts on the food supply chain. While advances in plant and crop science are relevant to this challenge, the context of application is crucial here and this will mean that many other considerations, discussed below, will potentially moderate the impact on crop growth and yield of what could be the introduction of very significant breakthroughs in genetic gain This paper considers opportunities for plant scientists seeking to address the world's growing food security challenge by exploiting new understanding of the basis of crop stress resilience. Ultimately the local challenge is to increase the resilience of cropping systems and rural communities.

Even though advances in plant and crop science understanding have helped us make considerable progress toward meeting the food-related Millennium Development Goals and the more recent Sustainable Development Goals, there is still a very significant "Global Food Security Challenge." This is a multidisciplinary challenge which depressingly now also involves a necessity to address the fact that for the first time in history, there are more obese people in the world than there are hungry people. We recognize that both hunger and obesity are promoting significant health problems associated with unhealthy and/or inadequate diets. While stress resilience is of less relevance to those addressing this set of issues, stress effects on crop and food quality can be appreciable and there are opportunities here for crop science to deliver change for the better.

We need to increase the availability of food in many regions of the world and also increase peoples' access to this food but the food should also be healthy. There are many social cultural and economic considerations that contribute to local differences in food availability. These considerations can be captured effectively in the following identity which describes major influences which can determine the impact of a change in a food production system:

$$G \times E \times M \times S \,(\text{Genetics} \times \text{Environment} \times \text{Management} \times \text{People/Society})$$

This interaction between a multitude of factors effectively tells us that a "local" approach to addressing many food challenges must be important. Crop science is well aware of the importance of G × E interactions in determining how effective new traits may be in particular locations/environments. Probably not surprisingly, some traits can have very positive effects on crop yield in some stress environments but the same traits can have neutral or even negative effects when environmental conditions are varied (Tardieu 2012; Bonneau et al. 2013). Often crop production is most profitable in good years (optimal conditions) and it is these profitable years that help to sustain farmers through suboptimal years when different stresses are present. Breeding for resilience, requires assessment of performance under optimal and suboptimal conditions to ensure that genetic gain under abiotic stress is not associated with a yield penalty in the absence of stress (Ribaut 2006). One of the major consequence of climate changes is the increasing unpredictability of climatic conditions and an increase in the stress intensity. As a result improved rice cultivars in some regions of southeast Asia need to be resistant to flooding during the first part of the crop cycle, but at the same time being drought tolerant as water limited conditions might occur during flowering or grain filling stages; the good news is that surprisingly those "opposite" stresses might have some common genetic basis (Fukao et al. 2011;. Rubaiyath et al. 2016).

Recent work by agronomists at CSIRO (Kirkegaard and Hunt 2010) in collaboration with breeders in the same organization shows the importance of even the most basic of crop management options (M in above equation) and many other studies show that social considerations (S) are also very important in determining whether an innovation is taken up and whether it impacts on peoples' lives. Even in the most general consideration of the Food Security challenge it is apparent that peoples' access to diets dominated by poorly nutritious, often unsafe food can cause massive health problems for many. Price et al. (2013) show how novel plant stress biology implemented through genetics and crop management can have very beneficial effects on the safety of food but this crop-specific challenge requires a local "solution."

Some Targets for Plant Scientists in the Delivery of "Sustainable Intensification"

Crop scientists who focus on the interaction between the genetic basis of their crop of choice and the environment are mostly concerned with the impact of the environment (stress) on the genetic potential yield. Increasingly however we are concerned with the impact of agriculture (the crop/food production process) on the environment. There is particular concern for the overuse of the input resources required for crop production and excessive water use is a major problem on several continents, with falling water tables due to over extraction of water for irrigation having a particularly significant effect on natural vegetation and ultimately promoting desertification (Kang et al. 2008). Overuse of fertilizer impacts adversely on soil quality (e.g., Guo 2010) and on quality of ground water and surface water which can create important health risks (e.g., Campbell et al. 2016)). The stress biology at issue here is variation in water and nutrient availability and there is now much information to show how these stress variables can be exploited to the benefit of both resource use and crop production. Stress is effectively being used as a crop growth regulator. Among the best example is alternate wetting and drying irrigation (AWD) which saves water while sustaining yield and can have beneficial effects on greenhouse gas emissions and crop quality (Yang and Zhang 2010)

It goes without saying that we should seek wherever possible to minimize the damaging effects of agriculture such as those detailed above, while still seeking new ways of increasing productivity. Exploitation of understanding of the genetic basis of crop stress resilience, or how to mitigate it such as through crop diversification (Lin 2011), can be key here. International Initiatives such as the Generation Challenge Programme (GCP) have demonstrated that translational research in crop improvement is not only achievable but can be highly successful with the right combination of technical and "soft" science skills and expertise. The GCP was able to demonstrate that harnessing plant genetic diversity and applying modern biology to the development of new crop varieties that meet the needs of smallholder farmers is both an efficient and effective means of conducting translational research. This Programme promoted a way of working based on "true" partnerships by assembling the right combination of expertise into teams, by providing these teams with adequate resources- including budget- and managing their evolution toward synergy and delivery of outputs while, at the same time, encouraging and enforcing information sharing (Ribaut 2014).

Recently, the term "sustainable intensification" has been coined to describe a target for future food production methodology. This may be a useful development but most are well aware that this term is highly location-specific and even in meta-environments, techniques for sustainable use of water and nutrients in agriculture will be context-specific, depending on for example the nature of the soils and the hydrology of the region. Local "solutions" need to consider agricultural, environmental and social factors which will differ in importance, again with location and land use objective. Pollock (2016) has highlighted the fact that the preservation of viable rural communities intimately linked to local agricultural needs to be given more attention if we are to also preserve/achieve rural social stability. We will see below how crop genetics and management techniques based on understanding of the basis of crop resilience can be influential in climate-stressed communities.

Crop Science to Ameliorate the Impact of a 4 Degree World on Food Production

Projections of climate change impacts produced by a number of different modeling approaches indicate near certainty that global crop production will be negatively affected by climate change (Challinor et al. 2014). Most predictions also suggest reduced crop quality and nutritional value (i.e., decreases in leaf and grain N, protein and nutrient (Fe, Zn, Mn, Cu) concentrations) associated with warmer climates and increased CO_2 levels. (Stress effects that need to be overcome).

To date, only a relatively few studies have delivered estimates of climate change effects for different regions of the world. Lobell et al. (2011) have identified South Asia and southern Africa as two regions that, in the absence of significant crop adaptation, would suffer the most negative impacts on important food crops (some of which have received little attention from stress biologists). The expectation is that future climate will be on average both warmer and wetter. Crop seasonality is affected by both the intensity and the distribution of the rains over time and both are affected by climate change (Feng et al. 2013). Increases in the inter-annual variability of yields are also likely to become more pronounced and will potentially affect stability of food availability and access (Porter et al. 2014).

Hochman et al. at CSIRO (2017) analyzed data from 50 weather stations located throughout Australia's wheat-growing areas and found that, on average, the amount of rain falling on growing crops declined by 2.8 mm per season, or 28% over 26 years, while maximum daily temperatures increased by an average of 1.05°C. By modeling these data using APSIM they calculate that the national wheat yield will fall from the recent average of 1.74 tonnes per hectare to 1.55 tonnes per hectare in 2041.

Plant science now has the capacity to develop crop varieties that are better suited to contrasting and new climatic conditions more rapidly than has previously been the case. Increases in the incidence of water deficits,

chronically high temperatures and an increase generally in mean temperature can sensitively affect different stages of reproductive crop development while also accelerating crop development, resulting in shorter crop durations and reduced time to accumulate biomass and grain yield. The time from trait identification, through breeding, local availability and adoption of a new variety can be up to 30 years and although revised breeding strategies and new methodologies, such as double haploids or genomics (Varshney et al. 2012), can reduce the cycle significantly, there are many other factors that determine the adoption of new varieties by farmers. In addition to market demands that might determine profitability, new varieties require efficient regulatory processes and distribution networks and will likely be accompanied by improved management practices that enhance yield and quality potential.

Challinor et al. (2016) have identified this chain of developments through to impact as the BDA process (Breeding, Development, Adoption). These authors show that for maize in Africa both adaptation and mitigation can reduce loss of yield due to shortening cropping duration and they argue that climate projections have the potential to provide target elevated temperatures for regional breeding operations. They also stress that while options for reducing BDA time are highly context-dependent, there are common threads.

Many recent reports on the global food security challenge have stressed the need for enhanced knowledge exchange strategies in many parts of the world, including the developed world (e.g., UK Foresight). This may particularly be the case in the developing world as highlighted by Challinor et al. (2016). As many of those living in poverty in the developing world depend on agriculture for their income, vibrant agricultural systems are the key to development. The five countries in the world with the greatest problems with agricultural production and hence the greatest food and nutrition needs are all found in sub-Saharan Africa. Agricultural development can feed more people in the region and can also link to more general economic growth and reduction of poverty by generating employment. GPC (http://globalplantcouncil.org/) can help focus the attention of plant science and scientists in the developed world on this region of the developing world.

In recent years, crop yields in many African countries have begun to rise and this is early evidence, that African agriculture may now be generating its own "Green Revolution." Progress has been driven by a number of factors, including increased investment in infrastructure, introduction of policies to enhance both local and international markets, and some development of extension programs to help farmers take profit from new knowledge which can enhance crop productivity (Foresight Africa). As is the case with many aspects of food systems around the world, there is no single silver bullet which will "solve" the problem of food and nutrition insecurity. There is, however, a general view that with appropriate focus upon regional constraints, capacity development, investment and partnerships, many African countries have the potential to address the problem of substantial crop yield gaps that historically have held back development on the continent (Van Ittersum et al. 2016).

Evidence for the considerable potential of African agriculture may be found by looking at recent or intended investments by the African Development Bank. Africa currently imports one-third of all calories consumed (USD 77 Billion pa) and with widespread poverty (49% of the population in Africa lives on <USD 1.25/day) and high youth unemployment (40–60%), the imperative for an agricultural transformation that will result in broader impacts is very obvious (Chianu, 2016).

The challenges are many. Up to 60% of all famers are non-commercial or semi-commercial. Markets are under-developed and in many instances value chains are very weak. However, the Feed Africa Initiative has set ambitious goals for the period to 2025. It will aim to substantially eliminate extreme poverty, end hunger and malnutrition, enhance the performance of value chains in agriculture and turn Africa into a net food exporter.

To achieve these ambitious aims will require a commitment by governments and many others, especially to invest in human capital; the researchers and practitioners who will drive the development and sustainability of agricultural commodities and processes. A key challenge will be to retain the best and brightest young minds and to create a cadre of innovative scientists, including plant breeders, who see a future in African agriculture. This will not be easy. Budding young scientists often see a future in developed countries or international agencies where their talents will be well-rewarded. However, we are optimistic, the potential is there (Diop et al. 2013). We see a future where agriculture and agricultural research play an important part in national economies; where science and education will be key to economic development and resourced accordingly; and, where regional initiatives and international organizations all have a role to play in creating an enabling and rewarding environment for young African researchers.

The development of African agriculture will be both global and local; globally, the biophysical potential is huge- about 60% of the world's un-utilized but potentially available cropland is in Africa. Locally, the vast migration of populations from rural to urban areas is creating new market opportunities.

New developments in KE with small holder farmers that might be applied globally with regional tuning have recently been described by Zhang et al. (2016). Here agronomy students from a range of regional Universities and from China Agricultural University (the project co-ordinator) are assigned to "Science and Technology Backyards" (STBs) in rural China. Often these are single villages or groups of small communities where the students work to develop farmers' co-operatives and to introduce new technology and changed farming practice. Increases in water and nutrient use productivity and yielding that have been achieved in these villages are impressive (Zhang et al. 2016).

Campbell et al. (2016) have recently argued that given the serious threats to food security posed by climate change, attention should shift to an action-oriented research agenda. He and co-authors see four key challenges:

(a) changing the culture of research;
(b) deriving stakeholder-driven portfolios of options for farmers, communities and countries;
(c) ensuring that adaptation actions are relevant to those most vulnerable to climate change;
(d) combining adaptation and mitigation strategies.

The emphasis here is to increase stakeholder engagement in research and by definition general principles and strategies to mitigate climate change impact must be implemented at the local level. In reality the BDA catena defined by Challinor et al. (2016), also termed the research to implementation gap, or the science-policy gap, is often substantial. Action is needed to address this shortcoming and GPC may have a role to play here.

Adoption rate of technologies with the potential to reduce risks in agriculture has traditionally been slow. For example, despite a global shortage of water for most purposes, the adoption of improved water management practices has been slow, even in agriculture, where around 70% of the world's available fresh water is used. There seems to be a clear case here for enhanced knowledge exchange between farmers, scientists and regional policy makers. How can stress resilience biology help us produce 'more crop per drop'?

Three Examples of Possible Local Interventions to Increase Food Security, Health and Well-being at Decreasing Scale of Operation

(a) The Community Scale: Eco- and Climate-Smart Villages

Some years ago, the EU funded the development of so-called eco-villages in different regions of sub-Saharan Africa. Introduction of technological innovation on a village scale resulted in enhancement of social sustainability of the communities as a component of enhanced environmental sustainability, the importance of which was highlighted by Pollock (2016). In particular, introduction of solar arrays generated significant increases in health and well-being of children as a result of phasing out of kerosene-based lighting system and their adverse effect on air quality in the home. Energy was also used to great effect for water pumping for irrigation and deficit irrigation techniques were applied. In the Chinese STB communities described above, crop scientists have shown villagers how to grow crops with reduced nutrient and water input. Crop geneticists have also played a part.

In what appears to be a very successful collaboration between CGIAR-CCAFS and several national programmes in Africa, rural communities are encouraged to develop Climate Sensitive Villages (CSVs) as platforms where researchers, local partners, farmers' groups and policymakers collaborate to select and trial a portfolio of technologies and institutional interventions. The focus is on the objectives of climate-smart agriculture (Campbell et al. 2016): namely, enhancing productivity, incomes, climate resilience and mitigation. Importantly, context-specific objectives are established by the stakeholders.

The Campbell paper notes that a broad range of adaptation technologies are introduced into the CSVs. These include water-smart practices, weather-smart activities, nutrient-smart practices, carbon-and energy-smart practices and knowledge-smart activities, all of which have been discussed above.

(b) The Farming System Scale: Conservation Agriculture

Conservation Agriculture (CA) has been widely adopted with some success throughout the Americas, where the effects of tillage had previously resulted in loss of soil structure, soil erosion with the loss of large quantities of good quality soil. CA is said to increase yields, to improve soil fertility, reduce soil water loss, control weed growth and reduce erosion. There may also be savings on use of tractor fuel and reduced C emissions, all changes resulting in a much more stress resilient agricultural system.

However, Giller et al. (2009) have suggested that CA can leave farmers with a heavy dependence on herbicides and fertilizers. The same group has highlighted particular concerns for use of conservation agriculture in Africa. These include: decreased yields often observed with CA, increased labor requirements when herbicides are not used, an important gender shift of the labor burden to women and a lack of mulch due to poor productivity and due to the priority given to feeding of livestock with crop residues. This appears to be an excellent example of different regional manifestations of the interaction between $G \times E \times M \times S$ (above).

(c) The Crop Scale: Putting Nitrogen Fixation to Work for Smallholder Farmers in Africa (N2Africa) http://www.n2africa.org/

Here, the crop stress which is a major problem in much of sub Saharan Africa, is a shortage of nitrogen for crops. N2Africa, a Gates-funded long term project directed by Ken Giller at Wageningen University, is focused on enabling African smallholder farmers to benefit more fully from symbiotic N2-fixation by grain legumes. The thrust of the project is a locally-focused knowledge exchange and capacity-building effort and the development of effective production technologies including inoculants and fertilizers. The capacity that is built will sustain the pipeline and deliver continuous improvement in legume production technologies tailored to local settings.

Discovery research is aimed at the identification of new elite strains of rhizobium for the several major grain legumes other than soybean – common bean, cowpea and groundnut. New elite strains will be made available to inoculant producers for scaling up the technology. The project website stresses that delivery and dissemination approaches will be tailored to local needs. New, innovative tools for monitoring and evaluation will allow "best fit technologies" to be developed at the field and farm-scale to be translated into "best-fit approaches" at the country or regional scale. In the first phase, N2Africa reached more than 230,000 farmers who evaluated and employed improved grain legume varieties, rhizobium inoculants and basal (P) fertilizers. The impact on the family of the increased utilization of legumes is particularly large as the crop is largely grown by women and used within the home.

Introduction of N fixation biology into non-legume crops may also be a game-changer if these new seeds can be made available to the very large numbers of smallholders in developing countries who can benefit from this stress resilience technology (Charpentier and Oldroyd, 2010).

It is clear from the above examples that there is much action-orientated research underway in farming communities around the world. It is equally clear that there is much still to do within the framework of the BDA pipeline (above) or the research to outcome catena. One size interventions will not "fit all" across the globe and we ask now what the Global Plant Council can do to facilitate progress in implementation as plant science and scientists seek to address a mounting number of global food challenges.

Food security is a global issue; by 2050 food production must increase by at least 60% to meet the demands of a growing population and changing diets. Meeting this challenge will require global and strategic thinking and planning. We have outlined some of the challenges to be addressed and presented examples of models that work. The key is stakeholder engagement at all levels and, in this regard, we submit that challenges for crop production will be best addressed at the local level either through the adoption and adaptation of generic solutions or through the development of local solutions through knowledge exchange and with the benefit of indigenous knowledge. There are encouraging signs that governments and regional bodies understand the importance of increasing agricultural productivity to meet the growing demands and we highlight the importance of international collaboration as a major element in increasing crop productivity and food production.

Actions for the Global Plant Council

1. Help facilitate partnerships in research to implementation projects across disciplinary boundaries and geographic borders (the right technology in the right place)
2. Help develop partnerships with international agencies
3. Promote sharing of data and working practices
4. Promote development of Knowledge Exchange resources and international training courses (novel science must be freely available to policy makers and importantly to the large numbers of practitioners producing food in the developing world)
5. Lead in the provision of advocacy for policy, practice, funding change
6. Lead in reducing the science-policy gap
7. Encourage a "bottom-up" approach to intervention
8. Lead in promoting regionally relevant interventions at a range of scales (understand the local landscape)
9. Encourage introduction of initiatives along the delivery chain.

Acknowledgments

This paper is based on outcomes from a Stress Resilience Symposium held in Brazil in October 2015 organized by the Global Plant Council and Society for Experimental Biology. The authors would like to thank the Society for Experimental Biology for funding support for this symposium.

Conflict of Interest

None declared.

References

Bonneau, J., J. Taylor, B. Parent, et al. 2013. Multi-environment analysis and improved mapping of a yield-related QTL on chromosome 3B of wheat. Theor. Appl. Genet. 126:747–761.

Campbell, B. M., S. J. Vermeulen, P. K. Agarwal, C. Corner-Dolloff, E. Girvetz, A. M. Loboguerrero, et al., 2016. Reducing risks to food security from climate change. Global Food Sec. 11:34–43.

Challinor, A., P. Martre, S. Asseng, P. Thornton, and F. Ewert. 2014. Making the most of climate impact ensembles. Nat. Clim. Change 4:77–80.

Challinor, A. J., A. K. Koehler, J. Ramirez-Villegas, S. Whitfield, and B. Das. 2016. Current warming will reduce yields unless maize breeding and seed systems adapt immediately. Nat. Clim. Chang. 6:954–958.

Charpentier, M., and G. Oldroyd. 2010. How close are we to nitrogen fixing cereals? Curr Opin Plant Biol. 13:556–564.

Chianu, J. 2016. Technologies for African Agricultural Transformation (TAAT) International Institute for Tropical Agriculture. Available via https://issuu.com/iita/docs/bulletin_taat_2341_special

Diop, N. N., F. Okono, and J.-M. Ribaut. 2013. Evaluating human resource capacity for crop breeding in national programs in Africa and South and Southeast Asia. Creat. Educ. 4:72–81.

Feng, X., A. Porporato, and I. Rodriguez-Iturbe. 2013. Changes in rainfall seasonality in the tropics. Nature Climate Change 3:811–815.

Foresight Africa. Available at https://www.brookings.edu/research/foresight-africa-top-priorities-for-the-continent-in-2016-2/. (accessed 17 March 2017).

Fukao, T., E. Yeung, and J. Bailey-Serres. 2011. The submergence tolerance regulator SUB1A mediates crosstalk between submergence and drought tolerance in rice. Plant Cell 23:412–427.

Giller, K. E., E. Witter, M. Corbeels, and P. Tittonell. 2009. Conservation agriculture and smallholder farming in Africa: the heretics' view. Field. Crop. Res. 114: 23–34.

Guo, J. H. 2010. Significant acidification in major Chinese croplands. Science 327:1008–1010.

Hochman, Z., D. L. Gobbett, and H. Horan. 2017. Changing climate has stalled Australian wheat yields: study. The conversation, Environment + Energy. Available at https://theconversation.com/changing-climate-has-stalled-australian-wheat-yields-study-71411. (accessed 17 March 2017).

Lin, B. B. 2011. Resilience in agriculture through crop diversification: adaptive management for environmental change. BioScience 61:183–193.

Kang, S. Z., X. L. Su, L. Tong, J. H. Zhang, L. Zhang, and W. J. Davies. 2008. A warning from an ancient oasis: intensive human activities are leading to potential ecological and social catastrophe. Int. J. Sustain. Dev. World Ecol. 15:440–447, 9.

Kirkegaard, J. A., and J. R. Hunt. 2010. Increasing productivity by matching farming system management and genotype in water-limited environments. J. Exp. Bot. 61:4129–4143.

Lobell, D. B., W. Schlenker, and J. Costa-Roberts. 2011. Climate trends and global crop production since 1980. Science 333:616–620.

Pollock, C. J. 2016. Sustainable Farming: chasing a mirage? Food Energy Secur. 5:205–209.

Porter, J. R., L. Xie, J. Andrew, K. Cochrane, S. M. Howden, M. M. Iqbal, D. B. Lobell, and M. Travasso (2014) Food security and food production systems. In: Climate Change 2014: Impacts, Adaptation, and Vulnerability. Part A: Global and Sectoral Aspects. Contribution of Working Group II to the Fifth Assessment Report of the. Intergovernmental Panel on Climate Change. Available via http://www.ipcc.ch/pdf/assessment-report/ar5/wg2/WGIIAR5-Chap7_FINAL.pdf (accessed 17 March 2017).

Price, A. H., G. J. Norton, D. E. Salt, O. Ebenhöh, A. A. Meharg, C. Meharg, M. R. Islam, R. N. Sarma, T. Dasgupta, A. M. Ismail, K. L. McNally, H. Zhang, I. C. Dodd, and W. J. Davies 2013. Alternate wetting and drying irrigation for rice in Bangladesh: is it sustainable and has plant breeding something to offer? Food Energy Secur. 2:120–129.

Ribaut, J.-M. 2006. Drought adaptation in cereals, 642 pp. Haworth's Food Products Press, New York.

Ribaut, J.-M. 2014. How to build research partnership that benefit farmers. SciDev.Net. Available at http://www.scidev.net/global/r-d/opinion/build-research-partnerships-benefit-farmers.html. (accessed 17 March 2017).

Rubaiyath, A., A. N. M. Bin Rahman, and J. Zhang. 2016. Flood and drought tolerance in rice: opposite but may coexist. Food Energy Secur. 5:76–88.

Tardieu, F. 2012. Any trait or trait-related allele can confer drought tolerance: just design the right drought scenario. J. Exp. Bot. 63:25–31.

van Ittersum, M. K., L. G. J. van Bussel, J. Wolf, P. Grassini, J. van Wart, N. Guilpart, et al. 2016. Can sub-Saharan Africa feed itself? PNAS 113:14964–14969.

Varshney, R., J.-M. Ribaut, E. S. Buckler, R. Tuberosa, J. A. Rafalski, and P. Langridge. 2012. Can genomics boost productivity of orphan crops. Nat. Biotech. 30:1172–1176.

Xue, F., A. Porporato, and I. Rodriguez-Iturbe. 2013. Changes in rainfall seasonality in the tropics. Nat. Clim. Chang. 3:811–815.

Yang, J., and J. Zhang. 2010. Crop management techniques to enhance harvest index in rice. J. Exp. Bot. 61:3177–3189.

Zhang, W., G. Cao, X. Li, H. Zhang, C. C. Wang, Q. Liu, et al. 2016. Closing yield gaps in China by empowering smallholder farmers. Nature 537:671–674.

15

Photobiology in protected horticulture

Phillip A. Davis & Claire Burns

Stockbridge Technology Centre, Cawood, Selby, North Yorkshire YO8 3TZ, UK

Keywords

Horticulture, Light emitting diodes,
crop production, photobiology,
photomorphogenesis, photosynthesis.

Correspondence

Dr Phillip A. Davis, Stockbridge Technology
Centre, Cawood, Selby, North Yorkshire YO8
3TZ, UK.
E-mail: phillip.davis@STC-nyorks.co.uk

Funding Information

Dr Davis' contribution to this review was
funded by AHDB Horticulture Fellowship
(CP085).

Abstract

The introduction of high power LED lighting systems for horticulture has
stimulated substantial interest from both the research community and the
protected horticulture industry. LED lighting systems have the potential to
reduce electrical energy consumption compared to conventional high pressure
sodium lights and their energy efficiency continues to improve. In addition
to the potential of LEDs to reduce carbon footprints and reduce running
costs, LED lighting also provides considerable opportunities to exploit the
wealth of photobiological knowledge to produce horticultural benefits. The
narrow emission spectra of LEDs allows lighting systems to be tightly designed
to stimulate specific plant photoreceptors, allowing plants to be manipulated
to produce desirable characteristics. Lighting systems can be designed to
maximize growth, control morphology, and optimize flavor and pigmentation.
This review outlines how the light spectrum influences photosynthesis and
how plant photoreceptors sense light and control growth. The review then
discusses the ways in which this knowledge is being implemented in commercial horticulture to improve factors such as yield, flavor, color, plant
growth, and flowering as well as pest and pathogen management and control.
Research in this area is moving rapidly as the LED systems improve and
increase in efficiency and as the range of novel horticultural applications
expands.

Glossary

Cryptochrome	A photoreceptor that is sensitive to blue and UVA light.
DE	Day-length-extension lighting. Light treatments provided to extend the length of the photoperiod.
EOD	End-of-day lighting. Light treatments provided for a short period at the end of the day to manipulation plant light responses.
HPS	High pressure sodium lighting.
LED	Light-emitting diodes.
NB	Night-break lighting. Light treatments provided during the middle of the night.
PAR	Photosynthetically active radiation (PAR) is light with wavelengths in the range 400–700 nm that can be used by plants for the process of photosynthesis.
PGR	Plant growth regulators.
Photomorphogenesis	The processes that causes plant morphology and pigmentation to change following exposure to light. These processes are activated and controlled by several photoreceptors
Photon irradiance	A measurement of the number of photons incident on a surface, which has units of $\mu mol[photons]/m^2/s$.
Photoreceptor	Light-sensitive proteins that initiate light responses.
Phototropin	A photoreceptor that detects blue and UVA light.
Phytochrome	A photoreceptor that can sense the red:far-red ratio of light.
UVR8	A photoreceptor that is able to detect UVB light

Introduction

Advances in technology are rapidly exploited by the protected horticulture industry, and have expedited the progression from structures that provide simple frost protection to sophisticated automated plant factories in which all environmental parameters are carefully regulated. Such advances have allowed crops to be produced year-round and can facilitate substantial yield increases. For example, lettuce production in plant factories increased crop yields by up to 100 times per unit area of land compared to conventional outdoor practices (Kozai 2013). As with glasshouse production in light-limited conditions such as winter months in northern latitudes, crop production in plant factories requires artificial lighting. In large-scale plant production systems, the efficiency of the lighting system is key to maintaining profitability. The introduction of LED lighting systems for horticultural use has attracted considerable attention for their potential to reduce electrical energy inputs, making winter crop production more financially viable. Horticultural LED lighting systems exhibit considerable diversity in their design, control systems, and the light spectra produced. LED lighting systems can be highly energy efficient; however, not all LEDs are more energy efficient than standard high-pressure sodium (HPS) lights (Pearson et al. 2015), and care is needed during design and implementation to ensure that the installed systems meet the needs of the crop production system.

The potential reduction in energy consumption that LED systems can deliver is highly desirable to the horticultural industry. However, the greatest potential of LED lighting systems to alter and improve crop production comes from the tight spectral control that LEDs provide. Light quality influences all aspects of plant biology, and much is known regarding the photobiological processes by which plants sense and respond to the light environment. Manipulation of the light spectrum using LED lighting allows the exploitation of this substantial body of knowledge to achieve improved crop production systems. Optimal lighting regimes have the potential to increase yields and improve plant quality, nutritional value, and flavor. In addition to enhancement of plant health and quality, the impacts of pests and pathogens can be reduced both as a result of elevated plant resistance and also by direct disruption of pest/pathogen biology. In this review, we briefly outline how plants use and respond to light before examining how LEDs are being used to manipulate crop photobiology and improve protected horticulture.

Optimizing the Use of Light to Maximize Photosynthesis

Plants are able to use light with wavelengths in the range 400–700 nm for photosynthesis. This waveband, which is often referred to as photosynthetically active radiation (PAR), contains 26% of the photons and 42% of the energy reaching the Earth's surface (calculated from ASTM G173-03 reference spectra). Plants must receive sufficient light to drive active growth and maintain plant quality and productivity. To maximize productivity and minimize energy inputs, artificially-supplied light must provide wavelengths that are used efficiently and meet plant needs. The light action spectrum for photosynthesis was first described by McCree (1971), who showed that red light was optimal for light-limited photosynthesis. More recently, research showed that changes in the concentration of plant accessory pigments such as carotenoids, which absorb predominantly in the blue region of the PAR spectrum, are responsible for the differences in light-use efficiency between a) the red and blue regions of the spectrum, and b) leaves grown under different conditions (Hogewoning et al. 2012). Although such pigments reduce the light-use efficiency of blue light, they are vital for protecting the photosynthetic machinery against UV damage (Middleton and Teramura 1993). While red light is utilized most efficiently for photosynthesis, red light alone is not sufficient to maximize photosynthesis. Blue light is required to prevent 'red light syndrome' (Trouwborst et al. 2016), which is characterized by suboptimal morphology and aberrant gene expression and biochemistry. Blue light is also needed to promote stomatal opening, improving access to CO_2, and driving transpiration and nutrient uptake (Hogewoning et al. 2010; van Ieperen et al. 2012; Nanya et al. 2012; Savvides et al. 2012). Other light wavelengths can further increase photosynthesis under certain circumstances. For example, green light can penetrate further into both canopies and individual leaves than red or blue light, and can drive photosynthesis in cells/leaves that are not reached by red and blue light (Sun et al. 1998; Terashima et al. 2009; Paradiso et al. 2011). The results of Terashima et al. (2009) also demonstrated the interactions of light quality and intensity with the greatest benefit of green light occurring at intermediate light intensities. The benefits of including green light in a customized spectrum for plant production would need to be evaluated with regard to the energy required to generate green wavelengths. Currently, red and blue LEDs are more energy efficient than green LEDs.

Unlike traditional lighting systems, LEDs can be turned on and off rapidly (hundreds of times per second). This creates the opportunity to potentially maximize the photosynthetic performance of crops while minimizing the energy inputs. In theory, it would be possible to pulse the light in such a way as to deliver the correct amount of light energy to excite every photosystem in a leaf without inducing the array of energy dissipation mechanisms that help protect plants from damage under natural conditions. This would help to maximize the light-use efficiency of plants. Tennessen et al. (1995) demonstrated that, provided intervals between light pulses were less than 200 μs, the amount of photosynthesis was proportional to the total amount of light provided to the plants. Jao and Fang (2004) observed that Potato plantlets grew fastest when light was pulsed at 720 Hz. and noted that 180 Hz provided the most energy-efficient system and would be appropriate where reduction in energy consumption was paramount. Shimada and Taniguchi (2011) found that photosynthetic rate and plant morphology were adversely affected in plants exposed to out-of-phase red and blue light pulses compared to plants exposed to in-phase light pulses. While this experiment provided interesting results from the perspective of how plants sense and use light, the most pronounced physiological effect observed was an increased shade avoidance response, which is of no benefit for most horticulture applications. In addition, pulsed light may not provide the desired effects in a glasshouse setting where natural light is present. Several research groups are developing sensor-controlled lighting systems that modulate the light regime to match ongoing plant needs. These technologies have the potential to maintain plant growth rates and quality during variable weather conditions while minimizing energy consumption.

Mobile lighting systems, which use fewer lamps and have correspondingly lower costs, have been trialed for crop production. Mobile systems have two major limitations: a) the amount of light supplied to plants is usually lower than with a static system, and b) systems to move lamps are needed, with associated installation and maintenance requirements. However, costs for mobile systems could be minimal if existing mobile irrigation booms were used to mount lamps. Li et al. (2014) showed that lettuce plants could be grown under mobile lights. However, this mobile system used half the number of lamps as were used for the fixed LED control treatment, and a more substantial reduction in lights would be needed to achieve viable savings in a commercial system. Due to the design of mobile lighting systems plants receive a variable intensity as the lights pass over the crop. However, plants can take up to 45 minutes of constant light to achieve maximum photosynthetic rates (Kirschbaum and Pearcy 1988), and much of the light provided during passage of a mobile light would therefore not be used for photosynthesis. Mobile lighting is therefore unlikely to be suitable for the majority of applications. However, in instances where only low doses of light are required (for example, end-of-day light treatments, or UVC/UVB treatments), mobile lights mounted on irrigation booms may provide an economically viable way of installing lamps.

Improving Yields with Interlighting

The light intensity within plant canopies decreases with depth as leaves absorb the light. Due to the light gradients within canopies leaves at the top may be light saturated when the canopy as a whole is light limited. Under these conditions adding artificial light at the top of the canopy can increase yield but any light absorbed by leaves that are already light saturated will provide no additional growth potential and those leaves may even become light stressed. By directly providing supplemental light to leaves that are shaded lower down in the canopy (interlighting) a greater proportion of the light can be used for photosynthesis without exceeding the point of photosynthetic light saturation. LEDs have made interlighting systems practical in commercial settings as they are cool to touch and can be placed close to crops without burning the plants. LEDs located within a cowpea (*Vigna unguicultata* L. Walp.) canopy were able to improve biomass production as well as reducing the senescence of older leaves deep in the canopy (Massa et al. 2008). Trouwborst et al. 2010 found that interlighting in cucumber crops increased leaf photosynthetic rate and photosynthetic potential of leaves lower in the canopy. However, the interlighting treatments caused extensive leaf curling. This led to a reduction in light interception by the canopy and prevented the interlighting treatment from increasing crop yields. Hao et al. 2012 also had mixed results when using interlighting with cucumber. Over the first two weeks of the experiment, visual quality and yield increased by more than the increase in total photon irradiance; however, these gains declined as the experiment progressed, especially in the blue interlighting treatment where some leaf curling was observed. Guo et al. (2016a) found that LED interlighting plus HPS top lighting increased the yield and concentration of health providing compounds of sweat peppers in comparison to HPS only light treatments. However, it should be noted that the total amount of light supplied in the interlighting treatments was greater than that provided by the HPS only treatments. Interlighting in tomato crops has proved highly successful and there are now a growing number of commercial installations, all of which are reporting significant increases in yields. Interlighting trials have also investigated the use of different light spectra at different locations with canopies. The addition of increasing amounts of blue light within the canopy was found to increase yields of cucumber but not tomato plants, though the blue light reduced the internode lengths of both species (Ménard et al. 2006). Guo et al. (2016b) found that applying far-red at the top of a cucumber canopy was able to increase the yield while providing blue at the top reduced yield. The best yields were achieved with far-red at the top of the canopy and blue light at the bottom. Many novel approaches to lighting crops with LEDs are now possible but trialing all these approaches would be costly and time consuming. To aid the development of lighting systems, modeling approaches will be increasingly important for identifying designs that should be trialed in the real world. de Visser et al. (2014) used 3D ray tracing models to assess the light interception of tomato canopies when illuminated with interlighting set at different angles of incidence. Their models predicted that light interception and photosynthesis could be maximized by positioning the interlights so they shone slightly upwards compared to horizontally as is the case in most commercial systems.

The Impact of LEDs of Overall Glasshouse Energy Budgets

One potential negative impact of LED lighting in glasshouses is the lack of radiative heat that is produced by LEDs. In experiments where the energy consumption of glasshouses has been monitored, LED-lit compartments required higher air temperatures to counteract the loss of radiative heat. This reduced the overall energy saving as there was a greater heating demand. Dueck et al. (2012) reported that the use of LEDs for tomato production increased energy consumption; however, this was attributed to the energy demands of the water cooling systems of the LEDs used in that particular system. The majority of current commercially available LEDs do not require water cooling systems. Gómez and Mitchell (2013) examined the use of LED towers in comparison to standard HPS lighting for tomato production. Their results indicated that the LEDs provided a significant energy saving but provided similar yields as the HPS lighting systems. In a more detailed analysis of the system, Gómez et al. 2013 measured the efficiency of electrical conversion into fruit biomass to be 75% greater for the LED lights compared to sodium lamps; however, this did not take heating requirements into consideration. In all experiments that compare HPS and LED light there is a need to assess the differences in plant temperature to ensure that any effect of temperature can be separated from the effects of light on plants responses. HPS light can increase leaf temperature by several degrees and this can increase plant growth rates. While the drop in crop temperature may have negative effects on crops in the colder months of the year, the lower temperature will benefit crops on warm days with low light levels. As with any significant change in crop environment, the switch from HPS to LED lighting will require a period of learning to develop protocols for correct management of plant irrigation and growth.

Overview of Plant Photoreceptors and Photomorphogenesis

Photomorphogenesis is the process by which photoreceptors drive changes in seedling morphology in response to light exposure after germination. Several processes responsible for efficient, healthy growth are stimulated during this process. For example, one of the prime ways in which photomorphogenesis optimizes plant performance is by stimulating seedlings to orient toward light, thereby maximizing light capture and photosynthesis. Photomorphogenesis is mediated by several types of photoreceptor, each of which is sensitive to distinct parts of the light spectrum. The different photoreceptors control several photomorphogenic processes that are important for plant survival, growth, and development.

The UVR8 photoreceptor responds to UVB light, with peak sensitivity at ~290 nm (Brown et al. 2009). Plants produce a range of pigments and other secondary metabolites in the presence of UVB light, and these act as sunscreens to provide protection against UV light damage (Chalker-Scott 1999). In particular, production of flavonoids and anthocyanins increases after exposure to UVB (Tevini and Iwanzik Wm Thoma 1981; Beggs and Wellman 1985), and the visual appeal of leaves and flowers is enhanced (Paul et al. 2006). Plants also retain a compact shape when exposed to UVB light (Gardner et al. 2009), and form tougher, more robust, leaves (Wargent et al. 2009). A further benefit of UVB light exposure is an increase in the concentration of essential oils in herbs (Kumari et al. 2009; Hikosaka et al. 2010).

Plants grown in the absence of blue light become etiolated and leaves tend to hang downwards and remain curled. Several photoreceptor families are responsible for sensing blue and UVA light, including the phototropins and cryptochromes. Phototropins control a wide range of plant responses such as stomatal opening, phototropism (bending toward light), chloroplast movement (Briggs and Christie 2002), leaf flattening (de Carbonnel et al. 2010), and de-etiolation of the hypocotyl (Folta and Spalding 2001). Phototropins regulate cellular processes while leaving gene expression unchanged. By contrast, cryptochromes regulate gene expression, resulting in downstream effects on secondary metabolism and pigment synthesis (Vlohr and Drumm-Herrel 1983), flowering (Giliberto et al. 2005), and inhibition of hypocotyl elongation (Folta and Spalding 2001). Cryptochromes also function to entrain circadian rhythms (Cashmore 2003). Phototropins and cryptochromes are similarly sensitive to UVA and blue light (Briggs and Christie 2002) but there is some evidence that cryptochromes can also be partially inactivated by green light (Sellaro et al. 2010). For a more detailed overview of the blue/ UV light signaling networks see Huché-Thélier et al. (2016).

Red and far-red light are sensed by another family of photoreceptors, the phytochromes, which function similarly to the cryptochromes in that they mediate their effects by altering gene expression. Phytochromes are involved in the control of a number of photobiological responses such as germination (photoblasty), inhibition of hypocotyl elongation, apical hook straightening, leaf expansion, flowering time, circadian rhythm entrainment, and chlorophyll biosynthesis. Plants possess several phytochromes (e.g., arabidopsis has five phytochromes, rice has three, and maize has six), each of which has a different functional range. Two types of phytochrome have received extensive investigation: phytochrome A (phyA) and phytochrome B (phyB). PhyB is activated by red light (peak absorbance, 666 nm) and deactivated by far-red light (peak absorbance, 730 nm). A short pulse of red light is sufficient to activate phyB; however, if a red pulse is followed by a far-red pulse the red light responses do not occur. This is termed red, far-red reversibility. In contrast with phyB, phyA can be activated by both far-red and red light (Shinomura et al. 1996). However, in the light, phyA is downregulated both transcriptionally and post-transcriptionally (Chen and Chory 2011) and PHYA primarily accumulates in plant tissues during periods of darkness. Although phytochromes are mainly considered with regard to their ability to detect red:far-red ratios, phytochromes absorb light wavelengths from the full spectrum, including blue light. Green light, for example, can stimulate phyA- and phyB-mediated germination (Shinomura et al. 1996). For a more detailed overview of red:far-red responses in plants, see Demotes-Mainard et al. (2016).

No photoreceptor with specificity for green light has been identified to date. Inclusion of green light in illumination mixes has been reported to increase plant growth rates; however, it remains to be seen whether this occurs as a result of the direct effect of green light on photosynthesis or via some other photomorphogenic effects. For a detailed review of the influence of green light on plant production, see Wang and Folta (2013).

The cellular and plant physiological responses induced by different photoreceptors overlap to some extent, with many of these responses caused by alterations to the synthesis and transport of plant hormones (Lau and Deng 2010). The tight spectral control provided by LED lighting allows different photoreceptors to be selectively activated, permitting the light environment to be tuned to manipulate plant responses and enhance desirable plant qualities. Such manipulation is impossible with traditional light technology, and it is through the precise activation of different photoreceptors in diverse crops that LED lighting has the potential to revolutionize horticultural practices.

The Effect of Light Quality on Propagation

Plant propagation is the first stage of horticultural crop production, and maximizing the efficiency of seed germination and rooting of cuttings can have huge impacts on overall yields and profitability. Light is an important cue both for seed germination and for root development of cuttings and providing optimal lighting conditions can greatly improve both the speed and success of propagation. Jankowska-Blaszczuk and Daws (2007) found that seeds were less dependent on light for germination as they increased in size. Seed size was also found to correlate with the tolerance of germination to shade conditions, as determined by the response to red:far-red light ratios (high red:far-red corresponds to full sun conditions with little shade). Smaller seed needed higher red:far-red ratios for germination than larger seeds. Germination efficiency can also be tailored using other regions of the light spectrum. For example, germination of Chinese ladder brake fern (*Pteris vittata*) spores can be inhibited by blue light. Interestingly, Sommer and Franke (2006) observed that exposing seeds of cress, radish, and carrots to bright green laser light caused the plants to grow considerably larger. No biological explanation for this observation has been elucidated, but further investigation may identify some useful practical applications.

Many important horticultural crops are propagated from cuttings or using micropropagation techniques. Using spectral manipulation to improve propagation efficiency is of particular interest for high-value crops that are challenging to root. One of the challenges of taking cuttings is preventing dehydration. Plastic sheeting and fogging helps to reduce transpiration, but can also be reduced by manipulation of the light spectrum. Blue light drives stomatal opening, so removing blue light from the spectrum helps to reduce transpiration and improve cutting survival. Red light has been shown to be beneficial in promoting root development in several species. Rooting was improved in two of three varieties of grape (*Vitis ficifolia*) when illuminated with red light compared to fluorescent or blue light. In the third grape variety, rooting levels were high and similar in all light treatments (Poudel et al. 2008). When Wu and Lin (2012) propagated *Protea cynarodies* plantlets under red LED light, 67% rooted compared to 7% under conventional fluorescent tubes, and 13% rooted under blue light or a red:blue (50:50%) LED light combination. Root development was also found to be more extensive under the red light treatments. In *Protea cynarodies* cuttings, Wu (2006) observed that the concentration of phenolic compounds increased over time and that root development only occurred after their concentration reached a certain level. Further investigation demonstrated that the phenolic compound 3,4-dihydroxybenzoic acid could promote root formation up to a concentration of 100 mg/L but inhibited root formation at higher concentrations (Wu et al. 2007). The use of red light in the propagation phase caused the plants to generate phenolic compound concentrations favorable to rooting, while the inclusion of blue light elevated concentrations further, thus inhibiting rooting. Similar effects may be occurring in other species, though the active compounds are likely to vary between species.

Limiting the spectrum to 100% red light is not always optimal for propagation, and a range of red:blue light mixtures have been found to produce optimal rooting in several species. *In vitro* propagation of banana plantlets and subsequent transfer to a soil based growing substrate was found to be best when performed under 80% red: 20% blue light (Nhut et al. 2002). A 50% red: 50% blue light treatment was most effective for the propagation of Cotton plants (Li et al. 2010). Strawberry plantlets performed best when propagated under 70% red: 30% blue light (Nhut et al. 2003). In climbing Gentian, red light was found to promote rooting while blue light inhibited rooting, with optimal rooting observed using a 70% red: 30% blue mixture (Moon et al. 2006). Even in species where 100% red light treatments are thought to provide optimal rooting, plants must still be moved to a light treatment containing some blue light after the critical stage of root initiation in order to prevent etiolation of the young plants and help ongoing root development: blue light enhances both root and shoot development (Nhut et al. 2003).

Light treatments provided to parental stock plants prior to removal of cuttings may also influence rooting success rates. *Eucalyptus grandis* cuttings had greater rooting success when the stock plants were grown under low red:far-red ratios (Hoad and Leakey 1996). Cutting success is closely linked to cutting quality, and improving the quality of stock plants through changes to lighting or with spectral filters is expected to provide significant benefits, especially if combined with optimal postcutting light treatments.

The Influence of Light Quality on Plant Growth

The first attempts at producing food crops under LED lighting systems were limited by the intensity and color of the LED lamps (only red LEDs were available), and fluorescent tubes were used to provide the blue light required to maintain plant health and growth rates (Bula et al. 1991; Barta et al. 1992; Tennessen et al. 1994). LED technologies have advanced greatly since these early attempts, and LED lighting systems that provide multiple colors of light are now available for horticultural

production. The benefits of including blue light in the spectrum have been demonstrated on numerous occasions. Wheat and Arabidopsis plants produced more seeds (Goins et al. 1997, 1998), lettuce, radish, and spinach produced more biomass (Yorio et al. 2001), and frigo strawberries produced more fruit with higher sugar contents (Samuoliene et al. 2010) when grown in the presence of blue light.

While the need for blue light in artificial growth systems is clear, there is less consensus regarding optimal red:blue mixtures. Part of this uncertainty is due to differences in stomatal light responses between species. Ouzounis et al. (2014) showed that stomatal conductance in campanula showed no response to supplemental LED lighting, whereas stomatal conductance increased in both roses and chrysanthemums. Light quality affected stomatal development in *Withania somnifera* plantlets, with monochromatic light resulting in impaired stomatal development (Lee et al. 2007). Stomatal development can also be influenced by UVB light. For example, soybean produced fewer stomata after UVB exposure and, although this could improve drought tolerance, photosynthetic performance could be adversely affected (Gitz et al. 2005). Although stomata routinely open and close in response to light to regulate water use and CO_2 uptake, any influence of light quality on the development and density of stomata during leaf growth will have long-term impacts on stomatal conductance, photosynthetic performance, and water-use efficiency.

In addition to its effects on stomata, light quality also influences many other responses required for healthy plant growth through regulation of plant metabolism and morphology. Cucumber plants grown under blue and UVA light were found to have both higher photosynthetic potential and increased transcription of the genes required for carbon fixation compared to plants grown under red, green, or yellow light (Wang et al. 2009). In rice, addition of blue light to a red background led to higher photosynthetic and stomatal conductance rates and was associated with higher chlorophyll and Rubisco contents (Matsuda et al. 2004). In lettuce, growth rates (measured as biomass accumulation) decreased as UVA and blue light increased (Li and Kubota 2009; Son and Oh 2013). In contrast, an increase in rapeseed growth rate was observed as blue light percentage increased from 0% to 75% (in a red: blue mix; Li et al. 2013). Folta and Childers (2008) observed the greatest growth of strawberry plants under 34% blue light. However, Yoshida et al. (2012) found that Strawberry fruit yield was greatest in plants grown under continuous blue light, and that red light inhibited flowering.

The inclusion of green light from LEDs increased fresh and dry weight biomass accumulation in lettuce when green light replaced some of the blue or red light in a light treatment mixture (Kim et al. 2004a; Stutte et al. 2009). There are two possible reasons for these observed increases in growth rate. First, green light can penetrate deeper into the plant canopy and, therefore, drive more photosynthesis. Second, reducing the amount of blue reduces the restriction on leaf expansion imposed by plant photoreceptors (Dougher and Bugbee 2004), thus increasing leaf area, light capture and growth. Research to date suggests that addition of green light to the growth spectrum does not enhance crop performance in all cases. Contrasting with the study noted above, Li and Kubota (2009) found that addition of green light caused no increase in lettuce biomass, but that plant morphology was affected and that an increase in stem and leaf elongation was observed. Early in lettuce crop development, larger leaves may benefit crop performance by allowing greater light capture; however, larger leaves later in the crop cycle may reduce plant quality, especially if combined with stem extension. Some of the discrepancies between the different sets of published results may be due to differences in the total amount of light provided as well as the proportions of green light provided. Kim et al. (2004b) found that 24% green light boosted lettuce yields; however, yields were reduced when greater than 50% green light was used, probably as a result of lower overall photosynthetic rates. Kim et al. (2004a) also found that in lettuce green light could cause stomatal closure and that stomatal opening was greatest under broad spectrum lighting (suggesting that white light may be better in this case). In Tomato transplants, the addition of small amounts of green (520 nm), orange (622 nm), or yellow (595 nm) LED light was found to reduce plant growth rates (Brazaitytė et al. 2010), and some of the negative impacts on plant growth could still be observed one month after exposure to the different light treatments (Brazaitytė et al. 2009). Yellow light was also found to suppress the growth of lettuce plants (Dougher and Bugbee 2001); however, it should be noted that these experiments were not performed with LED lighting and spectral assessments were complex. Lu et al. (2012) examined the effect of supplemental LED light on Tomato production on the single truss system and concluded that white light would be more effective at driving canopy photosynthesis in dense canopies than red or blue light, because the green light component of the white light spectrum penetrates further into the canopy than red or blue.

Far-red light is important for plant development and performance throughout the life of the crop. For example, far-red light can inhibit germination of lettuce seeds (Borthwick et al. 1952; Shinomura et al. 1996) and generally counteracts the influence of red and blue light on plant morphology. This results in plant stretching and reduced chlorophyll content (Li and Kubota 2009), which can also reduce photosynthetic rates. Far-red can, however, be beneficial to crop growth by increasing leaf area in lettuce (Li and Kubota 2009; Stutte et al. 2009), potentially allowing greater light capture. Far-red light can also increase the yields of green beans (Davis 2013). Due to the sensing mechanisms of the phytochromes, the effects of far-red light can be achieved by providing light at the end of day (EOD), or during the night period with day-length-extension (DE) or night break (NB) lighting techniques, rather than during the day. For example, EOD-far-red treatments are effective in encouraging tomato plants to grow taller, which could be used optimize the production of seedlings for grafting, and EOD-red treatments can improve plant compactness (Kubota et al. 2012). These lighting techniques generally require lower light intensities (1–5 μmol/m^2/s can provide strong influences on crop responses) and durations than daytime far-red light treatments and can therefore be used to influence morphology with lower energy investments.

Improving Crop Morphology and Reducing the Use of Plant Growth Regulators

Spectral manipulation can maximize biomass production but, depending on the particular conditions and crop, larger plants may not be desirable. Morphology and quality may be negatively affected if plants are grown 'too soft'. Several methods are used in the industry to control plant morphology during crop production, including reduced irrigation, increased electrical conductivity (EC) of irrigation solutions, application of plant growth regulators (PGRs), and altering temperature profiles (e.g., negative DIF). The use of PGRs in the ornamentals sectors is particularly widely used to help maintain crop compactness during periods of low light. The ability to control the light spectrum with LEDs, or with spectral filters, provides the potential to manipulate plant morphology, reducing the need for chemical intervention. Poinsettias grown under 80% red: 20% blue supplemental LED lighting were 20-34% shorter than those grown under HPS (5% blue) lamps (Islam et al. 2012). Although leaves were smaller and plants accumulated less dry matter, there was no delay in bract color formation or postproduction performance, indicating that LEDs could be useful for reducing the use of PGRs for Poinsettia production. An increase in the blue light proportion of supplemental light was also found to cause roses and chrysanthemum to remain more compact during production: the most compact roses were observed under 40% supplemental blue light (the highest proportion examined in the study; Ouzounis et al. 2014). The quality of the supplemental light was also found to strongly influence leaf morphology, with 100% red light treatments causing rose leaves to become curled. In many species, stem elongation decreases as the proportion of blue light increases (Moon et al. 2006; Nanya et al. 2012). While higher percentages of blue light reduce plant height, the concomitant reduction in leaf size may have negative influences on growth and development that could influence production periods. The red:blue ratio, while important, is not solely sufficient to control plant morphology: light intensity is also critical. In tomato plants, the absolute blue light intensity rather than the blue percentage in the light recipe controlled hypocotyl length and stem extension (Nanya et al. 2012). While stem elongation was controlled by blue light, the position of the first flower truss developed in proportion to the total photosynthetic rate of the plant (more photosynthesis = earlier truss development). In principle, this would mean that a plant grown in 75% blue light of 100 μmol/m^2/s would have the same internode size as a plant grown in 38% blue light at 200 μmol/m^2/s, though it should be noted that plants grown at the higher light intensity would grow more quickly and flower earlier. Higher light intensities can reduce crop production time which, after an initial capital investment in lamps, has the potential to reduce production costs.

Other colors of light also influence plant morphologies. In chrysanthemums, blue light reduced leaf mass, green light reduced stem mass, and red and far-red light caused a reduction in root mass (Jeong et al. 2012). In contrast to roses and chrysanthemums, campanula height was unaffected by supplemental blue light and, in this case, the addition of red light provided the greatest effect on reducing plant height (Ouzounis et al. 2014). The addition of red light likely reduced plant height by changing the red:far-ratio, and reducing far-red light with spectral filters could have a similar influence on plant morphology. Differences between plant morphological responses to red/far-red and blue light are associated with differences in the relative contributions of phytochromes and blue-sensitive photoreceptors (cryptochromes and phototropins) to inhibition of stem extension. A better understanding of the light regulatory pathways between species will help further improve lighting strategies but may also highlight new areas for detailed scientific investigation.

The Use of Far-red Light in Control of Flowering

The effects of far-red light on plant morphology can be substantial; however, the greatest potential for far-red light in horticultural applications is in the control of flowering time. Runkle and Heins (2001) demonstrated that far-red light promotes flowering in several long-day ornamental

species and that an absence of far-red light can even prevent flowering. This work examined plant light responses using spectral filters. With the use of LED light treatments, plants can be grown using even more extreme ranges of red:far-red ratios, providing the potential for either delaying or advancing flowering still further than has been seen with spectral filters. In petunias and pansies grown under red:blue:far-red light mixtures, flowering was induced up to two weeks earlier in plants treated with far-red light compared to plants grown without far-red light (Davis et al. 2015).

Adams et al. (2012) investigated the use of several types of LED with different spectra for use in NB and DE lighting to promote and delay flowering in several long- and short-day species. Spectral quality of the lights was found to have a significant impact on their effectiveness in controlling flowering. Far-red-only and red + white + far-red lamps promoted flowering at levels similar to incandescent lamps, while red + white lamps were less effective than incandescent light. For example, in short days, chrysanthemum flowering was delayed by NB and DE illumination with red + white and red + white + far-red lamps. Far-red-only lamps had no effect on any of the short-day plants. None of the LED light combinations were found to be as effective as incandescent lamps at delaying flowering in Christmas cactus. Begonia and poinsettia flowering times were advanced in response to red+white + far-red light treatments. LED treatments affected morphology as well as flowering time and plants grew taller as the amount of far-red in the treatments increased. Craig and Runkle (2016) showed that NB lighting had its greatest effect on flowering in several long-day plants when an intermediate red:far-red ratio was applied. If red:far-red ratios were too high or too low, flowering promotion was reduced. The influence on plant stretching was also greatest under the light treatments that had the greatest influence on flowering. Chrysanthemums normally flower when days are shorter than 13.5 h, but Jeong et al. (2012) observed that a 4 h day DE blue light treatment prompted flowering during a 16 h day. These experiments demonstrate that LEDs can be used effectively to control plant flowering and also highlight the importance of the spectral composition of lights used for this application.

Secondary Metabolites: Pigmentation, Flavor, and Aroma

Primary metabolites are the chemicals that are directly involved in normal growth, development, and reproduction, and loss of these compounds results in death. Plants also produce many other compounds, known as secondary metabolites, that act to improve the fitness of an organism and help it acclimate to a changeable environment (Lambers et al. 1998). Many of these compounds convey qualities that are desirable by humans such as color, flavor, and aroma. The production of many secondary metabolites is regulated by light (Samuoliene et al. 2013).

Red, far-red, and blue light have all been implicated in driving synthesis of the pigments required for photosynthesis (Tripathy and Brown 1995; Miyashita et al. 1997; Tanaka et al. 1998; Huq et al. 2004; Kim et al. 2004a; Moon et al. 2006; Li et al. 2010). Blue and red light cause an increase in chlorophyll levels, whereas far-red results in reduced chlorophyll contents. As well as influencing the appearance of plants, these changes can also alter the rate of photosynthesis and therefore impact plant growth rates. The link between secondary metabolites and photosynthesis comprises an additional layer of complexity that should be considered when designing light recipes.

Many crops have red-colored leaves or flowers that are distinctive and desirable, and maximizing pigmentation is important to retain quality for customers. Red pigmentation is mainly provided by two types of compound: anthocyanins and betacyanins. Anthocyanin synthesis is regulated by many different biochemical pathways, but blue-light via the cryptochromes (Ninu et al. 1999) is an important signal for driving synthesis. In lettuce, supplying supplemental LED lighting of different colors against a background of fluorescent white light resulted in increases in leaf anthocyanin, xanthophyll, and β-carotene concentrations (Li and Kubota 2009). UV-A and blue light both increased the anthocyanin concentration, with blue light prompting the largest increase. In contrast, far-red light and green light reduced anthocyanin concentration, and far-red light also reduced chlorophyll, xanthophyll, and β-carotene content. UVB was also shown to be a potent stimulator of anthocyanin production in lettuce (Park et al. 2007), and UV transparent spectral filters increased plant and flower pigmentation (Paul et al. 2006). Carotenoid concentration was found to be greater in buckwheat seedlings grown under white light compared to those grown with 100% blue or red light (Tuan et al. 2013). It should be noted that few plants perform well under 100% red or blue light, and a combination of red and blue light may produce carotenoids in similar quantities as white light. Polyphenol concentrations in chrysanthemum were at their highest levels when grown with red or green supplemental lighting and at their lowest levels when grown with blue supplemental lighting. However, the plants grown under blue light flowered, which may have influenced the production of secondary metabolites.

Betacyanins have replaced anthocyanins as red pigments in the Caryophyllales Order (excluding the families Caryphyllaceae, which contains *Dianthus*, and

Molluginaceae; Sakuta 2014). The Caryophyllales contains 6% of all eudicotes (~11,155 species) and includes the Amaranthaceae, (the family that contains spinach, swiss chard, and beetroot). Unlike anthocyanins, betacyanins do not appear to increase in concentration in response to blue/UVB light. These pigments instead appear to accumulate in response to red light and their synthesis is thought to be controlled by the phytochromes (Elliott 1979). This means that, while the red pigmentation of many species could be improved by increasing the amount of blue light, it is probable that plants in the Caryophyllales Order will improve their red pigmentation on provision of more red light.

Large differences in pigment contents can alter the flavor of crops, but light is also important in regulating the biosynthesis of many of the volatile compounds that create the flavor and aroma of leaves, fruits, and flowers. UVB light exposure has been linked to increased oil and volatile contents in a range of herb species including sweet flag (*Acorus calamus* L.; Kumari et al. 2009), japanese mint (*Metha arvensis* L var. piperascens; Hikosaka et al. 2010), lemon balm, sage, lemon catmint (Manukyan 2013), *Cynbopogon citratus* (Kumari and Agrawal 2010), and basil (Bertoli et al. 2013). In basil plants, blue light was also found to increase the oil content of leaves in comparison to white light treatments (Amaki et al. 2011). In the same study, green and red light were found to have little effect on oil contents, although green light was shown to increase crop biomass production compared to other light treatments. While more blue light can increase oil and other secondary metabolite contents, it is not always sufficient to simply provide more blue light. In basil plants grown under 100% blue light, rosmaric acid (RA) levels were 3 mg/L but under 100% red or white light the RA concentration reached 6 mg/L (Shiga et al. 2009). A possible reason for the lower level of secondary metabolite production observed in this study was that the photosynthetic rate under blue light was lower than under red or white light. Data from Manukyan (2013) indicated that increasing PAR led to an increase in production of secondary metabolites. It is important to provide plants with sufficient light to drive enough photosynthesis as this provides the metabolic building blocks for the various biosynthetic pathways as well as stimulating the biosynthetic pathways to maximize production of desirable compounds.

More recently, the effect of postharvest light treatments on secondary metabolites has been considered. Postharvest light treatments provide the potential to enhance crop qualities during transport, prior to sale or to delay the onset of senescence, thus extending shelf life. Costa et al. (2013) found that exposure to 2 hours of low intensity red light (30-37 μmol/m^2/s) delayed senescence of basil leaves for 2 days during storage at 20°C in the dark. The authors concluded that the effects were due to changes in gene expression mediated by phytochromes, which mediate cell senescence in low light conditions, rather than via photosynthetic carbon gain. Colquhoun et al. (2013) showed that postharvest light treatments in petunia, tomatoes, blueberries, and strawberries could alter the volatile compounds produced by the different crops. Eight-hour red and far-red light treatments increased levels of several volatile compounds in petunia that are known to be important components of flower scent. Fewer compounds were examined in strawberries and tomatoes, but both large increases and decreases were observed in amounts of volatile compounds in these crops following exposure to different light treatments.

Changing the light spectrum can cause amounts of some compounds to increase while others may decrease. While it is apparent that light treatments increase the concentration of certain compounds, it is not always understood how these changes may impact crop flavor. Many of the studies focus on just one or two compounds, but flavor is influenced by a large range of compounds. Due to the limits of our understanding regarding the ways in which secondary metabolites are influenced by light and how these influence flavor, it is currently more efficient to develop light treatments for improved flavor by trial-and-error. The compounds of importance, and their synthesis in response to light, can subsequently be elucidated.

Insect Management Under LEDs

Light is a highly important environmental cue for all insect species. Several aspects of the light environment influence insects, such as daylength, intensity, direction, polarization, spectrum, and contrast. These environmental cues influence many insect biological and behavioral responses including the circadian rhythm, host identification, take-off and landing frequency, reproductive success, phototaxis (movement toward or away from light), and feeding frequency. Improving our understanding of insect light responses will be important to ensure both pollination and pest control can be maintained under LED light sources. There are two main aspects to consider: 1) the direct effect of light quality on insect responses, and 2) the effect of host species responses to light quality in the insect of interest.

Our knowledge of the spectral sensitivity of insect vision is limited to a few species, but the diversity between species is considerable. For example, bees are able to see UV (peak absorbance ~350 nm), blue (peak absorbance ~450 nm), and green (peak absorbance ~550 nm) light but have low sensitivity for red light (Backhaus 1993). Many insects are also able to detect the polarization of light and use this information to navigate (Rossel 1993; Reppert et al. 2004). The pest *Caliothips phaseoli*, a thrip

species that attacks soy, has one photoreceptor that can only detect UV (UVA and UVB) light, and these insects are therefore blind to PAR. However, spectral sensitivity in this species is enhanced: some of the ommatidia of their compound eyes contain pigments that fluoresce under UVA light, and this acts as a UVA filter so those eyes only detect UVB light. This allows the insects to distinguish UVB from UVA even though they only have one photoreceptor (Mazza et al. 2010). Western flower thrips (*Frankliniella occidentalis*) see both visible and UV light. Males and females have similar visual responses but have different swarming behaviors, with males more likely to gather on flowers than females (Matteson et al. 1992). Behavioral responses to light, both innate and learned, provide added complexity to insect light responses that will provide added challenges to understanding and manipulating insect light responses.

A better understanding of pest light responses can be used to improve traps for monitoring insect populations. Making traps more attractive to insects can render them more effective, which enables earlier identification of pest issues. Green LEDs have been used to increase trap effectiveness for West Indian sweet potato weevils (*Eusceoes postfasciatus*; Nakamoto and Kuba 2004), Whitefly (*Bemisia tabaci*), Greenhouse whitefly (*Trialeurodes vaporarioum*), Fungus gnats (*Bradysia coprophila*), and Aphids (*Aphis gossypii*; Chu et al. 2004). In environments with no natural light, trap effectiveness will be more strongly influenced by the color of the traps and the color of the LEDs than in glasshouses where natural light dominates.

Indirect effects of light quality on pests are caused by plant responses to light. In a species of wild tomato (*Lycopersicon hirsutum*), seasonal changes in day length and quantity cause large changes in synthesis of 2-tridecanone, resulting in much greater concentrations in June than in January (Kennedy et al. 1981). When caterpillars of *Manduca sexta* were fed on tissue from plants grown in January, 8% died compared to 87% that perished when fed on plants grown in June. More subtle effects are likely to influence pest performance on crops grown in different light conditions. Plants produce a range of volatile organic compounds (VOCs) that act as attractants to both pests and beneficial insects. Changes in light quantity (Paré and Tumlinson 1999) and quality (Kegge et al. 2013) alter the production of VOCs and spectral manipulation may help enhance VOC production to maximize crop protection.

Plant Pathogens and Their Interactions with Light

The interactions between plants and their pathogens are also influenced by the light environment. Light affects many aspects of plant biology and many of these responses influence plant resistance to disease. The red:far-red ratio in particular has been shown to influence the expression of many genes, via the phytochromes, that are involved in disease resistance (Griebel and Zeier 2008). Low red:far-red ratios decrease the production of many secondary metabolites involved in disease resistance and thus reduce resistance (Ballaré et al. 2012). Salicylic acid (SA) and jasmonic acid (JA) both play important roles in mediating defenses against pathogens and low red:far-red ratios have been shown to reduce the response of both pathways to disease attack (de Wit et al. 2013).

Light will also have direct effects on fungal pathogens as they also possess an array of photoreceptors that modulate their gene expression (Corrochano 2007). Fungi have circadian rhythms (Liu and Bell-Pedersen 2006) and certain species sporulate at specific times of day to coincide with events that enable them to infect plants, such as during times when leaves are likely to be wet. Rose powdery mildew (*Podosphaera pannosa*) was found to release spores during the day and more spores were released with brighter light (Suthaparan et al. 2010). Light color was also important: compared to white light, more spores were released under blue and far-red light, and fewer spores were released under red light. Both day extension and night break light treatments with red light greatly reduced the release of mildew conidia, and such treatments may be useful in reducing the intensity and spread of mildew in crops. As powdery mildews are obligate pathogens, it is not possible to determine if the effect of the light treatments occurs as a direct effect on the pathogen or as a result of the plant responses. However, red light treatments have also been found to increase occurrence of two diseases in broad bean: *Alternaria tenuissima* and *Botrytis cinerea* (Islam et al. 1998). Spore germination rates were also affected by light color, with blue light reducing germination by 16.5% compared to other treatments. The spores of many plant pathogens are killed by exposure to solar radiation (Kanetis et al. 2010), with the UVB component of solar radiation being the most likely to cause spore death. Models of spore germination could be used to define the best time of day to provide a pulse of UV light that would maximize effectiveness. It may also be possible to use novel light strategies to increase disease control. Blue light inhibits spore germination, so if red light only is provided early in the day, spores will germinate and they can then be more easily killed with a UV pulse before the blue light is again turned on. UVC light has also been trialed for the control of plant diseases. For these treatments to be effective, it is important to make sure that treatments are applied when the pathogens are vulnerable. If the UV light is provided before sporulation or after infection then it will be ineffective at providing protection. If applied during the germination of the spores, then UVC can be effective at preventing infection. Designing the light scheme to co-ordinate UVC application with spore release or germination may be an effective method for controlling disease in controlled environment chambers. There are, however, health and safety issues regarding the use of UVC light in commercial settings where staff can be exposed.

Just as different plant species have different light responses, the light responses of different pathogens vary, as do the interactions in different plant/pathogen systems in response to light. Schuerger and Brown (1997) observed that in tomatoes infected with bacterial wilt (*Pseudomonas solanacearum*) and cucumber plants infected with powdery mildew (*Sphaerotheca fuliginea*), disease symptoms were at their lowest in plants grown under 100% red light. By contrast, for tomato mosaic virus (ToMV) on pepper plants, disease symptoms were slower to develop and less severe in plants grown in the presence of blue/UVA light. These data indicate that spectral modification could be used as part of an integrated disease management system, with the caveat that care must be paid to the development and achievement of appropriate light treatments.

Conclusions

The introduction of LEDs for use in horticulture is facilitating the application of photobiology at all stages of crop production from propagation to postharvest quality control. The diversity of LED applications is expected to increase in the near future as our understanding of plants, pest, and pathogens increases but also as LED technologies improve in efficiency and capital costs decrease. This technology has the potential to improve food quality, reduce energy consumption, and increase food security. This use of LEDs in horticulture is a rapidly evolving field and is expected to gradually revolutionize commercial crop production.

References

Adams, S., S. Jackson, V. Valdes, J. Akehurst, A. Hambidge, D. Fuller, et al. 2012. Protected ornamental: Assessing the suitability of energy saving bulbs for day extension and night break lighting. *HDC Project PC 296 Final report.*

Amaki, W., N. Yamazaki, M. Ichimura, and H. Watanabe. 2011. Effects of light quality on the growth and essential oil content in sweet basil. Acta Hortic. 907:91–94.

Backhaus, W. 1993. Color vision and color choice behavior of the honey bee. Apidologie 24:309.

Ballaré, C. L., C. A. Mazza, A. T. Austin, and R. Pierik. 2012. Canopy light and plant health. Plant Physiol. 160:145–155.

Barta, D. J., T. W. Tibbitts, R. J. Bula, and R. C. Morrow. 1992. Evaluation of light emitting diode characteristics for a space-based plant irradiation source. Space 12:141–149.

Beggs, C., and E. Wellman. 1985. Analysis of light-controlled anthocyanin formation in coleoptiles of *Zea mays* L.: The role of UV-B, blue, red and far-red light. Photochem. Photobiol. 41:481–486.

Bertoli, A., M. Lucchesini, A. Menuali-Sodi, M. Leonardi, S. Doveri, A. Magnabosco, et al. 2013. Aroma characterization and UV elicitation of purple basil from different plant tissue cultures. Food Chem. 141:776–787.

Borthwick, H. A., S. B. Hendricks, M. W. Parker, E. H. Toole, and V. K. Toole. 1952. A reversible photoreaction controlling seed germination. PNAS 38:662–666.

Brazaitytė, A., P. Duchovskis, A. Urbonavičiūtė, G. Samuolienė, J. Jankauskienė, V. Kazėnas, et al. 2009. After-effect of light-emitting diodes lighting on tomato growth and yield in greenhouse. Sci. Works Lithuanian Institute Hortic. Lithuanian University Agric. 28:115–126.

Brazaitytė, A., P. Duchovskis, A. Urbonavičiūtė, G. Samuolienė, J. Jankauskienė, J. Sakalauskaite, et al. 2010. The effect of light-emitting diodes lighting on the growth of tomato transplants. Zemdirbyste Agric. 97:89–97.

Briggs, W. R., and J. M. Christie. 2002. Phototropins 1 & 2: versatile plant blue-light receptors. Trends Plant Sci. 7:204–210.

Brown, B. A., L. R. Headland, and G. I. Jenkins. 2009. UV-B Action Spectrum for UVR8-Mediated HY5 Transcript Accumulation in Arabidopsis. Photochem. Photobiol. 85:1147–1155.

Bula, R. J., R. C. Morrow, T. W. Tibbitts, D. J. Barta, R. W. Ignatius, and T. S. Martin. 1991. Light-emitting Diodes as a Radiation Source for Plants. HortScience 26:203–205.

de Carbonnel, M., P. A. Davis, M. Rob, G. Roelfsema, S. Inoue, I. Schepens, et al. 2010. The *Arabidopsis* PHYTOCHROME KINASE SUBSTRATE 2 protein is a phototropin signalling element that regulates leaf flattening and leaf positioning. Plant Physiol. 152:1391–1405.

Cashmore, A. R. 2003. Cryptochromes: enabling plants and animals to determine circadian time. Cell 114:537–543.

Chalker-Scott, L. 1999. Environmental significance of anthocyanins in plant stress responses. Photochem. Photobiol. 70:1–9.

Chen, M., and J. Chory. 2011. Phytochrome signaling mechanisms and the control of plant development. Trends Cell Biol. 21:664–671.

Chu, C.-C., A. M. Simmons, T.-Y. Chen, A. P. Alexander, and T. J. Henneberry. 2004. Lime green light-emitting diode equipped yellow sticky card traps for monitoring whiteflies, aphids and fungus gnats in greenhouses. Entomologia Sinica 11:125–133.

Colquhoun, T. A., M. L. Schwieterman, J. L. Gilbert, E. A. Jaworski, K. M. Langer, C. R. Jones, et al. 2013. Light

modulation of volatile organic compounds from petunia flowers and select fruits. Postharvest Biol. Technol. 86:37–44.

Corrochano, L. M. 2007. Fungal photoreceptors: sensory molecules for fungal development ad behavior. Photochem. Photobiol. Sci. 6:725–736.

Costa, L., Y. M. Montano, C. Carrióna, N. Rolnya, and J. J. Guiamet. 2013. Application of low intensity light pulses to delay postharvest senescence of *Ocimum basilicum* leaves. Postharvest Biol. Technol. 86:181–191.

Craig, D. S., and E. S. Runkle. 2016. An intermediate phytochrome photoequilibria from night-interruption lighting optimally promotes flowering of several long-day plants. Environ. Exp. Bot. 121:132–138.

Davis, P. A. (2013) Securing skills and expertise in crop light responses for UK protected horticulture, with specific reference to exploitation of LED technology. AHBD Project CP085 Annual Report 2013.

Davis, P. A., R. Beynon-Davies, G. M. McPherson, J. Banfield-Zanin, D. George, C. O. Ottosen, et al. 2015. Understanding crop and pest responses to LED lighting to maximise horticultural crop quality and reduce the use of PGRs. AHDB Project CP125 Year One report

Demotes-Mainard, S., T. Péron, A. Corot, J. Bertheloot, J. Le Gourrierec, S. Pelleschi-Travier, et al. 2016. Plant responses to red and far-red lights, applications in horticulture. Environ. Exp. Bot. 121:4–21.

De Visser, P. H. B., G. H. Buck-Sorlin, and G. W. A. M. van der Heijden. 2014. Optimizing illumination in the greenhouse using a 3D model of tomato and a ray tracer. Front. Plant Sci. 5:48. doi:10.3389/fpls.2014.00048.

Dougher, T. A. O., and B. Bugbee. 2001. Differences in the response of Wheat, Soybean, and Lettuce to reduced blue radiation. Photochem. Photobiol. 73:199–207.

Dougher, T. A. O., and B. Bugbee. 2004. Long-term blue light effects on the histology of lettuce and soybean leaves and stems. J. Am. Soc. Hortic. Sci. 129:467–472.

Dueck, T. A., J. Janse, B. A. Eveleens, F. L. K. Kempkes, and L. F. M. Marcelis. 2012. Growth of tomatoes under hybrid LED and HPS lighting. Acta Hortic. 952:335–342.

Elliott, D. C. 1979. Temperature-sensitive responses of red light-dependent Betacyanin Synthesis. Plant Physiol. 64:521–524.

Folta, K. M., and K. S. Childers. 2008. Light as a growth regulator: controlling plant biology with narrow-bandwidth solid-state lighting systems. HortScience 43:1957–1964.

Folta, K. M., and E. P. Spalding. 2001. Unexpected roles for cryptochrome 2 and phototropin revealed by high-resolution analysis of blue light-mediated hypocotyl growth inhibition. Plant J. 26:471–478.

Gardner, G., C. Lin, E. M. Tobin, H. Loehrer, and D. Brinkman. 2009. Photobiological properties of the inhibition of etiolated Arabidopsis seedling growth by ultraviolet-B irradiation. Plant, Cell Environ. 32:1573–1583.

Giliberto, L., G. Perrotta, P. Pallara, J. L. Weller, P. D. Fraser, P. M. Bramley, et al. 2005. Manipulation of the blue light photoreceptor cryptochrome 2 in tomato affects vegetative development, flowering time, and fruit antioxidant content. Plant Physiol. 137:199–208.

Gitz, D. C., L. Liu-Gitz, S. J. Britz, and J. H. Sullivan. 2005. Ultraviolet-B effects on stomatal density, water-use efficiency, and stable carbon isotope discrimination in four glasshouse-grown soybean (*Glycine max*) cultivars. Environ. Exp. Bot. 53:343–355.

Goins, G. D., N. C. Yorio, M. M. Sanwo, and C. S. Brown. 1997. Photomorphogenesis, photosynthesis, and seed yield of wheat plants grown under red light-emitting diodes (LEDs) with and without supplemental blue lighting. J. Exp. Bot. 48:1470–1413.

Goins, G. D., N. C. Yorio, M. M. Sanwo-Lewandowski, and C. S. Brown. 1998. Life Cycle experiments with Arabidopsis grown under red light-emitting diodes (LEDs). Life Support Biosph. Sci. 52:143–149.

Gómez, C., and C. A. Mitchell. 2013. Supplemental lighting for greenhouse-grown tomatoes: intracanopy LED towers vs. overhead HPS lamps. Acta Hortic. 1037:855–862.

Gómez, C., M. C. Morrow, C. M. Bourget, G. D. Massa, and C. A. Mitchell. 2013. Comparison of intracanopy light-emitting diode towers and overhead high-pressure sodium lamps for supplemental lighting of greenhouse-grown tomatoes. Horttechnology 23:93–98.

Griebel, T., and J. Zeier. 2008. Light regulation and daytime dependency of inducible plant defenses in *Arabidopsis*: phytochrome signaling controls systemic acquired resistance rather than local defense. Plant Physiol. 147:790–801.

Guo, X., X. Hao, S. Khosa, K. G. S. Kumar, R. Cao, and N. Bennett. 2016a. Effect of LED interlighting combined with overhead HPS light on fruit yield and quality of year-round sweet pepper in commercial greenhouse. Acta Hortic. 1134:71–78.

Guo, X., X. Hao, J. M. Zheng, C. Little, and S. Khosa. 2016b. Response of greenhouse mini-cucumber to different vertical spectra of LED lighting under overhead high pressure sodium and plasma lighting. Acta Hortic. 1134:87–94.

Hao, X., J. M. Zheng, C. Little, and S. Khosl. 2012. LED inter-lighting in year-round greenhouse mini-cucumber production. Acta Hortic. 956:335–340.

Hikosaka, S., K. Ito, and E. Goto. 2010. Effects of Ultraviolet Light on Growth, Essential Oil Concentration, and Total Antioxidant Capacity of Japanese Mint. Environ. Control. Biol. 48:185–190.

Hoad, S. P., and R. R. B. Leakey. 1996. Effects of pre-severance light quality on the vegetative propagation of *Eucalyptus grandis* W. Hill ex maiden. Trees 10:317–324.

Hogewoning, S. W., G. Trouwborst, H. Poorter, W. van Ieperen, and J. Harbinson. 2010. Blue light dose–responses of leaf photosynthesis, morphology, and chemical composition of *Cucumis sativus* grown under different combinations of red and blue light. J. Exp. Bot. 61:3107–3117.

Hogewoning, S. W., E. Wientjes, P. Douwstra, G. Trouwborst, W. van Ieperen, R. Croce, et al. 2012. Photosynthetic quantum yield dynamics: from photosystems to leaves. Plant Cell 24:1921–1935.

Huché-Thélier, L., L. Crespel, J. Le Gourrierec, P. Morel, S. Sakr, and N. Leduc. 2016. Light signaling and plant responses to blue and UV radiations—Perspectives for applications in horticulture. Environ. Exp. Bot. 121:22–38.

Huq, E., B. Al-Sady, M. E. Hudson, C. Kim, K. Apel, and P. H. Quail. 2004. PHYTOCHROME-INTERACTING FACTOR 1 is a critical bHLH regulator of chlorophyll biosynthesis. Science 305:1937–1941.

van Ieperen, W., A. Savvides, and D. Fanourakis. 2012. Red and blue light effects during growth on hydraulic and stomatal conductance in leaves of young cucumber plants. Acta Hortic. 956:223–230.

Islam, S. Z., Y. Honda, and S. Arase. 1998. Light-induced resistance of broad bean against *Botrytis cinerea*. J. Phytopathol. 146:479–485.

Islam, M. A., G. Kuwar, J. L. Jihong, R. D. Blystad, H. R. Gislerød, J. E. Olsen, et al. 2012. Artificial light from light emitting diodes (LEDs) with a high portion of blue light results in shorter poinsettias compared to high pressure sodium (HPS) lamp. Sci. Hortic. 147:136–143.

Jankowska-Blaszczuk, M., and M. I. Daws. 2007. Impact of red: far red ratios on germination of temperate forest herbs in relation to shade tolerance, seed mass and persistence in the soil. Funct. Ecol. 21:1055–1062.

Jao, R. C., and W. Fang. 2004. Effects of frequency and duty ratio on the growth of potato plantlets in vitro using light emitting diodes. HortScience 39:375–379.

Jeong, SW, S Park, JS Jin, ON Seo, G-S Kim, H Bae, et al. 2012. Influences of four different light-emitting diode lights on flowering and polyphenol variations in the leaves of Chrysanthemum (*Chrysanthemum morifolium*). J. Agric. Food Chem. 60:9793–9800.

Kanetis, L., G. J. Holmes, and P. S. Ojiambo. 2010. Survival of *Pseudoperonospora cubensis* sporangia exposed to solar radiation. Plant. Pathol. 59:313–323.

Kegge, W., B. T. Weldegergis, R. Solerm, M. Vergeer-Van Eijk, M. Dicke, L. A. Voesenek, et al. 2013. Canopy light cues affect emission of constitutive and methyl jasmonate induce volatile organic compounds in *Arabidopsis thaliana*. New Phytol. 200:861–874.

Kennedy, G. G., R. T. Yamamoto, M. B. Dimock, W. G. Williams, and J. Bordner. 1981. Effect of day length and light intensity on 2-tridecanone levels and resistance in *Lycopersicon hirsutm* f. *glabratum* to *Maduca sexta*. J. Chem. Ecol. 7:707–716.

Kim, H. H., G. D. Goins, R. M. Wheeler, and J. C. Sager. 2004a. Green light supplementation for enhanced lettuce growth under red and blue light emitting diodes. HortScience 39:1617–1622.

Kim, H. H., G. D. Goins, R. M. Wheeler, and J. C. Sager. 2004b. Stomatal conductance of lettuce grown under or exposed to different light qualities. Ann. Bot. 94:691–697.

Kirschbaum, M. U. F., and R. W. Pearcy. 1988. Gas Exchange Analysis of the Relative Importance of Stomatal and Biochemical Factors in Photosynthetic Induction in *Alocasia macrorrhiza*. Plant Physiol. 86:782–785.

Kozai, T. 2013. Sustainable plant factory: closed plant production systems with artificial light for high resource use efficiencies and quality produce. Acta Hortic. 1004:27–40.

Kubota, C., P. Chia, Z. Yang, and Q. Li. 2012. Applications of far-red light emitting diodes in plant production under controlled environments. Acta Hortic. 952:59–66.

Kumari, R., and S. B. Agrawal. 2010. Supplemental UV-B induced changes in leaf morphology, physiology and secondary metabolites of an Indian aromatic plant Cymbopogon citratus (D.C.) Staph under natural field conditions. Int. J. Environ. Stud. 67:655–675.

Kumari, R., S. B. Agrawal, S. Singh, and N. K. Dubey. 2009. Supplemental ultraviolet-B induced changes in essential oil composition and total phenolics of *Acorus calamus* L. (sweet flag). Ecotoxicol. Environ. Saf. 72:2013–2019.

Lambers, H., F. S. III Chaplin, and T. L. Pons. 1998. Plant physiological ecology. Pp. 413–436. Springer, New York.

Lau, O. S., and X. W. Deng. 2010. Plant hormone signalling lightens up: integrators of light and hormones. Curr. Opin. Plant Biol. 13:571–577.

Lee, S.-H., R. K. Tewari, E.-J. Hahn, and K.-Y. Paek. 2007. Photon flux density and light quality induces changes in growth, stomatal development, photosynthesis and transpiration of *Withania somnifera* (L.) Dunal. plantlets. Plant Cell, Tissue Organ Cult. 90:141–15.

Li, Q., and C. Kubota. 2009. Effects of supplemental light quality on growth and phytochemicals of baby leaf lettuce. Environ. Exp. Bot. 67:59–64.

Li, H., Z. Xu, and C. Tang. 2010. Effect of light-emitting diodes on growth and morphogenesis of upland cotton (Gossypium hirsutum L.) plantlets in vitro. Plant Cell, Tissue Organ Cult. 103:155–163.

Li, H., C. Tang, and Z. Xu. 2013. The effects of different light qualities on rapeseed (*Brassica napus* L.) plantlet growth and morphogenesis in vitro. Sci. Hortic. 150:117–124.

Li, K., Q.-C. Yang, Y.-X. Tong, and R. Cheng. 2014. Using movable light-emitting diodes for electrical saving in a plant factory growing lettuce. Horttechnology 24:546–553.

Liu, Y., and D. Bell-Pedersen. 2006. Circadian Rhythms in Neurospora crassa and Other Filamentous Fungi. Eukaryot. Cell 5:1184–1193.

Lu, N., T. Maruo, M. Johkan, M. Hohjo, S. Tsukagoshi, Y. Ito, et al. 2012. Effects of supplemental lighting with light-emitting diodes (LEDs) on tomato yield and quality of single-truss tomato plants grown at high planting density. Environ. Control. Biol. 50:63–74.

Manukyan, A. 2013. Effects of PAR and UV-B Radiation on Herbal Yield, Bioactive Compounds and Their Antioxidant Capacity of Some Medicinal Plants Under Controlled Environmental Conditions. Photochem. Photobiol. 89:406–414.

Massa, G. D., H.-H. Kim, R. M. Wheeler, and C. A. Mitchell. 2008. Plant Productivity in Response to LED Lighting. HortScience 43:1951–1955.

Matsuda, R., K. Ohashi-Kaneko, K. Fujiwara, E. Goto, and K. Kurata. 2004. Photosynthetic characteristics of rice leaves grown under red light with or without supplemental blue light. Plant Cell Physiol. 45:1870–1874.

Matteson, N., I. Terry, A. Ascoli-Christensen, and C. Gilbert. 1992. Spectral efficiency of the western flower thrips, *Franklinella occidentalis*. J. Insect Physiol. 38:453–459.

Mazza, C. A., M. M. Izaguirre, J. Curiale, and C. L. Ballaré. 2010. A look into the invisible: ultraviolet-B sensitivity in an insect (*Caliothrips phaseoli*) revealed through a behavioural action spectrum. Proc. R. Soc. B 277:367–373.

McCree, KJ. 1971. The action spectrum, absorptance and quantum yield of photosynthesis in crop plants. Agric. Meteorol. 9:191–216.

Ménard, C, M Dorais, T Hovi, and A Gosselin. 2005. Developmental and physiological responses of tomato and cucumber to additional blue light. In V International Symposium on Artificial Lighting in Horticulture 711: 291–296.

Middleton, E. M., and A. H. Teramura. 1993. The role of flavonol glycosides and carotenoids in protecting soybean from ultraviolet-B damage. Plant Physiol. 103:741–752.

Miyashita, Y. T., Y. Kinura, C. Kitaya, C. Kubota, and T. Kozai. 1997. Effects of red light on the growth and morphology of potato plantlets in vitro using light emitting diodes (LEDs) as a light source for micropropagation. Acta Hortic. 418:169–173.

Moon, H. K., S.-Y. Park, Y. W. Kim, and C. S. Kim. 2006. Growth of Tsuru-rindo (*Tripterospermum japonicum*) cultured in vitro under various source of light-emitting diode (LED) irradiation. J. Plant Biol. 49:174–179.

Nakamoto, Y., and H. Kuba. 2004. The effectiveness of a green light emitting diode (LED) trap at capturing the West Indian sweet potato weevil, *Euscepes postfasciatus* (Fairmaire) (Coleoptera: Curculionidae) in a sweet potato field. Appl. Entomol. Zool. 39:491–495.

Nanya, K., Y. Ishigami, S. Hikosaka, and E. Goto. 2012. Effects of blue and red light on stem elongation and flowering of tomato seedlings. Acta Hortic. 956:264–266.

Nhut, D. T., L. T. A. Hong, H. Watanabe, M. Goi, and M. Tanaka. 2002. Growth of banana plantlets cultured in vitro under red and blue light-emitting diode (LED) irradiation source. Acta Hortic. 575:117–124.

Nhut, D. T., T. Takamura, H. Watanabe, K. Okamoto, and M. Tanaka. 2003. Responses of strawberry plantlets cultured in vitro under superbright red and blue light-emitting diodes (LEDs). Plant Cell, Tissue Organ Cult. 73:43–52.

Ninu, L., M. Ahmad, C. Miarelli, A. R. Cashmore, and G. Giuliano. 1999. Cryptochrome 1 controls tomato development in response to blue light. Plant J. 18:551–556.

Ouzounis, T., X. Fretté, E. Rosenqvist, and C. O. Ottosen. 2014. Spectral effects of supplementary lighting on the secondary metabolites in roses, chrysanthemums, and campanulas. J. Plant Physiol. 171:1491–1499.

Paradiso, R., E. Meinen, J. F. H. Snel, P. De Visser, W. Van Leperen, S. W. Hogewoning, et al. 2011. Spectral dependence of photosynthesis and light absorptance in single leaves and canopy in rose. Sci. Hortic. 127:548–554.

Paré, P. W., and J. H. Tumlinson. 1999. Plant volatiles as a defence against insect and herbivores. Plant Physiol. 121:325–331.

Park, J.-S., M.-G. Choung, J.-B. Kim, et al. 2007. Gene up-regulated during red colouration in UV-B irradiated lettuce leaves. Plant Cell Rep. 26:507–516.

Paul, N. D., J. M. Moore, and M. Huey. 2006. The potential benefits of three modified plastic crop covers in Hardy Ornamental Nursery Stock production: initial investigations on a grower holding (Garden Centre Plants, Preston). HDC Project PC19a Final report.

Pearson, S., P. A. Davis, and H. Kitchener 2015. Commercial review of lighting systems for UK Horticulture. AHDB Horticulture CP 139 Final Report.

Poudel, P. R., I. Kataoka, and R. Mochioka. 2008. Effect of red- and blue-light-emitting diode on growth and morphogenesis of grapes. Plant Cell, Tissue Organ Cult. 92:147–153.

Reppert, S. M., H. Zhu, and R. H. White. 2004. Polarized light helps monarch butterflies navigate. Curr. Biol. 14:155–158.

Rossel, S. 1993. Navigation by bees using polarized skylight. Comp. Biochem. Physiol. A Mol. Integr. Physiol. 104:695–708.

Runkle, E. S., and R. D. Heins. 2001. Specific functions of red, far red, and blue light in flowering and stem extension of long-day plants. J. Am. Soc. Hortic. 126:275–282.

Sakuta, M. 2014. Diversity in plant red pigments: anthocyanins and betacyanins. Plant Biotechnol. Rep. 8:37–48.

Samuolienė G, Brazaitytė A, Urbonavičiūtė A, Šabajevienė G and Duchovskis P (2010) The effect of red and blue light component on the growth and development of frigo strawberries, Zemdirbyste-Agriculture 97: 99–104.

Samuoliene, G., A. Brazaitytė, R. Sirtautas, A. Visile, J. Sakalauskaite, S. Sakalauskaite, et al. 2013. LED illumination affects bioactive compounds in romaine baby leaf lettuce. J. Sci. Food Agric. 93:3286–3291.

Savvides, A., D. Fanourakis, and W. van Ieperen. 2012. Co-ordination of hydraulic and stomatal conductance across light qualities in cucumber leaves. J. Exp. Bot. 63:1135–1143.

Schuerger, A. C., and C. S. Brown. 1997. Spectral quality affects disease development of three pathogens on hydroponically grown plants. HortScience 32:96–100.

Sellaro, P., M. Crepy, S. A. Trupkin, E. Karayekov, A. S. Buchovsky, C. Constanza Rossi, et al. 2010. Cryptochrome as a sensor of the blue/green ratio of natural radiation in Arabidopsis. Plant Physiol. 154:401–409.

Shiga, T., K. Shoji, H. Shimada, S. Hashida, F. Goto, and T. Yoshihara. 2009. Effect of light quality on rosmarinic acid content and antioxidant activity of sweet basil, Ocimum basilicum L. Plant Biotechnol. 26:255–259.

Shimada, A., and Y. Taniguchi. 2011. Red and blue pulse timing for pulse width modulation light dimming of light emitting diodes for plant cultivation. J. Photochem. Photobiol., B 104:399–404.

Shinomura, T., A. Nagatani, H. Hanzawa, M. Kubota, M. Watanabe, and M. Furuya. 1996. Action spectra for phytochrome A- and B-specific photoinduction of seed germination in Arabidopsis thaliana. PNAS 93:8129–8133.

Sommer, A. P., and R.-P. Franke. 2006. Plants grow better if seeds see green. Naturwissenschaften 93:334–337.

Son, K.-H., and M.-M. Oh. 2013. Leaf shape, growth, and antioxidant phenolic compounds of two lettuce cultivars grown under various combinations of blue and red light-emitting diodes. HortScience 48:988–995.

Stutte, G. W., S. Edney, and T. Skerritt. 2009. Photoregulation of bioprotectant content of red leaf lettuce with light-emitting diodes. HortScience 44:79–82.

Sun, J., J. N. Nishio, and T. C. Vogelmann. 1998. Green Light Drives CO2 Fixation Deep within Leaves. Plant Cell Physiol. 39:1020–1026.

Suthaparan, A., S. Torre, A. Stensvand, M. L. Herrero, R. I. Pettersen, D. M. Gadoury, et al. 2010. Specific Light-emitting diodes can suppress sporulation of Podosphaera pannosa on greenhouse roses. Plant Dis. 94:1105–1110.

Tanaka, M., T. Takamura, H. Watanabe, M. Endo, T. Yanagi, and K. Okamoto. 1998. In vitro growth of Cymbidium plantlets cultured under super bright and blue light emitting diodes (LEDs). J. Hortic. Sci. Biotechnol. 73:39–44.

Tennessen, D. J., E. L. Singsaas, and T. D. Sharkey. 1994. Light-emitting diodes as a light source for photosynthesis research. Photosynth. Res. 39:85–92.

Tennessen, D. J., R. J. Bula, and T. D. Sharkey. 1995. Efficiency of photosynthesis in continuous and pulsed light emitting diode irradiation. Photosynth. Res. 44:261–269.

Terashima, I., T. Fujita, T. Inoue, W. S. Chow, and R. Oguchi. 2009. Green light drives leaf photosynthesis more efficiently than red light in strong white light: Revisiting the enigmatic question of why leaves are green *Plant Cell*. Physiology 50:684–697.

Tevini, M., and U. Iwanzik Wm Thoma. 1981. Some effects of enhanced UV-B irradiation on the growth and composition of plants. Planta 153:388–394.

Tripathy, B. C., and C. S. Brown. 1995. Root-shoot interaction in the greening of wheat seedlings grown under red light. Plant Physiol. 107:407–411.

Trouwborst, S. W., G. Trouwborst, H. Maljaars Hm Poorter, W. van Ieperen, and J. Harbinson. 2010. Blue light dose-responses of leaf photosynthesis, morphology and chemical composition of *Cucumis sativus* grown under difference combination of red and blue light. J. Exp. Bot. 121:75–82.

Trouwborst, G., S. W. Hogewoning, O. van Kooten, and J. Harbinson. 2016. Plasticity of photosynthesis after the 'red light syndrome' in cucumber. Environ. Exp. Bot. 61:3107–3117.

Tuan, P. A., A. A. Thwe, Y. B. Kim, J. K. Kim, S.-J. Kim, S. Lee, et al. 2013. Effects of white, blue, and red light-emitting diodes on carotenoid biosynthetic gene expression levels and carotenoid accumulation in sprouts of Tartary Buckwheat (*Fagopyrum tataricum* Gaertn.). Journal of Agricultural and Food Chemistry 61:12356–12361.

Vlohr, H., and H. Drumm-Herrel. 1983. Coaction between phytochrome and blue/UV light anthocyanin synthesis in seedlings. Physiol. Plant. 58:408–414.

Wang, Y., and K. M. Folta. 2013. Contributions of green light to plant growth and development. Am. J. Bot. 100:70–78.

Wang, G., M. Gu, J. Cui, K. Shi, Y. Zhou, and J. Yu. 2009. Effects of light quality on CO_2 assimilation, chlorophyll-fluorescence quenching, expression of Calvin cycle genes and carbohydrate accumulation in *Cucumis sativus*. J. Photochem. Photobiol., B 96:30–37.

Wargent, J. J., J. P. Moore, A. R. Ennos, and N. D. Paul. 2009. Ultraviolet radiation as a limiting factor in leaf expansion and development. Photochem. Photobiol. 85:279–286.

de Wit, M., S. H. Spoel, G. F. Sanchez-Perez, C. M.M. Gommers, C. M. J. Pieterse, L. A. C. J. Voesenek, et al. 2013. Perception of low red:far-red ratio compromises both salicylic acid- and jasmonic acid-dependent pathogen defences in Arabidopsis. Plant J. 75:90–103.

Wu, H. C. 2006. Improving in vitro propagation of *Protea cynaro*ides L. (King Protea) and the roles of starch and phenolic compounds in the rooting of cuttings. PhD thesis. University of Pretoria, Pretoria.

Wu, H.-C., and C.-C. Lin. 2012. Red light-emitting diode light irradiation improves root and leaf formation in difficult-to-propagate *Protea cynaroides* L. plantlets in vitro. HortScience 47:1490–1494.

Wu, H. C., E. S. du Toit, C. F. Reinhardt, A. M. Rimando, F. van der Kooy, and J. J. M. Meyer. 2007. The phenolic, 3,4-dihydroxybenzoic acid, is an endogenous regulator of rooting in *Protea cynaroides*. Plant Growth Regul. 52:207–215.

Yorio, N. C., G. D. Goins, H. R. Kagie, R. M. Wheeler, and J. C. Sager. 2001. Improving spinach, radish, and lettuce growth under red light emitting diodes (LEDs) with blue light supplementation. HortScience 36:380–383.

Yoshida, H., S. Hikosaka, E. Goto, H. Takasuna, and T. Kudou. 2012. Effects of light quality and light period on flowering of everbearing strawberry in a closed plant production system. Acta Hortic. 956:107–112.

16

Diversification and use of bioenergy to maintain future grasslands

Iain S. Donnison & Mariecia D. Fraser

Institute of Biological, Environmental & Rural Sciences, Aberystwyth University, Gogerddan Campus, Aberystwyth SY23 3EE, UK

Keywords
Bioenergy, biorefining, conservation, grasslands, Miscanthus, pasture

Correspondence
Iain S. Donnison, Institute of Biological, Environmental & Rural Sciences, Aberystwyth University, Gogerddan Campus, Aberystwyth, SY23 3EE, UK.
E-mail: isd@aber.ac.uk

Funding Information
The Biotechnology and Biological Sciences Research Council (BBSRC) for long-term strategic funding of the authors and grassland research at the Institute of Biological, Environmental & Rural Sciences, Aberystwyth University. The Engineering and Physical Sciences Research Council (EPSRC), Natural Environment Research Council, Department for Environment, Food and Rural Affairs (Defra), Energy Technologies Institute and Welsh Government (Low Carbon, Energy & Environment National Research Network) and European Commission for additional research funding.

Abstract

Grassland agriculture is experiencing a number of threats including declining profitability and loss of area to other land uses including expansion of the built environment as well as from cropland and forestry. The use of grassland as a natural resource either in terms of existing vegetation and land cover or planting of new species for bioenergy and other nonfood applications presents an opportunity, and potential solution, to maintain the broader ecosystem services that perennial grasslands provide as well as to improve the options for grassland farmers and their communities. This paper brings together different grass or grassland-based studies and considers them as part of a continuum of strategies that, when also combined with improvements in grassland production systems, will improve the overall efficiency of grasslands as an important natural resource and enable a greater area to be managed, replanted or conserved. These diversification options relate to those most likely to be available to farmers and land owners in the marginally economic or uneconomic grasslands of middle to northern Europe and specifically in the UK. Grasslands represent the predominant global land use and so these strategies are likely to be relevant to other areas although the grass species used may vary. The options covered include the use of biomass derived from the management of grasses in the urban and semi urban environment, semi-natural grassland systems as part of ecosystem management, pasture in addition to livestock production, and the planting and cropping of dedicated energy grasses. The adoption of such approaches would not only increase income from economically marginal grasslands, but would also mitigate greenhouse gas emissions from livestock production and help fund conservation of these valuable grassland ecosystems and landscapes, which is increasingly becoming a challenge.

Introduction

Humankind has been changing the planet and especially its vegetation in a significant way for millennia. This includes the domestication of crops and animals, which enabled early civilization to evolve. Many of these crops and animal species, including, for example, cereals, forage grasses, pigs, and cattle, were native to grasslands and make up modern intensive agriculture. More extensive grasslands, made up of pasture and rangeland, underwent more indirect selection pressure and still represent the dominant global land use type. Since the industrial revolution, humankind has increasingly moved from living in rural-based societies and economies to urban-based ones. Our relationship to the countryside has changed and in many parts of the world concerns have moved from productivity to environmental stewardship. This has presented grassland farmers and their communities with significant challenges as incomes fall and rural infrastructure declines. Grassland agriculture is also experiencing the biggest threat to date in terms of loss of land area to other uses, including expansion of the built environment as well as from cropland, forestry, and energy (solar, wind, biofuels). The use of grasslands as a natural resource either in terms of existing vegetation and land cover or planting of new species for bioenergy present an opportunity, and potential solution, to maintaining the broader ecosystem services that perennial grasslands provide as well as improving the options for grassland farmers and their communities. A number of older studies exist on the processing and fractionation of biomass for feed, food, energy, and diversification of grassland products. However, in this paper, the focus is on the more recent literature, and in particular over the last decade, on the use of grasses and grasslands for energy and other nonfood applications by the authors and others. The paper brings together different grass or grassland-based studies and considers them as part of a continuum of strategies that when also combined with improvements in grassland production systems (Gerssen-Gondelach et al. 2015) will improve the overall efficiency of grasslands as an important natural resource and enable a greater area to be managed, replanted, or conserved. All these grassland systems are capable of delivering conservation, grazing and/or energy but each system has a different set of features and properties which make them more or less suited to certain locations, farming systems, and end uses (Table 1). The combination of several of these strategies enables farmers and regions to provide energy feedstocks from more marginal land, and even has the potential through remediation to increase food production. This paper therefore explores a number of approaches which seek to maximize individually or collectively the use of grassland as a natural resource in addition to its traditional uses and other ecosystem services. In other words, bioenergy has the potential to be a land management tool for grasslands, helping to protect habitats, livelihoods, as well as contributing to national and global renewable energy and greenhouse gas emission targets.

Use of Natural and Semi-natural Grasslands

Natural and semi-natural grasslands systems, sometimes referred to as rough grazings or rangeland are important habitats providing a resource for extensive livestock production and a wide number of ecosystem services, including carbon sequestration, biodiversity, water management, landscape, recreation, and leisure. However, reductions in livestock numbers across Europe in response to changes in support mechanisms and other economic pressures have led to abandonment of large areas of poorer quality native plant communities as farmers focus their activities on improved pastures and crops. Left unmanaged, this vegetation becomes characterized by a build-up of mature and senescent plant material with a high fiber content and relatively low nutrient value. Many of these communities are dominated by plant species rejected by stock (e.g., *Juncus* spp., *Molinia caerulea*, *Deschampsia cespitosa*, *Juniperus* spp., *Genista* spp.) or have become invaded by undesirable or alien species (e.g., *Pteridium*, *Ulex* spp., *Calamagrostis epigejos*, *Solidago* spp., *Lupinus polyphyllus*). However, many of these grasslands by their nature are important habitats and often form part of the Natura 2000 network of protected areas consisting of sites designated by EU Members under the Habitats and Birds Directives (Melts et al. 2014). Increasingly these areas are

Table 1. Categorization of grassland into types based on species content, location or function; the grassland major and minor outputs; and the challenges; and potential solutions for each type.

Grassland type	Major outputs	Minor outputs	Challenges	Solutions
Natural/semi-natural	Biodiversity and other ecosystem services, landscape	Energy, livestock forage	Abandonment, cost of environmental management	Develop diversification technology options
Urban	Landscape, recreation	Energy, ecosystem function	Cost of establishment and management, pollution and loss through building	Develop diversification technology options
Extensive pasture	Livestock forage, biodiversity and other ecosystem services	Landscape, energy	Environmental impact of ruminant production	New grass varieties and legume varieties, selected breeding of livestock
Intensive pasture	Livestock forage	Landscape, ecosystem function, energy	Environmental impacts of agronomic inputs and ruminant production	New grass and legume varieties, dietary supplements
Energy	Energy and other nonfood uses	Ecosystem function	Cost of establishment and lack of knowledge of the crop	New establishment methods (seed rather than vegetative propagation), agronomy support

managed not by grazing but by cutting excess biomass late in the season to maintain species diversity and avoid dominance by a small number of species. In the US bioenergy production from feedstocks grown on marginal or underutilized land, such as those enrolled in the Conservation Reserve Program, has also been shown to provide greenhouse gas benefits and supplement government subsidies (Gelfand et al. 2011; Jungers et al. 2013). This movement is also increasing more globally, with trends in urbanization and rationalization, away from the more isolated rural lifestyles and living. Habitat management is also often important to protect birds or animals, which may require areas of grassland with a short sward height as feeding sites. Conservation organizations which own or manage land are therefore often active in making interventions such as cutting in these habitats that are no longer managed by agriculture. A further challenge is that even when cut as hay or ensiled, this material is of such low nutritional value to animals that it remains unused and is dumped. Therefore, there is an opportunity for bioenergy production from a genuine waste product which if it is not used presents a management and potential environmental risk. In turn it helps avoid the controversy that there is not enough land to grow energy with an increasing world population, and expectations of that population (Ghose 2014).

The value of low-input high-diversity mixtures of native grassland perennials for biofuels has also been championed on the basis of greater greenhouse gas reductions and less agrichemical pollution per hectare than can be achieved from corn grain ethanol or soybean biodiesel (Tilman et al. 2006). The material, however, is heterogeneous and very variable from site to site based on the individual species composition in addition to variability caused by local climate and soil types. Bioenergy strategies for the utilization of this resource therefore tend to focus on preprocessing and potentially fractionation. For example, one solution is the integrated generation of solid fuel and biogas from biomass (IBFT; Richter et al. 2010; Bühle et al. 2012) where a hot water pretreatment is used to wash the sugars and corrosive alkaline metals and chlorine concentrations from the biomass. The biomass is then squeezed through a screw press to generate a soluble fraction which can be fermented through anaerobic digestion to make biogas to power the hot water generation, and a press cake which can be further processed by pelletizing for combustion or pyrolysis to bio-oil and biochar. Fuel quality is significantly influenced by the botanical composition, but the quality can be improved and to an extent controlled by processing (Hengsen et al. 2012). An initial question on such approaches is whether they are predominantly an environmental problem that needs a solution or whether they can address renewable energy targets

and carbon dioxide emission targets. A study by Corton et al. (2013) predicted that conservation management in Wales could potentially generate one million tonnes of biomass annually. This equates to the equivalent of 38% of the Welsh transport sector's greenhouse gas (GHG) reduction target for 2020 or a reduction in GHG emissions by 11% of the domestic sector's reduction target for 2020, depending on conversion routes used. Not only do new applications for high diversity low input grasslands provide more options for those managing these established grasslands but also provides incentives to restore such grassland systems including on contaminated land (Zhou et al. 2009).

Historically, moderate levels of animal performance were achieved in many marginal grassland areas, but structural and compositional changes in vegetation following reductions in stocking rates have significantly lowered the production potential and related economic viability of many types of semi-natural grassland (Dumont et al. 2013). Furthermore, recent research has confirmed that enteric methane emission intensities are greater when animals consume poorer quality indigenous grassland (Fraser et al. 2014), and thus the carbon footprint per kilogram of output is considerably higher compared to more intensive production systems (Edwards-Jones et al. 2009; Gill et al. 2009). However, there are benefits both for human health and food security from grass-based meat production, particularly when forages from areas unsuitable for cultivation are turned into human-edible products (Lind et al. 2009). Mechanical removal of biomass avoided by grazers for bioenergy production should stimulate new plant growth with a higher nutritional value to livestock, in turn improving protein and energy supplies and related production efficiencies. In situations where competitive plant species that are avoided by grazing livestock have become dominant it should also facilitate changes in species balance which reduce foraging time and improve intake potential by increasing the availability of preferred food items. Thus, production of bioenergy and biorenewable products from underutilized native vegetation should complement rather than compete with traditional pastoral uses when undertaken as part of integrated multi-functional management systems. By stimulating utilization of abandoned or undergrazed grasslands, this approach would increase the land cover available for pastoral livestock systems and free up scarcer high-grade land for crop production. Moving livestock production back into areas which have seen agricultural abandonment would enhance food security and development while preserving rural communities and cultural heritage, including landscapes valuable for tourism and recreational activities. Management of areas of flammable dried grass and shrubby vegetation will also reduce fire risk, a constant hazard within certain regions, which carries environmental costs, for example, in terms of soil erosion and CO_2 emissions, and which has also been exacerbated by reductions in grazing livestock.

Use of Improved Grasslands, Forage, and Turf

The urban environment including parks, sports pitches, and transport verges are often planted with grasses and these are also cut. As above, this is regarded as a waste product, not so much because it is too indigestible to feed to livestock but because of the risk of contamination. Bioenergy is therefore an ideal technology to make use of this digestible feedstock. Likewise for livestock production, especially in grass-dominated feeding systems, there can be periods of the year when excess grass can be cut in addition to that needed so that farmers can produce both meat or milk, and energy. This can provide an incentive to optimize grassland system productivity and also help mitigate greenhouse gas emissions from other parts of the farming enterprise. For both urban green waste (Van Meerbeek et al. 2015) and surplus forage, fermentation-based conversion routes such as anaerobic digestion (Kyazze et al. 2008) and ethanol production (Martel et al. 2010; Farrar et al. 2012) are more likely to be suited to this largely green wet biomass. These feedstocks are also ideally suited for fermentation as a feedstock to a wider range of chemical building blocks or other industrial biotechnology targets (Hull et al. 2014). The high sugar perennial ryegrasses (*Lolium perenne*) have proved to be of particular interest, given the greater availability of readily fermented soluble sugars and principally fructan (Kyazze et al. 2008; Farrar et al. 2012). Once again exploitation of these grasses for bioenergy production could take place as part of management systems integrating this with livestock production. Recent research found that the voluntary intake of lambs was higher when offered silage prepared from grass fiber than when offered silage prepared from unprocessed grass. This was likely due to the physical damage to the grass during processing promoting a more rapid breakdown of the feed in the rumen, leading to a shorter retention time. Related performance figures suggest that these higher voluntary intakes could offset the relatively lower nutritional value of pressed grass, highlighting the potential for simultaneous production of bioenergy and meat/milk from grass, and offering another opportunity to reduce greenhouse gas burdens associated with ruminant production.

Being perennial, forage and turf grasses offer wider ecosystem benefits including carbon sequestration, water management, and soil structure. However, these grasses are often replanted on a medium term (5–10 year) basis and this provides opportunities for replanting with new varieties of forage grasses, such as high sugar types, or the planting of different species of grasses. Potentially these grasses could be nonnative which may provide new opportunities in relation to increases in productivity and also tolerance to changes in climate.

Use of Dedicated Energy Grasses

Studies on dedicated energy crops have highlighted and calculated: areas of underutilized land (Hastings et al. 2009); the opportunities to be created through plant breeding and agronomy to increase yields and reduce land take (Karp and Shield 2008); fossil carbon substitution and reduction in atmospheric CO_2 (Hughes et al. 2010); and the provision of wider ecosystem services (Hedde et al. 2013; Bourke et al. 2014). To date, much of the dedicated energy grass science and plant breeding has been focused on grasses such as Miscanthus, switchgrass, napier grass, arundo, and reed canary grass. The advantage of these grasses is that they grow on marginal land and produce high outputs from low inputs. They are typically harvested after senescence and so nitrogen has been remobilized to below ground rhizome and, or, rooting systems. These grasses are also often capable of growing in multiple environments and geographies thereby representing good targets for, and justification of, plant breeding and agronomy research as the markets for seeds or propagules are expanded. The choice of energy grass is largely based on the target environmental conditions (Clifton-Brown et al. 2011; Don et al. 2012). This applies both to temperatures required for germination, but also for crop growth, over wintering, and then ripening. Miscanthus in particular has received a lot of interest as an energy crop because it combines the high productivity and water and nutrient use efficiencies of a C4 tropical grass with remarkable cold adaptation, with its natural geographic range extending from the tropics in South East Asia up to northern China, Japan, and Siberia. Miscanthus was originally collected by botanists and horticulturalists and brought to Europe and North America from the 19th century onwards (Dougherty et al. 2014). These collections included the naturally occurring triploid hybrid between *M. sinensis* (2×) and *M. sacchariflorus* (4×), *M. x giganteus* (3×). The two parental species and the hybrid have been the subject of studies to understand this species and optimize it for bioenergy production. Research tends to focus on three main areas: (1) yield and yield maintenance; 2) biomass quality; and (3) sustainability and environmental impacts. The key advantage in the development of dedicated energy grasses is that plant breeding and agronomy can be targeted on their specific use for bioenergy, although with quality traits a wider number of markets are developing including for biomaterials, chemicals and even the use for winter animal bedding as a lower cost and local alternative to cereal straws. One major trait of interest has been flowering time because it impacts yield (Jensen et al. 2013) with delayed flowering extending vegetative growth into later in the season (Jensen et al. 2011a); quality through association with the start of senescence; and sustainability and the environment through remobilization of nitrogen following senescence (Mos et al. 2013) and the control of seed dispersal to minimize the risk of invasiveness (Dougherty et al. 2014). In addition, the ability to control flowering enables the crossing of different species that do not necessarily flower at the same time with *M. sacchariflorus* being a short day plant (Jensen et al. 2013) and *M. sinensis* flowering more according to accumulated temperature (Jensen et al. 2011a) with rainfall also having an effect (Jensen et al. 2011b). The generation of seed-based Miscanthus varieties also has the potential to significantly reduce financial and carbon establishment costs compared to rhizomes.

Strategies to increase yield, beyond the manipulation of flowering time, include extending crop canopy duration, improving net photosynthetic efficiency, incorporating disease, chilling and drought resistance, and improving nutrient and water use efficiencies. Extending the crop canopy duration through earlier emergence in spring and delayed senescence in the autumn, and an efficient crop canopy architecture, represent means to improve the proportion of radiant light intercepted (Robson et al. 2013a). However, extensions of crop canopy duration result in a greater risk of the plant experiencing lethal temperatures. In the case of Miscanthus, the impact of early season frosts can be responded to by the emergence of new shoots assuming sufficient resources have been build up in the previous season and so the negative impact is much lower than in many crop species where the entire crop would be lost because the crop is annual or because the harvested portion is a result of flowering and flower initials are laid down the previous year. Significant differences exist between Miscanthus species and genotypes for drought and freezing tolerance, thereby providing routes to improvement (Clifton-Brown et al. 2002; Purdy et al. 2013). A number of other single or compound traits have also been evaluated and canopy or plant height is one of the more highly correlated traits to yield (Robson et al. 2013b), whilst nonstructural carbohydrate profiles and ratios between soluble sugars and starch have also being shown to be predictive indicators of future productivity (Purdy et al. 2015).

Strategies to improve energy crop quality depend on the ultimate end use(s). For example at present Miscanthus is predominantly used commercially as a feedstock for heat and power either in large power stations or in smaller community or municipal heating systems. For the thermochemical conversion market, the ideotype is a crop which has low N and P to reduce future fertilizer inputs and low NOx emissions, low K, and Cl to reduce corrosion in boilers, low moisture to minimize drying and reduce spoilage during storage, low ash to reduce slagging and consequent operational downtime, and high processability and calorific value to increase energy density. There is additional interest in the use of Miscanthus as a feedstock for fast pyrolysis (Hodgson et al. 2011) or as a source of sugars for fermentation to transport fuels (Brosse et al. 2012) and through anaerobic digestion to biogas (Klimiuk et al. 2010). For fermentation, the ideotype is likely to include greater digestibility of the lignocellulose. Other uses such as a source of biomaterials for insulation (Uihlein et al. 2008) or fiberboard (Velasquez et al. 2003) may require an optimization of processability and fiber sizes. Comparison of biomass samples harvested from the same genotypes grown at five different locations in Europe indicate that biomass composition is reasonably stable across multiple environments (Hodgson et al. 2010a) and therefore composition represents a target for optimization for different uses through plant breeding. Therefore, variation exists within Miscanthus for the optimization of bioenergy and industrial end use quality traits (Allison et al. 2011) as markets develop and mature. Other studies have also examined the potential for identifying higher value compounds in Miscanthus to increase the value within the biomass chain (Parveen et al. 2011) but the challenge is whether the market size of such compounds will be sufficient to significantly impact a large enough area of the crop and therefore number of growers.

As energy grasses are a renewable source of energy, it is important that they are sustainable and that environmental impacts are predominantly positive. Miscanthus already exhibits efficient use of resources, especially nitrogen and does not require annual fertilizer application (McCalmont et al. 2015). Indeed there is evidence that high nitrogen treatments have negative impacts on biomass yield and quality. For example, nitrogen fertilization can have a negative impact on biomass quality for thermochemical conversion with a reduction in cell wall components and an increase in ash content (Hodgson et al. 2010b). The high nitrogen efficiency of Miscanthus comes from two routes, firstly the efficient recycling of nitrogen through senescence (Mos et al. 2013) and secondly through association with nitrogen fixing bacteria (Davis et al. 2010; Keymer and Kent 2014). Water use is also already efficient compared to many crops, however because of the high biomass production improvements in water use efficiency are also considered desirable. Once established and usually after the first season Miscanthus does not require herbicide treatment even when planting into preexisting grasslands. The crop, because of the dense and efficient canopy closure, is effective at suppressing weeds which makes it ideally suited for marginal land and the bioremediation of weed infestation. There is then a potential trade-off between productivity and suppression of long term problematic weeds and potential biodiversity of other species. Positive benefits tend to be reported during the early years of establishment, when there are gaps in the crop and also when compared to annual arable crops including those like wheat and oil seed rape which may be used for biofuels. Crop establishment was often patchy in the earlier plantings of Miscanthus, again beneficial for biodiversity, but is less common in more recent plantings as crop agronomic techniques have improved. The use of appropriate field margins may help to balance trade-offs between productivity and biodiversity, and be another source of biomass following cutting late in the season as described above for conservation grasslands. In comparison to annual cropping systems, higher densities and diversity of soil invertebrates were observed under Miscanthus planted onto contaminated land (Hedde et al. 2013). Likewise in a large multi-location study in Ireland differences between biodiversity in Miscanthus and conventional crops were mostly positive with higher vascular plant richness and higher solitary bee abundance and richness compared with conventional crops (Bourke et al. 2014). Perhaps, however not surprisingly, biodiversity benefits tend to be smaller and disbenefits can occur when comparisons are made to more natural or native grasslands (Dauber et al. 2015). The high biomass accumulation and leaf fall over winter help contribute to an accumulation of soil carbon. Several studies have determined soil carbon under Miscanthus to be similar to long-term grasslands and more recent studies have been looking at the transition (Harris et al., 2014). The Miscanthus canopy can also bring about positive changes in terms of albedo and therefore contribute to global cooling measures compared to other vegetation types (Jørgensen et al. 2014). When changing land use even between different grass species, a holistic approach is needed to ensure that the benefits exceed the disbenefits. Therefore, approaches such as scenario (Harvolk et al. 2014) and opportunity mapping (Lovett et al. 2014) represent important strategies for identifying those locations where land use change is most likely to bring about predominantly positive impacts, and therefore aids policymaking.

Concluding Remarks

Grasslands are often described as multifunctional because of the wider ecosystems services in addition to the food that they can provide, predominantly meat and milk from ruminants. The use of grassland biomass for energy and other bioproducts is an extension of this natural resource and can provide additional options for farmers and communities that are often looking for means to diversify. Moreover, large areas of poorer quality native plant grasslands are being abandoned as farmers focus their activities on improved pastures and crops, and the mechanical removal of biomass avoided by grazers for bioenergy production can be used to stimulate new plant growth with a higher nutritional value to livestock, as part of a long-term rotation strategy with the production of energy from the removed biomass and food from the new growth. This extension in use can therefore be through the use of existing grassland vegetation or from the planting of new grassland species which can fit in with existing farming machinery and practices (as described in Table 1). Energy and other industrial uses can then provide new market opportunities for grassland farmers including to diversify income streams, to provide a means to fund land management for areas of conservation value, and be part of the solution to reduce the overall greenhouse gas emissions from livestock farming enterprises. Grasses within the built environment, like conservation biomass, represents a currently underutilized and under valorized resource, and presents an opportunity for economic and environmental impacts. Collectively grasslands and forage, energy and turf grasses form a continuum of perennial systems which can provide multiple benefits in terms of productivity and ecosystem services for those that live, work or visit them. Improving the productivity and environmental sustainability of grassland agriculture through multiple strategies, is going to be needed over the coming decades to ensure the sustainability of these important habitats and landscapes. Embracing the diversification of grass-derived products will in many instances be an important step to ensuring grassland survival and protection from the onslaught of competition for land from urbanization and from other agricultural and forestry-based land uses.

Acknowledgments

The Biotechnology and Biological Sciences Research Council (BBSRC) for long-term strategic funding of the authors and grassland research at the Institute of Biological, Environmental & Rural Sciences, Aberystwyth University. The Engineering and Physical Sciences Research Council (EPSRC), Natural Environment Research Council, Department for Environment, Food and Rural Affairs (Defra), Energy Technologies Institute and Welsh Government (Low Carbon, Energy & Environment National Research Network) and European Commission for additional research funding.

Conflict of Interest

None declared.

References

Allison, G. G., C. Morris, J. Clifton-Brown, S. J. Lister, and I. S. Donnison. 2011. Genotypic variation in cell wall composition in a diverse set of 244 accessions of Miscanthus. Biomass Bioenergy 35:4740–4747.

Bourke, D., D. A. Stanley, E. O'Rourke, R. Thompson, T. Carnus, J. Dauber, et al. 2014. Response of farmland biodiversity to the introduction of bioenergy crops: effects of local-factors and surrounding landscape context. GCB Bioenergy 6:275–289.

Brosse, N., A. Dufour, X. Z. Meng, Q. N. Sun, and A. Ragauskas. 2012. Miscanthus: a fast-growing crop for biofuels and chemicals production. Biofuels Bioproducts & Biorefining-Biofpr 6:580–598.

Bühle, L., F. Hensgen, I. Donnison, K. Heinsoo, and M. Wachendorf. 2012. Life cycle assessment of the integrated generation of solid fuel and biogas from biomass (IFBB) in comparison to different energy recovery, animal-based and non-refining management systems. Bioresour. Technol. 111:230–239.

Clifton-Brown, J. C., I. Lewandowski, F. Bangerth, and M. B. Jones. 2002. Comparative responses to water stress in stay-green, rapid- and slow senescing genotypes of the biomass crop, Miscanthus. New Phytol. 154:335–345.

Clifton-Brown, J., P. Robson, R. Sanderson, A. Hastings, J. Valentine, and I. Donnison. 2011. Thermal requirements for seed germination in Miscanthus compared with Switchgrass (Panicum virgatum), Reed canary grass (Phalaris arundinaceae), Maize (Zea mays) and perennial ryegrass (Lolium perenne). GCB Bioenergy 3:375–386.

Corton, J., L. Bühle, M. Wachendorf, I. S. Donnison, and M. D. Fraser. 2013. Bioenergy as a biodiversity management tool and the potential of a mixed species feedstock for bioenergy production in Wales. Bioresour. Technol. 129:142–149.

Dauber, J., S. Cass, D. Gabriel, K. Harte, S. Aström, E. O'Rourke, et al. 2015. Yield-biodiversity trade-off in patchy fields of Miscanthus x giganteus. GCB Bioenergy 7:455–467.

Davis, S., W. Parton, F. Dohleman, C. Smith, S. Grosso, A. Kent, et al. 2010. Comparative biogeochemical cycles of bioenergy crops reveal nitrogen-fixation and low greenhouse gas emissions in a Miscanthus x giganteus agro-ecosystem. Ecosystems 13:144–156.

Don, A., B. Osborne, A. Hastings, U. Skiba, M. S. Carter, J. Drewer, et al. 2012. Land-use change to bioenergy production in Europe: implications for the greenhouse gas balance and soil carbon. GCB Bioenergy 4:372–391.

Dougherty, R. F., L. D. Quinn, A. B. Endres, T. B. Voigt, and J. N. Barney. 2014. Natural history survey of the ornamental grass Miscanthus sinensis in the introduced range. Invasive Plant Sci. Manag. 7:113–120.

Dumont, B., A. G. Thórhallsdóttir, A. Farruggia, and A. Norderhaug. 2013. Livestock grazing and biodiversity in semi-natural grasslands. Grass. Sci. Europe 18:314–326.

Edwards-Jones, G., K. Plassmann, and I. M. Harris. 2009. Carbon footprinting of lamb and beef production systems: insights from an empirical analysis of farms in Wales, UK. J. Agric. Sci. 147:707–719.

Farrar, K., D. N. Bryant, L. Turner, J. Gallagher, A. Thomas, M. Farrell, et al. 2012. Breeding for bioethanol production in L. perenne: selection for high water soluble carbohydrate content results in increased harvestable early spring biomass. Bioenergy Res. 5:149–157.

Fraser, M. D., H. R. Fleming, and J. M. Moorby. 2014. Traditional vs modern: role of breed type in determining enteric methane emissions from cattle grazing as part of contrasting grassland-based systems. PLoS One 9:e107861.

Gelfand, I., T. Zenone, P. Jasrotia, J. Chen, and S. K. Hamilton. 2011. Carbon debt of Conservation Reserve Program (CRP) grasslands converted to bioenergy production. Proc. Natl Acad. Sci. USA 108:13864–13869.

Gerssen-Gondelach, S., B. Wicke, and A. Faaij. 2015. Assessment of driving factors for yield and productivity developments in crop and cattle production as key to increasing sustainable biomass potentials. Food Energy Secur. 4:36–75.

Ghose, B. 2014. Food security and food self-sufficiency in China: from past to 2050. Food Energy Secur. 3:86–95.

Gill, M., P. Smith, and J. M. Wilkinson. 2009. Mitigating climate change: the role of domestic livestock. Animal 4:323–333.

Harris, Z. M., N. P. McNamara, R. Rowe, M. Dondini, J. Finch, M. Perks, et al., 2014. Research Spotlight: The ELUM project: Ecosystem Land-Use Modeling and Soil Carbon GHG Flux Trial. Biofuels 5:111–116.

Harvolk, S., P. Kornatz, A. Otte, and D. Simmering. 2014. Using existing landscape data to assess the ecological potential of Miscanthus cultivation in a marginal landscape. GCB Bioenergy 6:227–241.

Hastings, A., J. Clifton-Brown, M. Wattenbach, C. P. Mitchell, P. Stampfl, and P. Smith. 2009. Future energy potential of Miscanthus in Europe. GCB Bioenergy 1:180–196.

Hedde, M., F. van Oort, E. Boudon, F. Abonnel, and I. Lamy. 2013. Responses of soil macroinvertebrate communities to Miscanthus cropping in different trace metal contaminated soils. Biomass Bioenergy 55:122–129.

Hengsen, F., L. Bühle, I. Donnison, M. Frasier, J. Vale, J. Corton, et al. 2012. Mineral concentrations in solid fuels from European semi-natural grasslands after hydrothermal conditioning and subsequent mechanical dehydration. Bioresour. Technol. 118:332–342.

Hodgson, E. M., S. J. Lister, A. V. Bridgwater, J. Clifton-Brown, and I. S. Donnison. 2010a. Genotypic and environmentally derived variation in the cell wall composition of Miscanthus in relation to its use as a biomass feedstock. Biomass Bioenergy 34:652–660.

Hodgson, E. M., R. Fahmi, N. Yates, T. Barraclough, I. Shield, G. Allison, et al. 2010b. Miscanthus as a feedstock for fast-pyrolysis: does agronomic treatment affect quality? Bioresour. Technol. 101:6185–6191.

Hodgson, E. M., D. J. Nowakowski, I. Shield, A. Riche, A. V. Bridgwater, J. C. Clifton-Brown, et al. 2011. Variation in Miscanthus chemical composition and implications for conversion by pyrolysis and thermo-chemical bio-refining for fuels and chemicals. Bioresour. Technol. 102:3411–3418.

Hughes, J. K., A. J. Lloyd, C. Huntingford, J. W. Finch, and R. J. Harding. 2010. The impact of extensive planting of Miscanthus as an energy crop on future CO_2 atmospheric concentrations. GCB Bioenergy 2:79–88.

Hull, C. M., E. J. Loveridge, N. J. Rolley, I. S. Donnison, S. L. Kelly, and D. E. Kelly. 2014. Co-production of ethanol and squalene using a Saccharomyces cerevisiae ERG1 (squalene epoxidase) mutant and agro-industrial feedstock. Biotechnol. Biofuels 7:133.

Jensen, E., K. Farrar, S. Thomas-Jones, A. Hastings, I. Donnison, and J. Clifton-Brown. 2011a. Characterisation of flowering time diversity in 244 Miscanthus genotypes. GCB Bioenergy 3:387–400.

Jensen, E. F., M. Squance, A. Hastings, S. Thomas-Jones, K. Farrar, L. Huang, et al. 2011b. Understanding the value of hydrothermal time on flowering in Miscanthus species. In: biomass and Bioenergy Crops IV. Asp. Appl. Biol. 112:181–189.

Jensen, E., P. Robson, J. Norris, A. Cookson, K. Farrar, I. Donnison, et al. 2013. Flowering induction in the bioenergy grass Miscanthus sacchariflorus is a quantitative short-day response, whilst delayed flowering under long days increases biomass accumulation. J. Exp. Bot. 64:541–552.

Jørgensen, S. V., F. Cherubini, and O. Michelsen. 2014. Biogenic CO_2 fluxes, changes in surface albedo and biodiversity impacts from establishment of a miscanthus plantation. J. Environ. Manage. 146:346–354.

Jungers, J. M., J. E. Fargione, C. C. Sheaffer, D. L. Wyse, and C. Lehman. 2013. Energy potential of biomass from conservation grasslands in Minnesota, USA. PLoS One 8:e61209.

Karp K, and I Shield. 2008. Bioenergy from plants and the sustainable yield challenge. New Phytol. 179:15–32.

Keymer, D. P., and A. D. Kent. 2014. Contribution of nitrogen fixation to first year Miscanthus x giganteus. GCB Bioenergy 6:577–586.

Klimiuk, E., T. Pokoj, W. Budzynski, and B. Dubis. 2010. Theoretical and observed biogas production from plant biomass of different fibre contents. Bioresour. Technol. 101:9527–9535.

Kyazze, G., R. Dinsdale, F. R. Hawkes, A. J. Guwy, G. C. Premier, and I. Donnison. 2008. Direct fermentation of fodder maize, chicory fructans and perennial rye grass to hydrogen using mixed microflora. Bioresour. Technol. 99:8833–8839.

Lind, V., J. Berg, L. O. Eik, S. M. Eilertsen, J. Mølmann, M. Hersleth, et al. 2009. Effects of concentrate or ryegrass-based diets (Lolium multiflorum) on the meat quality of lambs grazing on semi-natural pastures. Acta Agric. Scand. A Anim. Sci. 59:230–238.

Lovett, A., G. Sünnenberg, and T. Dockerty. 2014. The availability of land for perennial energy crops in Great Britain. GCB Bioenergy 6:99–107.

Martel, C. M., A. G. S. Warrilow, C. O. Jackson, J. G. L. Mullins, R. C. Togawa, J. E. Parker, et al. 2010. Expression, purification and use of the soluble domain of Lactobacillus paracasei beta-fructosidase to optimise production of bioethanol from grass fructans. Bioresour. Technol. 101:4395–4402.

McCalmont, J. P., A. Hastings, N. P. McNamara, G. M. Richter, P. Robson, I. S. Donnison, et al. 2015. Environmental costs and benefits of growing Miscanthus for bioenergy in the UK. GCB Bioenergy. doi:10.1111/gcbb.12294.

Melts, I., K. Heinsoo, and M. Ivask. 2014. Herbage production and chemical characteristics for bioenergy production by plant functional groups from semi-natural grasslands. Biomass Bioenergy 67:160–166.

Mos, M., S. W. Banks, D. J. Nowakowski, P. R. H. Robson, A. V. Bridgwater, and I. S. Donnison. 2013. Impact of Miscanthus x giganteus senescence times on fast pyrolysis bio-oil quality. Bioresour. Technol. 129:335–342.

Parveen, I., M. D. Threadgill, B. Hauck, I. Donnison, and A. Winters. 2011. Isolation, identification and quantitation of hydroxycinnamic acid conjugates, potential platform chemicals, in the leaves and stems of Miscanthus x giganteus, using LC-ESI-MSn. Phytochemistry 72:2376–2384.

Purdy, S., A. Maddison, L. Jones, R. Webster, J. Andralojc, I. Donnison, et al. 2013. Characterization of chilling-shock responses in four genotypes of Miscanthus reveals the superior tolerance of M. x giganteus compared with M. sinensis and M. sacchariflorus. Ann. Bot. 111:999–1013.

Purdy, S. J., A. L. Maddison, J. Cunniff, I. Donnison, and J. Clifton-Brown. 2015. Non-structural carbohydrate profiles and ratios between soluble sugars and starch serve as indicators of productivity for a bioenergy grass. AoB Plants. plv032 doi:10.1093/aobpla/plv032.

Richter, F., T. Fricke, and M. Wachendorf. 2010. Utilization of semi-natural grassland through integrated generation of solid fuel and biogas from biomass. III. Effects of hydrothermal conditioning and mechanical dehydration on solid fuel properties and on energy and greenhouse gas balances. Grass Forage Sci. 65:185–199.

Robson, P. R. H., K. Farrar, A. P. Gay, E. F. Jensen, J. C. Clifton-Brown, and I. S. Donnison. 2013a. Variation in canopy duration in the perennial biofuel crop Miscanthus reveals complex associations with yield. J. Exp. Bot. 64:2373–2383.

Robson, P. R., E. F. Jensen, S. L. Hawkins, S. White, K. Kenobi, J. C. Clifton-Brown, et al. 2013b. Accelerating the domestication of a bioenergy crop: identifying and modelling morphological targets for sustainable yield increase in Miscanthus. J. Exp. Bot. 64:4143–4155.

Tilman, D., J. Hill, and C. Lehman. 2006. Carbon-negative biofuels from low-input high-diversity grassland biomass. Science 314:1598–1600.

Uihlein, A., S. Ehrenberger, and L. Schebek. 2008. Utilisation options of renewable resources: a life cycle assessment of selected products. J. Clean. Prod. 16:1306–1320.

Van Meerbeek, K., S. Ottoy, A. De Meyer, T. Van Schaeybroeck, J. Van Orshoven, B. Muys, et al. 2015. The bioenergy potential of conservation areas and roadsides for biogas in an urbanized region. Appl. Energy 154:742–751.

Velasquez, J. A., F. Ferrando, X. Farriol, and J. Salvado. 2003. Binderless fiberboard from steam exploded Miscanthus sinensis. Wood Sci. Technol. 37:269–278.

Zhou, X. P., B. Xiao, R. M. Ochieng, and J. K. Yang. 2009. Utilization of carbon-negative biofuels from low-input high-diversity grassland biomass for energy in China. Renew. Sustain. Energy Rev. 13:479–485.

Permissions

List of Contributors

Li-Min Zhang, Yan Xia, Yu-Miao Zhang, Zhi-Quan Liu and Hai-Chun Jing
Key Laboratory of Plant Resources, Institute of Botany, Chinese Academy of Sciences, Beijing 100093, China

Sylvester Elikana Anami
Key Laboratory of Plant Resources, Institute of Botany, Chinese Academy of Sciences, Beijing 100093, China
Institute of Biotechnology Research, Jomo Kenyatta University of Agriculture and Technology, Nairobi, Kenya

Yaghoob Jafari
Institute for Food and Resource Economics, University of Bonn, Bonn, Germany

Jamal Othman
National University of Malaysia, Bangi, Selangor, Malaysia

Rattan Lal
Carbon Management and Sequestration Center, The Ohio State University, Columbus, Ohio 43210

Carlos A. Salas Macías
Facultad de Ingeniería Agronómica, Universidad Técnica de Manabí, Lodana, Santa Ana, Ecuador

Julio C. Alegre Orihuela
Universidad Nacional Agraria La Molina, Lima, Perú

Sergio Iglesias Abad
Universidad Católica de Cuenca, Cuenca, Ecuador
Aklilu Bajigo Madalcho
Department of Natural Resource Management, College of Dryland Agriculture, Jigjiga University, 1020, Jigjiga, Ethiopia

Gareth J. Norton, Anthony J. Travis and Adam H. Price
Institute of Biological and Environmental Sciences, University of Aberdeen, Aberdeen AB24 3UU, UK

John M. C. Danku and David E. Salt
Institute of Biological and Environmental Sciences, University of Aberdeen, Aberdeen AB24 3UU, UK
Centre for Plant Integrative Biology, School of Biosciences, University of Nottingham, Sutton Bonington Campus, Loughborough LE12 5RD, UK

Mahmud Hossain and Md. Rafiqul Islam
Department of Soil Science, Bangladesh Agricultural University, Mymensingh, Bangladesh

Navin Sharma, Rodrigo Ciannella and Phil Dobie
World Agroforestry Centre, UN Gigiri, Nairobi, Kenya

Babita Bohra and Namita Pragya
World Agroforestry Centre, NASC Complex, New Delhi, India

Sarah Lehmann
Intern at GIZ, Untergasse 15, 65510 Idstein, Germany

Timothy A. Volk, Justin P. Heavey and Mark H. Eisenbies
State University of New York College of Environmental Science and Forestry (SUNY-ESF), Syracuse, New York

Sarah Gerssen-Gondelach and Birka Wicke
Copernicus Institute of Sustainable Development, Utrecht University, Heidelberglaan 2, 3584 CS Utrecht, The Netherlands

Andre Faaij
Energy and Sustainability Research Institute, University of Groningen, Nijenborg 4, 9747 AC Groningen, The Netherlands

Guang Chu, Zhiqin Wang, Hao Zhang, Lijun Liu and Jianchang Yang
Jiangsu Key Laboratory of Crop Genetics and Physiology/Co-Innovation Center for Modern Production Technology of Grain Crops, Yangzhou University, Yangzhou, Jiangsu, China

Jianhua Zhang
School of Life Sciences and State Key Laboratory of Agrobiotechnology, The Chinese University of Hong Kong, Hong Kong, China

Vicky Buchanan-Wollaston and Jim Beynon
School of Life Sciences, University of Warwick, Gibbet Hill, Coventry CV4 7AL, U.K

Zoe Wilson
School of Biosciences, University of Nottingham, Sutton Boningon Campus, Sutton Bonington, Leicestershire LE12 5RD, U.K

François Tardieu
INRA Laboratoire d'Ecophysiologie des Plantes sous Stress Environnementaux (LEPSE), Montpellier, France

Katherine Denby
Department of Biology, University of York, Heslington, York YO10 5DD, U.K

Tanya Y. Curtis and Nigel G. Halford
Plant Biology and Crop Science Department, Rothamsted Research, Harpenden, Hertfordshire AL5 2JQ, UK

William J. Davies
The Lancaster Environment Centre, Lancaster University, Bailrigg, Lancaster LA1 4YQ, UK

Jean-Marcel Ribaut
Generation Challenge Programme (GCP) c/o CIMMYT, Carretera Mexico-Veracruz, El Batan, Texcoco, Estado de Mexico, Mexico

Phillip A. Davis and Claire Burns
Stockbridge Technology Centre, Cawood, Selby, North Yorkshire YO8 3TZ, UK

Iain S. Donnison and Mariecia D. Fraser
Institute of Biological, Environmental & Rural Sciences, Aberystwyth University, Gogerddan Campus, Aberystwyth SY23 3EE, UK

Index

CPSIA information can be obtained
at www.ICGtesting.com
Printed in the USA
BVHW062000260822
645617BV00004B/170